Reporting Technical Information

Reporting
Technical
Information

5TH EDITION

Kenneth W. Houp

*Associate Professor
of English Composition
The Pennsylvania State University*

Thomas E. Pearsall

*Professor and Head,
Department of Rhetoric
University of Minnesota*

Macmillan Publishing Company
NEW YORK
Collier Macmillan Publishers
LONDON

Earlier editions copyright © 1968 by The Glencoe Press, copyright © 1973 and 1977 by Benziger Bruce and Glencoe, Inc. copyright © 1980 by Glencoe Publishing Co., Inc.

Macmillan Publishing Company
866 Third Avenue, New York, New York 10022

Collier Macmillan Canada, Inc.

Library of Congress Cataloging in Publication Data

Houp, Kenneth W.
  Reporting technical information.

  Bibliography: p.
  Includes index.
  1. Technical writing. I. Pearsall, Thomas E.
II. Title.
T11.H59 1984   808'.0666021   83-5479
ISBN 0-02-474920-6

Printing: 4 5 6 7 8   Year: 4 5 6 7 8 9 0 1 2

ISBN   0-02-474920-6

Preface

IN THE 1970s, the teaching of technical writing grew rapidly and steadily. Although exact figures are difficult to obtain, approximations based on the sale of textbooks in the field indicate that the number of students per year taking technical writing has jumped from about 25,000 in 1968 to over a quarter million in 1982. Another indicator is the growth of the Association of Teachers of Technical Writing from a nucleus of ten charter members in 1972 to an organization with over a thousand members. Currently, five summer workshops in technical writing enroll over 150 students a year. Ten years ago such workshops did not even exist. Numerous panel discussions about technical writing are now regular features at national meetings of English teachers. Why has this discipline come so far so fast?

We suggest that one answer lies in the nature of technical writing itself. If there is one thing that all technical writing practitioners and teachers can agree upon, it is that technical writing makes things happen. As Professor Thomas L. Warren puts it, "The purpose of any piece of technical writing is to inform so that the reader can act. . . ."[1] Certainly, there is a sense of both practicality and power in technical writing. Technical writing is practical because it is situational. It addresses itself to a particular purpose and audience. Technical writing is powerful because it makes things happen. Without manuals, plans, and instructions, our machines, factories, and ships would not get built. Without the instructions that come with a computer, it would be a collection of worthless boxes. Old scientific theories fall before newly published scientific theories. If our energy crisis and economic and environmental problems are one day resolved, it will be because the solutions were set down in technical writing. Students and teachers alike enjoy the power and technical writing's rational practicality.

Teachers and students alike see technical writing courses as being worthwhile because—like tennis and unlike football—they have a high carry-over value beyond the college days. A survey conducted by Professor Richard M. Davis among persons listed in *Engineers of Distinction* led him to the following facts and conclusions:

- The respondents spent an average of 24.35 percent of their time writing.
- The respondents spent 31 percent of their time working with other people's written materials.
- Most of the respondents felt that the ability to write effectively is either very important or of critical importance to them.
- Almost all reported that as their responsibility increased, so did their need to write.
- Almost all felt that the ability to write well had contributed to their advancement.
- In advancing subordinates, almost all considered their writing ability.[2]

Information from other fields, such as agriculture and business, provides similar direct evidence: graduates need to be able to communicate effectively with workers, supervisors, and the general public. Other evidence, more indirect but perhaps even more powerful, supports the need for good writing skills. Everywhere we turn these days, articles and ads proclaim we are entering "The Information Age." Is this merely hyperbole to sell us word processors and communication equipment, or are those of us in the developed countries truly entering an age in which information will be our major product? Evidence seems to indicate that we are, indeed, entering something very like an "information age."

In a recent talk, John Naisbitt, publisher of *Trend Report* and a perceptive observer of changes in society, stated that the developed countries are rapidly shifting "from a mass industrial society to an information society." In Naisbitt's words, "The final impact will be more profound than the 19th century shift from an agricultural to an industrial society." Naisbitt points out that in 1950, 65 percent of the work force of the United States was in industry; now that figure stands at 30 percent. In 1950, the proportion of people in information occupations was 17 percent; today the proportion is over 55 percent. Naisbitt defines information occupations as "those involved in the creating, processing, and distribution of information. . . ."[3]

Much supports Naisbitt's position. For example, the major resource for the entrepreneur in the past has been capital. Now it may be knowledge and data. Small firms are springing up all around us, many based on some form of microcomputer technology, itself a part of the information explosion. These small firms have as their major resource not money but rather the information in the heads of the young people who founded them. Because information about new technology has become so vital, a lack of information can be as harmful as a lack of machine tools. Recently a computer firm near one of us halted production on one of its new models because the needed user information for it had not yet been produced.

The growth in the numbers of students in our technical writing classes, then, would seem to be a sign of the times. We are filling a need that profound changes in our society are bringing about. As more and more people had technical information that they had to report, technical writing classes began to attract students from many disciplines—agriculture, business administration, economics, forestry, medical technology, pyschology—to join the engineers who once were the primary audience. With this change in the student body came a change in the teaching of technical writing. Once it had heavily depended on teaching from models of existing reports, primarily engineering reports. But the new student body, representing many disciplines with many models to choose from, caused the teachers of technical writing to look beyond the models for the process that produced the models.

We are happy to say that we anticipated this change in our 1972 second edition with our Part 1 "Process of Technical Reporting" and our introduction of audience analysis as part of the process. The models were by no means disregarded. They remained in a section called "Applications." However, in that section in the second edition, and now in this fifth edition, we used applications such as correspondence, instructions, proposals, and feasibility reports that seemed common to all the disciplines present in our classes. We want the readers of this book to understand the process that underlies good report writing, but we also want them to know that there are models that can be useful to them. Writers have always used models, whether the models are sonnets, heroic couplets, three-act plays, or well-constructed feasibility reports.

Our continuing interest in process has led in this edition to a new chapter called "Getting Started" (Chapter 2) that deals with the beginning of the process—that rather terrifying moment when the writer is trying to get from nowhere to somewhere. We show the writer how not only must topic be considered but audience, purpose, and writer's role as well. With these as the building blocks, we take the writer in later chapters through audience analysis, gathering and checking information, organizing, and writing and revising the report. We make it clear many times, through statement and examples, that the writing process is a back-and-forth process. Although the end result may be a piece of writing with a hierarchy of information that can be read in a linear fashion, the actual process is more like the putting together of a jigsaw puzzle. The parts are fitted together in little chunks here and there until finally, after a good deal of trial and error, a total picture falls into place. In other words, we introduce the reader to the world of the real writer who knows all too well that ideas do not usually come in a well-ordered 1–2–3 fashion.

What, then, will the reader find in this 5th edition?

Part 1, "Process of Technical Reporting," covers the basic process from getting started to organizing, writing, and revising the report. Interspersed with the process chapters are chapters on the report writer's tools such as the library, the rhetorical modes, graphics, and such formal elements of reports as abstracts, introductions, conclusions, tables of content, and headings.

Part 2, "Applications," covers advanced and extended applications of the basic principles. Here we show how to write correspondence, proposals, progress reports, feasibility reports, and the like.

Part 3, "Handbook," provides a ready reference when questions of grammatical usage, punctuation, and mechanics arise.

The appendixes include an extended student report, additional guides to library research and computerized information retrieval, and a bibliography of books that can lead the writer to other books about the many subjects we cover in this book.

What other changes have we made for this 5th edition?

- We have unified examples when this seemed to be a change that would aid the learning process. For example, we have created a new, contemporary, and, we hope, lively and realistic example for a report project that runs through both Chapter 2, "Getting Started" and Chapter 5, "Gathering and Checking Information." By using the same example we show how a report can grow from a small idea, barely conceived, to a fully grown body of material ready to be organized and written. In a similar way, we carry one project through Chapters 15, "Proposals," 16, "Progress Reports," and 17, "Feasibility Reports." Again, the reader should be able to see how the report grows from the original concept to the finished product.
- Our Chapter 9, "Achieving a Clear Style," has benefitted greatly from the significant research being carried on by Dr. Janice Redish and her colleagues at the American Institutes for Research in Washington, D.C. Their work has caused us in some matters to modify our views and in others to state previous positions even more strongly. In any case, as such research continues, we can all be a bit more confident about what makes one report more readable than another.
- We have combined two chapters, "Prose Elements" and "Mechanical Elements" into one, Chapter 11, "Formal Elements of Reports." By so doing, we can treat the elements in the order in which they appear in a

report without regard for whether they are a prose element such as an introduction or a mechanical element such as a table of contents. We think the information will be more useful in this form. Also, we have covered more thoroughly than ever the reasons underlying the use of a format. In both Chapter 8, "Organizing Your Report," and Chapter 11, we emphasize the importance óf using a format that allows for selective reading of a report.

- We have revised Chapter 15, "Proposals," to make it more useful for the student. We have added examples of student-sized proposals and proposals for reports that should be useful in the classroom.
- Our 4th edition chapter on "Physical Research Reports" is completely updated. Because the changes made should make the chapter useful to writers in thė social sciences as well as the physical sciences, we have changed the name of the chapter to "Empirical Research Reports."
- The University of Chicago Press style manual continues to be our guide. Consequently, we have modified slightly our own documentation practices and our instruction on documentation to conform to the 13th edition of this fine book, now called *The Chicago Manual of Style*. Also, we have pulled footnotes out of our chapters and gathered them all together in "Chapter Notes" on pages 527–533. We hope we have accomplished two things with this change: cleaned up the format of our pages and provided many note examples in one place for the student.

In addition, we have made numerous smaller changes. We have refreshed the book with many new examples. We now offer more in-depth explanations of certain writing practices, such as how to construct a report title and how to use tense properly in an empirical research report. Those of you who have used this book over the years know our touchstone, expressed in every preface, that all writing is subject to infinite improvement. We truly believe that statement about our own writing and welcome the opportunity that each new edition offers for improvement. However, in researching for this 5th edition, we came across a short passage that we feel perhaps cannot be improved. Its three short imperatives express the philosophy of this book and we would like to leave them with you:

1. Know your reader
2. Know your objective
3. Be simple, direct, and concise[4]

We acknowledge all the many sources we have drawn upon in writing this 5th edition. Detailed acknowledgments can be found in "Chapter Notes." However, we acknowledge here a number of people who have been particularly helpful. We are grateful to those reviewers who have read earlier editions of this book and this edition in manuscript and provided insights and suggestions for improvement, including Paul Anderson, Miami University; Carol M. Barnum, Southern Technical Institute; Virginia Book, University of Nebraska; C. William Brewer, Texas Tech University; Russell Briggs, Kalamazoo Valley Community College; Neal Duane, Northeastern University; Margaret Duggan, South Dakota University; Michael Fallet, Northeastern University; Richard Ferguson, University of Minnesota; Ralph E. Jenkins, Temple University; Perry D. Luckett, University of Colorado; Allison McCormack, Miami University—Hamilton; Michael G. Moran, Clemson University; Carol Pemberton, Normandale Junior College; Martha Satz, Southern Methodist University; Ed Stoddard, Northeastern University; and Thomas L. Warren, Oklahoma State University.

We thank our many friends in the business world and the Society for Technical Communication, in particular, J. Paul Blakely, Oak Ridge National Laboratory; Mary Fran Buehler, Jet Propulsion Laboratory, California Institute of Technology; Karen Bunting, The Trane Company; Gerald Cohen, IBM; James M. Lufkin and Cindy Urman, Honeywell; and Charles R. Pearsall, Deere and Company, who sent us example materials.

Donald J. Barrett, Chief Reference Librarian, United States Air Force Academy, has once again revised Appendix B, "Technical Reference Guides" for us, and Professor James Connolly, University of Minnesota, has again contributed the section on visuals found in Chapter 19, "Oral Reports and Group Conferences." For that we thank them deeply.

Finally we express our love and gratitude to our wives Lois and Anne who are as much a part of this book as we are.

KENNETH W. HOUP
THOMAS E. PEARSALL

Contents

19   Oral Reports and Group Conferences   416

PART 3
HANDBOOK   451

Reporting
Technical
Information

Part 1 Process of Technical Reporting

Part 1 serves as an introduction to the process of technical reporting. Following the overview of technical reporting given in Chapter 1, we introduce you to the process from the inception of an idea for a report all the way to the finished product. Interspersed with the chapters on process are chapters on the tools you will need for technical reporting, such as the library, the rhetorical modes, and graphics. If you wish to read the process chapters through before dealing with the tools, read in sequence Chapters 2, "Getting Started," 3, "Analyzing Your Audience," 5, "Gathering and Checking Information," 8, "Organizing Your Report," and 10, "Writing and Revising Your Report."

Chapter 1

An Overall View of Technical Reporting

THIS FIRST CHAPTER is purely introductory. It is intended to give you the broadest possible view of report writing. Beginning with Chapter 2 we go into details, but for the details to be most meaningful, they have to be seen against the background given here.

SOME MATTERS OF DEFINITION

What does the term *report writing* mean to you? Does the term *technical reporting* convey any clearer notion? In this book these terms are often used interchangeably. Therefore, we are compelled at the outset to provide you with a working definition, one that we expand and refine in later chapters.

This need for definition always arises when a novel and sometimes complex term is introduced. Suppose you were to learn from a bulletin board announcement that a lecture is to be given tomorrow evening on "operations research." If you didn't know the meaning of the term, would you attend the lecture anyway? Probably not.

> *operations research:* analysis, usually mathematical, to determine the effectiveness of a process, or the like, to increase efficiency.

This definition, brief as it is, might very well enable you to make up your mind.

Now let us turn specifically to the title and subject matter of this book: *Reporting Technical Information.*

reporting: providing an account or description of what has been learned by experience, observation, or investigation.

technical: peculiar to or characteristic of a particular art, science, trade, or profession.

information: a body of knowledge gained from experience, observation, or investigation.

The reporting of technical information is thus seen to involve three elements at one or more stages of the process:

1. A problem or subject matter that is not popular knowledge but, rather, is specialized in that it belongs to art, science, medicine, engineering, or the like.
2. Study, investigation, observation, analysis, experimentation, and measurement to obtain accurate and precise information about the problem or subject matter.
3. The organization and presentation of the information thus gained so that it will be clear and meaningful to the person or persons for whom it is intended.

The final product of this three-stage process is a technical report that may range in size and complexity from a simple memorandum to a stack of books. To expand our overall view of technical reporting, we elaborate our set of definitions under these five headings:

The Nature of Technical Reports
The Substance of Technical Reports
The Attributes of Good Report Writers
The Qualities of Good Reports
A Day in the Life of Two Report Writers

THE NATURE OF TECHNICAL REPORTS

Some people—especially those who have never tried their hand at it—believe that report writing is simply good writing. They believe that it has no qualities to distinguish it from other expository writing such as the essay, journal article, or news account. Some take the opposite stand: they hold that writers of technical reports must use a brand of writing so peculiar and specialized that it is essentially divorced from all other writing.

The truth, we believe, lies between these extremist attitudes: technical report writing is most certainly a specialty within the field of writing as a whole. Beginning report writers have to serve an apprenticeship. They must gain a working knowledge of their new subject matter and its terminology. They must learn to develop a prose style that is clear, objective, and economical. They must learn report types, variations in format, standards for abbreviations, the rules that govern the writing of numbers, the kinds of people who read reports and their expectations. That is, beginning report writers have to learn the whole special business of being report writers.

And yet a broad and sound foundation in other writing is a tremendous asset to report writers, for it gives them versatility both on and off the job. They can write a good letter, prepare a brochure, compose an essay. In this comprehensive sense, they are simply *writers*. As writers they understand, for example, that subordination and coordination, coherence and emphasis are important in *all* writing. They understand that not all writing is done in the same tone and style. As writers they have not one style but a battery of them:

Example	*Commentary*
. . . the very nice plant my mother had on her table in the front hall.	Everyday, homey diction; much depends on the reader's imagination.
. . . in a shaft of yellow sunlight, a white-flowering begonia in a red clay pot.	Pictorial, vivid, sensory; "shows" rather than "tells about."
. . . a twelve-inch begonia propagated from a three-inch cutting; age, 42 days.	Specific, "technical"; factually informative.

As a reporter of technical information, whether part-time or full-time, you may have to use all of these "languages," for your job will be to convey your message to your intended readers. By playing the right tune with these languages in different combinations, and by adding other writing skills in generous measure, you can produce leaflets, brochures, and sales literature; reports to stockholders; a great variety of letters; and articles for magazines and journals.

However, in much technical reporting, particularly when you are writing for your professional colleagues, you will be nearer to the closing begonia example than the first two. Your diction will be objective and accurate. By relying on this style you can produce operating manuals, feasibility reports, research reports, progress reports, and similar materials. When you choose this style, your reports are likely to show these characteristics:

- The writing is marked by a no-nonsense approach to the subject it treats. It is single-minded and earnest. Interesting points are seldom introduced for their interest value alone; they must also be pertinent.

- The purpose of the article or paper is usually spelled out in the opening paragraph or two. All included information bears upon the accomplishment of the stated purpose. For example, a technical paper on smoke detectors may set forth only one major objective: to determine the relative effectiveness of photoelectric and ionization chamber types in detecting smoldering fires, flaming fires, and high temperatures. Other major topics would be reserved for other papers.
- The vocabulary tends to be specialized. Some of the terms may not appear in general dictionaries. If the audience shares the writer's professional specialization, the specialized terms may not be defined within the text, on the assumption that professional colleagues will be familiar with them.
- Sentences are highly specific and fact filled.
- When appropriate to the material, numbers and dimensions are numerous. These are usually in arabic form and are exact rather than rounded to the nearest whole number.
- Signs, symbols, and formulas may pepper the prose. The terms may be listed and defined in accompanying glossaries (minidictionaries).
- Graphs and tables may substitute for prose or reinforce and expand upon the surrounding prose. Figures and illustrations of all sorts are widely used, sometimes to supplement prose, sometimes to replace it.
- Documentation and credits appear in notes and bibliographies.

Finally, as perhaps these listed characteristics make clear, audience analysis is tremendously important to successful technical reporting. In matters of definition, for example, terms are not normally defined if the audience can be expected to know them. But the indispensable corollary to that proposition is that terms *must be defined* when your audience, for whatever reason, cannot be expected to know them. Or using another characteristic, sentences can be fact filled when the audience is highly professional and highly motivated. However, when your readers do not share your motivation, profession, and enthusiasm, you may have to slow your pace and make your prose less dense. In technical reporting, you must know your audience as well as your objectives.

THE SUBSTANCE OF TECHNICAL REPORTS

Organizations produce technical reports for both internal and external use. Internally, reports such as feasibility studies, technical notes, and memorandums go from superiors to subordinates, from subordinates to superiors, and between colleagues at the same level. If reports move in more than one direction, they may have to be drafted in more than one version. Company policy, tact, and the need-to-know are important considerations for intracompany paperwork. Many examples come to mind. The director of information services has to study and report on the feasibility of providing middle management with personal computers. The research department has to report the results of tests on new products. The personnel department has to instruct new employees about company policies and procedures. In fact, the outsider cannot conceive of the amount and variety of paperwork a company must generate simply to keep its internal affairs in order.

Externally, letters and reports of many kinds go to other companies, to the government, and to the users of the company's products. Let us cite a few of the many possibilities: A computer company must prepare instructional manuals to accompany its computers. A university department prepares a proposal to a state government offering to provide research services. An architectural firm prepares progress reports to report to clients the status of contracted building programs. A firm must report on its accounting practices to the Internal Revenue Service.

As we pointed out in the preface, the manufacture of information is becoming a major industry in its own right. Much of that information is research related. Many government agencies, scientific laboratories, and commercial companies make research their principal business. They may undertake this research to satisfy their internal needs or the needs of related organizations. The people who conduct the research may include chemists, physicists, mathematicians, psychologists—the whole array of professional specialists. They record and transmit much of this research via reports. The clients for such research may be government agencies or other institutions that are inadequately equipped to do their own research. Reports may, in fact, be the only products of some companies and laboratories.

Much technical reporting goes on at universities and colleges. Professors have a personal or professional curiosity that entices them into research. If they believe that their findings are important, they publicize the information in various ways—books, journal articles, papers for professional societies. Students assigned research problems to further their academic training present what they have done and learned in laboratory reports, monographs, and theses.

Many reports are prepared for public consumption. For example, a state department of natural resources is entrusted not only with conserving our woodlands, wetlands, and wildlife but also with making the public aware of these resources. State and federally supported agricultural extension services have as a major responsibility the preparation and dissemination of agricultural information for interested users. Profit-earning companies must create and improve their public image and also attract customers and applicants for employment. Airlines, railroads, distributors of goods and services, all have to keep in the public view. Pamphlets, posted notices, and radio and television announcements are commonly used to meet all these needs.

Myriad applications such as these—company memos and reports, government publications, research reports, public relations releases—create a great flood of paperwork. Some of it is only of passing interest; some of it makes history. All of it is prepared by report writers: some are full-time report writers, but most are professionals who are reporting on their own work.

THE ATTRIBUTES OF GOOD REPORT WRITERS

To write clear and effective reports, you have to build upon the natural talents you have in communicating ideas to others. But how can you build successfully? What skills, characteristics, and attitudes are of most value to the report writer? From experience, we can summarize some of the major attributes that will stand you in good stead as a report writer:

- You have to be reasonably methodical and painstaking. Plan your work for the day and for the rest of the week. Look up from time to time to take stock of what you and the others are doing, so that you do not squander your time and energy on minor tasks that should be put off or dispensed with altogether. File your correspondence. Keep at your desk the supplies you need to do your work. Keep a clear head about ways and means for accomplishing your purpose.
- Be objective. Try not to get emotionally attached to anything you have written; be ready to chuck any or all of it into the wastebasket. While reading your own prose or that of your colleagues, do not ask whether you or they are to be pleased but whether the intended audience will be pleased, informed, and satisfied.
- As a professional, keep in mind that most of what you do will eventually have to be presented in writing. Do your work so that it will be honestly and effectively reportable. Keep a notebook or a deck of note cards. Record what you do and learn.
- Remind yourself every morning and afternoon that *clarity* is the most important single objective of writing. Until the sense of a piece of writing is made indisputably clear, until the intended reader can understand it, nothing else can profitably be done with it.
- As someone who must write, understand that writing is something that can be learned, even as chemistry, physics, and mathematics can be. The rules and formulas of writing are not as exact, perhaps, as those of science, but they can never be thrown overboard if you are to bring your substance home to your reader.

THE QUALITIES OF GOOD REPORTS

Good report writers turn out good reports. Because the qualities of good reports vary from report to report, depending upon audience and objective, we cannot offer you a list that applies equally to all reports. However, some qualities are apparent more often than not in good technical reports.

The good report . . .

- Arrives by the date it is due.
- Arrives in good physical condition—properly labeled and packaged, and without postage due.
- Makes a good impression when it is picked up, handled, and flipped through.
- Has the necessary identifiers on the cover and title page, so the librarian can log it in.
- Has the necessary preliminary or front matter to characterize the report and disclose its purpose and scope.
- Has a body that provides essential information and that is written clearly without jargon or padding.
- When they are needed, has a summary or set of conclusions to reveal the results obtained.
- Has been so designed that it can be read selectively: for instance, by some users, only the abstract; by other users, only the introduction and conclusions; by still other users, the entire report.
- Has a rational and readily discernible plan, such as may be revealed by the table of contents.
- Reads coherently and cumulatively from beginning to end.
- Answers readers' questions as these questions arise in their minds.
- Conveys an overall impression of authority, thoroughness, soundness, and honest work.

Beyond all these basic characteristics, the good report is free from typographical errors, grammatical slips, and misspelled words. Little flaws distract attention from the writer's main points.

A DAY IN THE LIFE OF TWO REPORT WRITERS

By way of summary, let us recreate two representative writers whom we shall identify as Marie Enderson and Ted Freedman.

Marie Enderson: Junior Engineer and Occasional Report Writer

Marie Enderson works as a junior engineer in the design and development department of a small electronics firm that employs some 700 persons. She still has 18 credits to earn before obtaining her B.S. degree in Electrical Engineering from State University, and is working off this requirement at the rate of 3 credit hours per term.

Enderson has been with the company about nine months. Since her early teens she has been recognized in her neighborhood as a whiz at building and repairing electronic gadgets. Therefore, she has practical experience not possessed by employees several years her senior. Her specialty is that of designing and "bread-boarding" novel electronic circuits. Her current project is to develop a solid-state frequency multiplier.

Marie is especially valued for her originality and drive. She is also great with schematics, pliers, and soldering iron.

Her first project with the company concerned bow-tie antennas. On the Monday following completion of the engineering work, Marie was reminded of the requirement to submit a full report on the six-month project. She had a ghastly time the next two weeks. She found, as do many novices at writing, that she knew *what* she wanted to say but not *where* or *how* to say it. But the 25-page report did somehow get written and, after a thorough overhaul by Ted Freedman (whom we'll meet next), went to publication.

Her first experience with on-the-job reporting taught Marie four unforgettable things: (1) even a junior engineer is not simply a fix-it person whose only product is a gadget that works; (2) things that go on in your head and hands are lost unless they are recorded; (3) reporting what you have thought and done is a recurring necessity; and (4) reporting, strange and difficult as it may seem at first, is something that can be learned by anyone of reasonable intelligence and perseverance.

Marie Enderson is now earning $9.60 per hour as a junior engineer, and she will be eligible for promotion the first of July or January after obtaining her B.S. degree in Electrical Engineering. At the moment, her biggest worry is whether her report on the frequency multiplier will get by Ted Freedman.

Ted Freedman: Technical Writer and Company Editor

Ted Freedman was hired as a technical writer/editor. Freedman is 27. He holds a B.S. degree in Technical Communication from a university in the Midwest.

His office is a sparsely furnished cubicle down the hall from the publications and mailing departments. His office possessions include an old typewriter, a brand-new word processing station, a three-foot shelf of dictionaries and reference manuals, including *The Chicago Manual of Style*, and an extra-large wastepaper basket.

At 8:45 this morning Ted is scheduled for a project review session in the company auditorium. He arrives at the auditorium with five minutes to spare. For the next hour he studies flip charts, slide projections on the huge screen, chalk-and-blackboard plans for company reorganization (minor), and staffing proposals for three new projects totaling $378,400. From the platform, Chief Scientist Muldoon requests that Ted develop research timetables and preview reporting needs.

At 10:20 Ted meets with a commercial printer to examine the artwork and layout for a plush report his company is preparing for a state commission. The work looks good, mighty good, but a little lacking in typographical variety, he suggests.

At 12:55, back from lunch in the company cafeteria, Ted glances over the memos that collected on his desk during the morning—nothing urgent. Then he opens the manila folder lying in his mail rack. It is the manuscript on Marie Enderson's frequency multiplier project.

At 1:30 he gets Marie on the telephone and arranges for a meeting at 3:00 so that they can run through the manuscript together. But he feels that he should do some work on it before the meeting to demonstrate the variety of things that need to be done.

He comes upon a disjointed, mixed-up paragraph. Turning to the typewriter he makes a complete rewrite of it. He detects a "hole" or discontinuity in the second chapter. Another telephone call to Marie confirms that a page somehow got lost and will have to be reinserted.

However, it is the two-page set of conclusions that really stirs Ted into action. The conclusions are a hodgepodge of preliminary intentions, procedures employed during the research, reservations about the quality of the work, and some unsupported speculation. Therefore, he pulls this section and attaches a covering critique for Marie's guidance. Thus the afternoon wears on.

At 3:00 Marie appears and the two confer, make changes, and plan later alterations in the manuscript. As always, they work amicably together. To relieve tension they intersperse their writing and editing with an occasional trip to the coffee urn, a chat with a department head, and a trip to the library to consult a specialized reference work.

For this kind of work Ted is paid $24,500 a year, with two weeks of paid vacation and ten days of sick leave. He is good at his work and considered to have a great future.

Enderson and Freedman are roughly representative of many thousands of technical report writers/editors, most of them, like Enderson, part-time as the need arises. Of course, to gain a more rounded understanding of their duties and behavior we would have to pay them many additional visits. It is evident, though, that much of the time they are not writing at all, in the popular sense. Some of the time they are simply listening hard to what people are saying to one another—trying to clarify, simplify, and translate into other terms. A generous portion of their time is spent on tasks that have little direct connection with writing but eventually provide grist for the writing mill. The skills that writers such as Enderson and Freedman need to accomplish their work are the subject matter of this book.

EXERCISES

1. As your instructor directs, bring to class one or more magazine or newspaper clippings that you believe to be technical. In what respects is the writing technical? Subject matter? Purpose? Tone? Vocabulary? Other?

2. Rewrite a brief paragraph of technical prose (perhaps a clipping submitted in Exercise 1) to substantially lower its technical level. Explain what you have done and why.

3. With the help of *Ulrich's International Periodicals Directory* (see page 64), find several periodicals in your professional field. Examine one or more copies of them. In what ways and to what extent does your examination of such periodicals confirm or change your first impressions of technical writing?

4. On a two-column page, list your present *assets* and *limitations* as a technical writer.

5. Turn to the job advertisements section of a large metropolitan newspaper such as the Sunday *New York Times*. What advertisements for technical writers do you find? What qualifications are demanded of them?

6. If you have the opportunity, talk with several professional people of your acquaintance. Ask them how much writing they do and what kinds. Ask them how much importance they attach to good writing in their profession.

Chapter 2

Getting Started

WE TEACH WRITING to both college students and to professionals already at work. A favorite beginning for the first meeting is to ask the class, "What bugs you about writing?"[1] The students' most frequent response is "Finding a topic." The professionals almost never mention topic selection. Their most frequent responses are "Organizing the report" and "Being concise." The reason for the different responses is probably rather obvious. Professionals write when they have a situation that calls for a piece of writing. Normally, the topic comes with the situation: *We have to explain to the clients our progress (or lack of progress) in installing the air conditioning system in their new plant.* Or, in another typical situation, *We have to provide instructions for the bank tellers who will use the computer consoles we have installed at their stations.* And so on. Not only the topic but the audience and purpose as well come with the situation. We hasten to add that the situation is not always this clear-cut, at least not to the writer. But, generally, professionals do find that their topics seek them out and not the other way around.

In professional writing, you will find that topic, purpose, audience, and—a fourth factor—writer's role are all closely interrelated. Also, the topic will probably be in a subject matter area where the writer has some competence. The purpose may be persuasive, instructional, analytical, or some combination thereof. Typically, the readers will be people who need to learn some procedure or who need information and analysis in order to reach or understand some decision or conclusion. The writer's role will be somewhere on a continuum from salesperson to scientist. That is, the writer may actively be selling a point of view, objectively presenting facts and their implications, or somewhere between these two positions. How well the writer handles all these variables will determine the success or failure of the communication. In getting started, then, the writer must begin to deal with these variables, some or all of which in a classroom setting may be given by the teacher.

SELECTING A TOPIC

Because professionals normally write in areas where they already have some competence, you, the student, should try to do the same. You may be lucky enough to have topics from your classes and experience to write about. If not, you can follow several paths that should lead to good topic selection. You can talk over topic possibilities with students and teachers in courses in your major field. You can go to the library and examine periodicals in your field. You can read scientific and technical magazines for items that may relate to your expert knowledge and that may trigger an idea for you. For example, here are some recent article titles from *Science 81* and *Science 82:*

"The Gene Machine"
"Rationing a River: The Tortuous Course of the Colorado"
"Pickleweed, Palmer's Grass, and Saltwort"
"Metals That Remember"
"Seven Wonders of the Solar System"
"Synfuels Stall Out"
"Micromemories"
"Roots of Madness"

In these eight articles are represented, at the least, eight major fields: biology, hydrology, horticulture, metallurgy, astronomy, chemistry, computer science, and psychology.

Even daily newspapers can help. They regularly carry articles on current problems either caused or subject to cure by technology. Such problems and their solutions lend themselves well to research and feasibility studies.

By the way, in looking for a topic don't be blind to the impact that information from other fields may have on your field. For instance, suppose you find an article in the daily paper that describes how, because of the declining birthrate and rising life expectancy, the average age in the United States is increasing. Think about that for a moment. What impact will this aging of America have on such fields as education, clothing, housing, agriculture, restaurant management, military defense, Social Security and other pension funds, recreation, energy, and medicine? The creative mind is always looking for the cross-relationships that both create and solve problems.

Once you have a topic, check it for usability. You would be unwise to commit yourself to researching a topic without observing these five safeguards:

1. You must have competence in the necessary technical background to handle the subject matter. Competence doesn't mean that you already possess all the information needed for the writing project. It does mean that you know where to find needed information and how to interpret it after you find it.
2. You must have access to the necessary means and facilities. These may include physical equipment, laboratory setups, time and money for travel, sites for inspection, and so on. Again, from the reporting view, you must restrict yourself to reasonable requirements.
3. You must be able to complete the project within the hours and calendar time available to you. If you have only 10 or 15 weeks (the usual length of a college term or semester), do not obligate yourself to grow a crop of corn, test some material under conditions of the four seasons, or record weather data for the coming 10 years. To do so would be to guarantee failure before you get started.
4. Insofar as your particular circumstances allow, choose a general subject and a specific problem of interest to you. Subjects that at the beginning are of only mild interest may become absorbing, but subjects that are genuinely distasteful tend to become intolerably so with the passage of time.
5. Because investigations ordinarily depend to some degree on information already provided by other people, go to your local libraries to run a check on the sources of information available in the form of periodical literature, standard reference works, previous reports, and pamphlets. If you can locate three or four potentially good sources within an hour, you can reasonably conclude that enough additional sources to make the project feasible will come to light if you spend several more hours of effort.

DETERMINING AUDIENCE, PURPOSE, AND WRITER'S ROLE

Once you have a topic that passes the usability criteria, you are ready to think about your audience, purpose, and role as a writer.

Audience

Because the audience for the writing project will often determine your purpose for you, perhaps it is best to think about that aspect of the problem first. As we explain in Chapter 3, "Analyzing Your Audience," at least four broad categories exist. The audience may be made up of lay people who are reading out of some combination of curiosity, self-interest, and desire for entertainment. Executives make up another possible audience, one of people who will use the information in decision making. Technicians read for "how to" information on everything from how to choose a variety of wheat for planting on the high plains to how to calculate what next year's inflation rate may be. Experts read to remain experts; alone of all the audiences, they want a good deal of information in a narrow area.

It is not too early at this beginning stage to think about audience knowledge and attitudes. Such knowledge may help to guide your research. Your readers, for instance, may possess a good deal of information about your subject or none at all. They may be well educated or not. They may be friendly, hostile, or apathetic to your work. Other audience factors to consider may be age, socioeconomic status, occupation, indeed anything that may have a bearing on the readers' acceptance and understanding of the presentation.

Purpose

As to your purpose, what are you hoping to accomplish? In a how-to, technician situation, your purpose may mesh perfectly with the readers' purposes. That is, the readers may need to learn a procedure, and your purpose is to teach it to them. But even in writing instructions and procedures, the situation can sometimes be more complicated than simply learning and teaching. It may be that the technicians already have a procedure that they like, and they may be skeptical about the procedure you wish to teach them. Now your purpose is persuasive as well as instructional. You have to convince them that the new procedure is better than the old. Therefore, you will need to gather and present information to justify the change.

You may have many purposes, from convincing a government agency to fund your research to intriguing a youngster with the wonders of geology. The possibilities are limitless, but be sure that without a clearly stated, well-thought-out purpose, you will not be able to limit or develop your topic properly. Your information gathering and ultimately your paper will wander aimlessly if you do not set your purpose early in the process.

Writer's Role

Finally, what is your role as a writer? If you are a student, play some role other than student. In school a student's primary purpose in writing is either to display knowledge for a teacher or to display writing ability. Both are legitimate reasons for writing, but as we hope our discussion to this point indicates, neither has much relationship to the purposes that will guide your writing as a working professional. Therefore, project yourself into a role that resembles the roles you will play on the job. Many possibilities exist: salesperson, researcher, engineer, teacher, junior executive, manual writer, or legislative aide, to name but a few. To see the importance of knowing the writer's role, imagine the difference in approach between that of a Chevrolet sales representative trying to sell a fleet of Chevrolets to a company and that of a young executive of the same company reporting to his or her superior that the results of a feasibility study demonstrate Chevrolets to be the best purchase. In the first instance, only the advantages of the Chevrolets are likely to be emphasized. In the second, a more balanced appraisal will be expected.

To illustrate all this, let's suppose for a moment that the topic of the aging of America interests you. Further imagine that your major is restaurant management. Because you are interested in management, you decide to write for an executive reader. You project yourself into the role of a young executive working for a small restaurant chain. The new owner of the chain, Jane Lewis, is also new to the restaurant field. She bought the chain after retiring from a successful career as an insurance sales representative. She is 45 years old with a college degree in history. You know she is intelligent, well educated, and understanding about business but not too well informed about the restaurant business. You cannot assume that she will possess a thorough grounding in restaurant management techniques. Therefore, you will need to justify fully any recommendations about restaurant management that you may make to her.

The restaurant chain caters to a young preteen and teenage clientele and serves mainly hot dogs, hamburgers, gigantic ice-cream concoctions, and the like. The food is served with a lot of hoopla, and the decor is bright and colorful. You know from a newspaper article you read that while in 1960 the 5- to 17-year-old group greatly outnumbered the 22- to 34-year-old group, by 1980 the two groups were almost equal. Furthermore, the older group is projected to soon outnumber the younger group. You wonder if your new boss might not be interested in this change. It could very well have considerable impact on the chain's menu, manner of service, and environment. You decide that, indeed, your boss would be interested and decide to do some further research and write a report for her. In the report, if your research uncovers enough information to warrant them, you may even make some tentative suggestions for change.

You now have a topic, a role as a writer, and an audience with a reason to be interested in your presentation. Your purpose is taking shape. First of all, you want to inform your boss of the changes taking place in America's age groups. You'll need to gather accurate and complete statistics to supplement the information in the newspaper story. You'll need to interpret the statistics so that you'll be able to predict with some hope of certainty whether the aging of America is likely to be a lasting phenomenon. This last means you'll need to have some understanding of why the change is coming about. You'll also have to answer management's favorite question when presented with facts—so what? What, that is, are the implications of your facts?

Here, in answering the so-what question, is where your expert knowledge in food management comes into play. Are there significant differences in food tastes between the two groups in question? Do 30-year-olds eat hamburgers and ice cream in the same quantity that 14-year-olds do? What about the environment? Will the 30-year-olds prefer softer lighting and more subdued colors? How about noise level? The 14-year-olds seem to enjoy the ear-shattering sound of the siren that the waiters turn on when a super-duper ice-cream confection is being served. Will the 30-year-olds enjoy the sound as well? And so on. Some of the questions you can answer right now; some you can't. But you know where to look for the answers. You have a good beginning on your writing project.

We hope our example illustrates the process for you. It is a thinking process. Like all thinking processes, it is likely not to happen in a neat, 1–2–3, linear way when you are doing it. You have many variables in your problem, and changing one variable may change other variables and cause you to rethink your solution. But by following the process of considering topic, audience, purpose, and writer's role as four closely related parts of one whole, you are much more likely to make a good beginning than if you do not. To help you with the process, as part of the exercise that follows, we provide you with a form that should help you get started.

EXERCISES

After discussion with your teacher as to what the writing project will be, get started on a project that will be due at some future date. Follow the steps we have outlined in this chapter to determine topic, audience, purpose, and role of writer. Using the form that follows,[2] record the results of your efforts. (You will find that nothing clarifies thought or reveals weakness in your ideas quite so well or so quickly as the exercise of writing them out. If you can't express something in words, it's highly likely that you haven't yet thought it through thoroughly.)

Topic
Reader(s)
   Technical level (education, existing knowledge of subject, experience, etc.)
   Relationship to writer (boss, client, subordinate, peer, lay reader, etc.)
   Attitude toward subject (neutral, friendly, skeptical, curious, uninterested, hostile, etc.)
   Other factors

Reader's Purpose
 Why will the reader(s) read the report
 What should the reader(s) know after reading the report
 What should the reader(s) be able to do after reading the report
Writer's Purpose
Writer's Role

Chapter **3**

Analyzing Your Audience

IN CHAPTER 2, "Getting Started," we intro-
duce you to audience analysis as a key part of the writing process. In
this chapter, we develop still further how you write not only to satisfy your own purposes but also the purposes of your readers. *Why*
do readers want to read what you write—what do they hope to learn
from you, and what do they intend to do with that knowledge?

You must understand not only the purpose but also the background of your audience. You must know who your readers are,
what they already know, and what they don't know. You must know
what your readers will understand without explanation and without
definitions. You must know what information you must elaborate,
perhaps with simple analogies. You must know when you can use a
specialized word and when you cannot. You must know when to
define a specialized word that you can't avoid using. All this requires
a lot of thought, but the successful writer always remembers that
readers bring their experience *and their experience only* to their
reading.

To be a good writer, then, you must know your audience—its purpose and knowledge. Perhaps in no other kind of writing is this business of matching a particular piece of writing to a particular audience as important as it is in technical writing.

Engineers of the nineteenth century were aware of the problems
facing audiences of different levels. In 1887, the American engineer
Arthur M. Wellington had different parts of his book *The Economic
Theory in the Location of Railways* set in three different type sizes:

large type for the lay reader, medium-sized type for the reader who could understand some technical data, and small type for the reader who needed the most detailed scientific data. Readers, knowing their own limitations and interests, had clear signals as to what to read and what not to read.[1]

In this century, Philip W. Swain, an engineering professor at Yale and later chief editor of *Power* magazine, expressed the distinctions that exist between audiences this way:

> Note the many languages within our language. The college freshman learns that "the moment of force about any specified axis is the product of the force and perpendicular distance from the axis to the line of action of the force." Viewing the same physical principle, the engineer says: "To lift a heavy weight with a lever, a man should apply his strength to the end of a long lever arm and work the weight on a short lever arm." Out on the factory floor the foreman shouts, "Shove that brick up snug under the crowbar and get a good purchase; the crate is heavy." The salesman says: "Why let your men kill themselves heaving those boxes all day long? The job's easy with this new long-handled pinch bar. With today's high wages you'll save the cost the first afternoon."[2]

Swain here defines essentially four audiences: the scientist, the engineer, the technician or the operator, and the executive. Each audience has different interests and understands the same problem through a different language and from a different perspective.

In this chapter we break down audiences a bit differently. We tell you how to deal with lay people, executives, experts, and technicians. We also tell you how to put together a combined report for an audience in which these groups are mixed. Before we discuss these four audiences, let us caution you that no audience is uniform, falling readily into a neat category. We speak of a lay audience, or an executive audience, but such audiences are by no means totally homogeneous units. An audience might be compared to an aggregate of rocks of all shapes and sizes as opposed to a mass of smooth marble.

This chapter, therefore, is not audience analysis made simple, foolproof, and mechanical. Rather, it contains a series of generalizations that should help you understand the process of audience analysis. But only you can analyze your own audience.

A LAY AUDIENCE

Who are lay people? They are fourth graders learning how the moon causes solar eclipses. They are the bank clerk reading a Sunday newspaper story about genetic engineering and the biologist with a Ph.D. reading an article in *Scientific American* entitled "The Nature of Metals." In short, we are all lay people once we are outside our own particular fields of specialization.

We can make only a few generalizations about lay people. They read for interest. They read to understand the world in which they live. They are not very expert in the field or they would not be reading an article written for a lay audience. In these days of environmental concern and consumerism, they may be reading as a prelude to action. Certainly their main reason for reading is practical. They are much more concerned with what things do than how they work. Their interest is personal. What impact will this new development have on them? They are more interested in the fact that widespread computer networks may invade their privacy than the fact that computers work on a binary number system. They are more interested in the efficiency and cost of automobile antipollution devices than the theory of such devices.

Many lay people are more attuned to fiction and television than to scientific exposition. They like drama. For this reason, narrative is often an effective device when writing for them. Anecdotes and incidents can be a useful way to illustrate what something is and what it does. Nonscientific lay people, in particular, need reminding that science and technology are human activities.

Beyond these generalizations, lay people present a bewildering complexity of interests, skills, educational levels, and prejudices. How then can we define exactly how to write for them? The truth is that we cannot—completely. But we can make some broad statements about their needs, interests, likes, and dislikes that—to paraphrase Lincoln—apply to all of the lay people some of the time, some of the lay people all of the time, but *not* all of the lay people all of the time.

To simplify matters a little, let's get a picture in our minds of typical lay people. For our present purposes, we are excluding specialists with advanced degrees, like the biologist mentioned earlier, from the "typical lay people" category—though, as noted, specialists may be defined as lay people when outside their particular fields. These typical lay people are fairly bright and somewhat interested in science and technology. They have at least a high school education. (In 1980, 68.6 percent of the United States population had at least four years of high school; 17 percent had at least four years of college.[3]) They read well and have a smattering of mathematics and science but are a little vague about both subjects. How do we treat them? What approaches are best when we write for them?

Human Interest

To begin with, you often have to motivate lay people to read what you have written. Frequently this means introducing human interest into your writing. Most of us, at any educational level, have an interest in other human beings and in human personalities. Most writers for lay audiences recognize this interest and use it to gain acceptability for their subject matter. For example, an article in *Time* about unemployment will give us statistical information about the numbers of people out of work. But the writer of the article knows that many of us do not relate very well to bare, abstract statistics, even when they affect us deeply, as do the facts about unemployment. Therefore, the *Time* writer will also usually introduce people into the article. Perhaps Bill and Mary Gould, a "typical" couple living in Detroit, Michigan, will be cited. We'll learn what effect Bill's being unemployed has had on their lives. We are interested in learning about what happens to real people, and through such knowledge we can better understand the problems unemployment creates.

In the following example, notice how the introduction to an article on hurricane prediction uses human drama to capture the reader's interest:

> Today, some 45 million Americans live in coastal areas vulnerable to hurricanes. Many of them have seen merely the fringes of a hurricane and believe it capable of nothing more dangerous than tearing off shingles or flattening large trees. Only a fraction have endured the unbelievable fury at the storm's center.
>
> How bad, in fact, can a hurricane be? Bob Sheets, a forecaster with the National Hurricane Center, presents visitors with two photographs of an apartment house in Pass Christian, Mississippi. The first shows a solid, three-story brick structure, separated from the Gulf of Mexico by an eight-foot seawall, a four-lane highway, a row of substantial oak trees, a generous front lawn, and a swimming pool. When Hurricane Camile was predicted in August 1969, 25 people felt confident enough to stage a "hurricane party" there. The second photo shows the same site, after Camille passed through. Only the swimming pool remained. Of the partygoers, one managed to cling to the upper branches of a tree. Another was swept out to sea and cast back 12 hours later and four miles down the beach, semi-conscious, but somehow still alive. The other 23 died.[4]

Even fairly complex scientific material for a high-level audience can benefit by the introduction of human interest. In a *Saturday Review* article on new fuel sources, the author tells about some complex technology but when possible introduces the human beings behind the technology, as in this excerpt:

> Among the newer systems are "hydrocarbon miscible flooding," "carbon dioxide flooding," and "miscellar polymer flooding." But while a number of systems have succeeded in varying degrees of getting at the oil, none has been able to do so with a sufficient net-energy gain or at an economic price.

Enter two Texans, Frank DeFalco and Charles McCoy, and a Hungarian-born scientist, George Merkl. In 1974 DeFalco and McCoy created the Molecular Energy Research Company (MERCO) for the purpose of finding a technologically efficient and economically advantageous method of capture.

Merkl, 48, has credentials in theoretical and solid-state physics, nuclear chemistry, petroleum geology, and electronics. The contribution he is making to the solution of America's energy problem is reminiscent of the role of other foreign-born scientists—Einstein, Szilard, Fermi, and Wigner among them—in helping the United States to use science as a decisive factor in the war against Nazi Germany. Merkl came to the United States in 1957 after advanced studies in Vienna. He worked in several American research laboratories and immersed himself in research on catalysts, agents that speed up chemical reactions.

Merkl teamed up with DeFalco and McCoy in 1976. Like many others, he was obsessed with the possibility that a way could be devised to get at the estimated 350 billion barrels of petroleum locked into America's natural underground vaults. Almost intuitively, he turned to inorganic polymers—large-moleculed hard substances, like silicon or graphite, with a very high melting point. The process involved a polarized hydrogen bond. The bonding releases atomic hydrogen, which cracks the light end of the oil, creating natural gas. This gas forces the trapped oil out of the rock and causes it to form a pool or reservoir, making it accessible to conventional collection methods.[5]

Because the article is written for a high-level lay audience, the writer can use a wide-ranging vocabulary. But really technical terms such as *catalysts* and *inorganic polymers* are defined. Most important for our purposes here, the writer whets our interest by telling us something about George Merkl. We learn about his education and experience and that he is "obsessed with the possibility" of getting at America's trapped petroleum. We are more interested in the technology because we see it as a human activity. (See also pages 125–126.)

While providing human interest and perhaps even human drama, be careful not to exaggerate scientific achievements. Newswriters, when writing stories about scientific achievements, sometimes forget this need for caution—to the dismay of the scientists involved. A newspaper story concerning research on the skin ailment psoriasis carried this headline: "Psoriasis Cure Breakthrough Seen." The lead of the story announced, "Scientists Wednesday announced a breakthrough in treating psoriasis, the skin disease which causes misery for about 6 million Americans." The scientist involved criticized the story, saying that nowhere in the report presented by the scientists was the word *breakthrough* used. He concluded by saying, "The last sentence of our writeup said cure of psoriasis is probably 50 years away. Yet the title of this article you sent says 'Psoriasis Cure Breakthrough Seen.' All I can say is [censored]!"[6]

Background

As a writer on a technical or scientific subject, you are likely to be an expert. Your biggest difficulty, perhaps, will be to remember how little you knew about your subject matter before you studied and experienced it. Do try to remember and to provide your lay audience with the background information they'll need to comprehend and absorb your material. Provide such information in terms your reader can understand. In the following example, the writer introduces his subject, dietary control of salt, by discussing the role of salt (sodium chloride) in the human body. Such information would be known to the nutritionist but not to the lay reader.

> Sodium chloride, most frequently encountered in the food supply as common table salt, is an essential part of the human diet. As it dissolves in water, it dissociates into two ions—one of sodium and the other of chloride. In all mammals, including humans, the sodium ion is required to maintain the pressure and volume of blood. It is also essential in controlling the passage of water into and out of the body's cells, and the relative volumes of fluids inside and outside those cells. In addition, sodium is needed for the transmission of nerve impulses and for the metabolism of carbohydrates and proteins.
>
> The chloride ion, too, is essential, and is involved in maintaining the acid-base balance in the blood, and in tissue osmolarity (the passage of water across cell walls to maintain proper concentrations of various chemical entities). It is necessary for activating certain essential enzymes, and for the formation of hydrochloric acid in the stomach, needed in the digestive process.
>
> Thus, both sodium and chloride are normal and necessary constituents of body tissues and fluids, and must be provided for in the diet.[7]

In providing background, the writer begins with the familiar, "common table salt," and works his way to the unfamiliar. Analogy is a special way of comparing the familiar to the unfamiliar and as such is often very useful in lay writing. Notice its use in this excerpt from an article on laser technology:

> The main laser bay is a giant clean room two stories high and nearly the size of a football field. To take the analogy a step further, a single low-power laser beam begins at one end zone, is amplified, and is divided by mirrors into six separate beams. Each beam is then amplified further and directed through one of six long tunnel-like chains of optical components. Just beyond the 50-yard line, each beam is divided again, this time into four beams, and all 24 are directed toward a four-foot stainless steel sphere right about where the opposite goal line would be.[8]

Analogy is a powerful device to help the lay reader. In the everyday world around us there are countless things—such as light bulbs, radios, garden hoses, faucets, windows, mirrors, trees, tennis rackets, baseballs, clay, loam, granite, ocean waves—known and somewhat understood by everyone that the writer can use to explain about every law of science. It's a question of using one's imagination and knowing how to talk in lay terms without being condescending. (See also pages 113–114.)

In addition to familiarizing the reader with the subject, background information is often used to stress the importance of the subject at hand. A *Harvard Medical School Health Letter* on alcoholism begins this way:

> Consider that half of all deaths in automobile accidents, half of all homocides, and a fourth of all suicides are related to alcohol abuse; that persons with a "drinking problem" are seven times more likely to be separated or divorced than those in the general population; that the total cost of alcohol abuse in this country may exceed 44 *billion* dollars; that an alcoholic's life span is shortened (on average) by 10–12 years; and that at least ten million persons in this country abuse alcohol. No wonder some have labeled alcoholism the most devastating socio-medical problem faced by human society short of war and malnutrition.[9]

Give your readers, then, a grounding in your subject. When possible use familiar things as points of comparison. If your background also wakens your readers' concerns and interest, so much the better.

Definitions

Although we learn our expert language long after we have acquired our common, everyday language, expert language becomes such a part of our life that we often forget that others don't share it with us. We forget that we are, in effect, bilingual, possessing both a common language that we share with others and an expert language that we share with a much smaller group. In reaching out to lay people, we must remember that they need specialized terms and words defined. You are the host when you write. You have invited your readers to come to you. You owe them every courtesy, and defining difficult terms is a courtesy. If you force your readers to the dictionary every fourth line, their interest will soon flag.

Depending upon the needs of both writer and audience, terms can be defined either briefly, usually by the substitution of a more familiar term, or at length. The use of brief definitions is seen in this segment of *The Harvard Medical School Health Letter* on alcoholism:

> Alcoholics have a much higher incidence of *peptic ulcers* and *pancreatitis* (inflammation of the pancreas) than non-alcoholics. In addition, many

suffer from repeated episodes of nausea, vomiting, and abdominal distress—most often due to superficial *gastritis* (inflammation of the lining of the stomach).[10]

Here, the lay definition substitutes simpler, more familiar language for the physician's technical terms. In the same article is this more extended definition:

> If a group of experts attempts a definition of "alcoholism," as many definitions as experts usually emerge, which, of course, points out the complexity of this particular "-ism." Some argue the relative merits of biological (it is a "disease") versus social-psychological (it is a "behavior disorder") theories, but most settle for descriptive definitions—such as the one from the Rutgers University Center of Alcohol Studies: "An alcoholic is one who is unable consistently to choose whether he shall drink or not, and who, if he drinks, is unable consistently to choose whether he shall stop or not." Ultimately, all will agree to descriptive definitions which point out that for the person with a serious drinking problem, tremendous disruption occurs in terms of health, interpersonal relationships, and the basic activities of life—eating, sleeping, and working.[11]

Notice how this definition is used to stress one of the major points of the article—that alcoholism, however caused, disrupts the lives of alcoholics and the lives of those around them. If the writer had left the readers to define *alcoholism* for themselves, the emphasis wanted by the writer would have been lost. Likewise, a dictionary definition may not emphasize the points important to the writer. Compare this definition of *alcoholism* from *The American Heritage Dictionary of the English Language* with the one from *The Harvard Medical School Health Letter:*

> 1. A chronic pathological condition, chiefly of the nervous and gastroenteric systems, caused by habitual excessive alcoholic consumption.
> 2. Temporary mental disturbance, muscular incoordination, and paresis caused by excessive alcoholic consumption.

The dictionary definition deals primarily with the physiological aspects of alcoholism and overlooks the social aspects that concern the writer.

Be careful not to distort the true meaning of terms if you substitute more common terms for technical language. One researcher, for example, felt his work was distorted by this lead in a newspaper story:

> A research group reported Friday that marijuana causes chimpanzees to overestimate the passage of time, and a single dose can keep them befuddled for up to three days.

The researcher commented:

> The term "befuddle" was not employed in our scientific report, and the statement in the news article "and a single dose can keep them befuddled for up to three days" is erroneous and misleading. Three days were required to recover normal baseline performance following administration of high doses.[12]

Scientists choose words very precisely and for good reason. Although their findings must be interpreted for lay readers, to distort or to sensationalize their work is a disservice both to them and to the reader. Define, then, for clarity and understanding and to aid your own exposition but do it with care. Refer also to the section on definition on pages 103–107.

Simplicity

There are several ways to keep an article simple for lay people. Two of them we have already discussed: give needed background and define those specialized terms that you must use. Most often, avoid specialized words for which you can find simple substitutes. Experts can read certain meanings into the word *homeostatis* and you should use it for them, but for a lay audience *stable state* or *equilibrium* will serve as well. But another caution here: most people like to enrich their vocabularies. So don't avoid technical terms altogether. Just don't put them together in incomprehensible strings with a reader-be-damned attitude.

Some scientific specialities are loaded with mathematics. Others, such as biochemistry, are full of formulas, complicated charts, and diagrams incomprehensible to lay people. Mathematics, formulas, and diagrams are useful shorthand expressions for experts. Through them, experts find a precision impossible to obtain in any other way. But what experts sometimes forget are the years spent learning how to handle such precise tools. The average person, lacking those years of training, rarely can handle them. When you write for lay people, you must force yourself to express your ideas in plain language. As we have seen in the background examples, it can be done.

Another way to achieve simplicity is in the way you handle language. Modern research confirms a fact that is perhaps intuitively obvious: sentences that are too long or too complex cause difficulties for readers.[13] A reader familiar with a subject can comprehend sentences of greater difficulty than can a reader unfamiliar with the subject. In effect, the unfamiliar reader is fighting two battles at one time: one with the complex sentence structure, another with the subject matter. Therefore, when dealing with lay readers be ever conscious of the need to keep your sentence structures, as well as your vocabulary, at a simpler level than might be appropriate for a more expert audience. In Chapter 9, "Achieving a Clear Style," we discuss how to achieve a style that aids rather than impedes understanding.

Graphics

You can use graphics in many ways in a report designed for a lay audience. You can use bar charts in place of equations to explain mathematical concepts or in place of formulas for chemical concepts. Or you can combine tables and pictographs that establish facts, formulas, or definitions quickly, clearly, and in a way that interests readers.

Figure 3-1 was used by the Jet Propulsion Laboratory to illustrate the paths of two unmanned space missions to Jupiter and then around the poles of the sun. It graphically illustrates what is difficult to express in words. The illustration is both clear and a bit dramatic—note the presentation of flares erupting from the surface of the sun.

Mary Fran Buehler, an editor, and Andrea Stein, a writer, of the Jet Propulsion Laboratory at the California Institute of Technology, believe that "people-related art" increases reader interest and understanding. They put their case this way:

> Science and technology can seem forbiddingly impersonal, especially to nonscientists. Hardware and mathematical equations and computer printouts can seem to have an existence of their own. The truth is, of course, that there would be no science and no technology without people. Somewhere in the dim past, people struggled with theory and concepts, or tinkered with Rube Goldberg-like contraptions, that made possible the scientific advances and the gleaming mechanical marvels we enjoy today.
>
> One of our responsibilities in technical communication is to help people understand the importance of science and technology in their lives so that they can participate—through understanding—in today's scientific advances. This paper describes some efforts to promote this understanding by using people-related art (in this case, historical artwork) to provide human interest in scientific publications that are designed for lay audiences. The authors believe that this interest, in turn, allows the reader to identify with and understand the scientific message more easily, since the experience of humanness is universal to humans, but the knowledge of science is not.[14]

Buehler and Stein put their ideas to work in many Jet Propulsion Laboratory publications. Figure 3-2 is an example of a sixteenth century woodcut used in a document for astronomers, both amateur and professional, who will be watching for Halley's Comet in 1986.

We have a good deal more to say about graphics in Chapter 12, "Graphical Elements of Reports."

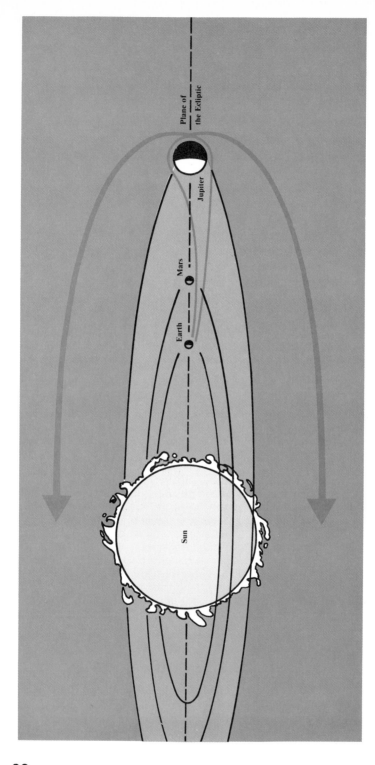

FIGURE 3-1 Paths of Two Unmanned Space Missions Around Jupiter and the Sun. [From U.S., National Aeronautics and Space Administration, *Solar Polar* (Washington, D.C., 1978).]

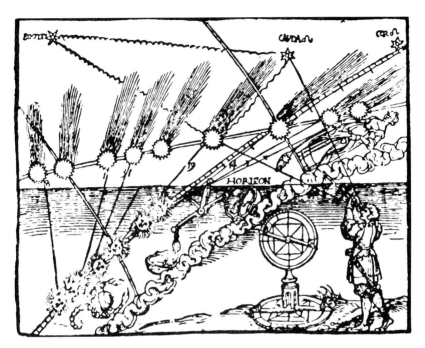

FIGURE 3-2   Woodcut showing P. Apian's Observations of Halley's Comet in August 1531 A.D. [From *The International Halley Watch* (Washington, D.C.: National Aeronautics and Space Administration, 1980), 5.]

A Challenging Audience

Lay people present a challenging audience to write for because their needs and capabilities are so difficult to pin down and define. We can say, in general, that in writing for them we should provide ample background (without mathematics or complicated formulas), definitions, illustrations, and simple charts. Remember, lay people are reading mainly for interest and their interest is mainly practical: How will this scientific development affect our lives? So your cardinal rule in writing for lay people should be to keep things uncomplicated, interesting, human, practical, and personal; and, if possible, to provide a touch of drama.

Examples of good writing for lay people are easy to find. Magazines such as *Time, Newsweek,* and *National Geographic* regularly print articles on scientific subjects for their readers. The new science magazines designed specifically for lay people by the American Association for the Advancement of Science and designated by their years, as in *Science 81* and *Science 82,* provide excellent examples of lay writing. Reading and analyzing such articles is the best school we know for learning how to write for a lay audience.

EXECUTIVES

Much of what we have said about lay people also applies to executives. You cannot assume that executives possess very much knowledge in the field you are writing about. While most executives have college degrees and many have technical experience, they represent many disciplines—not necessarily including the one you are writing about. Some executives may have training in management, accounting, a social science, or the humanities, but little or no technical background.

Even more than lay people, executives' chief concerns are with practical matters—what things do as opposed to how they work. Executives want to know how technological development will affect the development of their companies, and they probably can use more technical background than lay people. Both of these attributes are evident in this passage from the *Bell Laboratories Record*, a journal published, as it states, "for the management and technical staff of the Bell System":

> The SDP [a new computer] components were chosen to be not only fast but flexible, so the processor is programmable. The same computer, therefore, can be used for other tasks that require fast processing and great memory capacity. It is already being used in several Navy applications where custom-designed equipment would otherwise have had to be developed. The resultant cost savings, and the logistics advantage of having the same spare parts inventory for a variety of applications, are two of the attractive features of the SDP.
>
> Quite different applications are possible. For example, the SDP could also be used to process signals from radio telescopes. Because these signals have a low signal-to-noise ratio, analysis requires the processing of a great deal of data collected over a long period of time. Millions of data from several hours may have to be stored and repetitively processed.
>
> The SDP might also be used in telecommunication systems where large quantities of data must be processed quickly. It could be used to recognize speech and synthesize spoken answers, cancel echoes, or multiplex signals by time assignment speech interpolation. Its ability to process many signals simultaneously might be useful in a variety of applications in central offices.[15]

Notice that throughout the passage the emphasis is on *what* the computer does, its functions and potential capabilities, not on *how* it does what it does.

James W. Souther, who has made an extensive study of executives and their needs, suggests that writing intended for them should approximate the level found in *Scientific American*.[16] Use and define important technical terms, but avoid shop jargon entirely. Executives are busy people; don't force them to use a dictionary any more than you would lay people.

When writing for executives, write in plain language, avoiding undue sentence complexity. Do not use mathematics beyond their grasp. Use simple graphics of the types suitable for lay people: bar graphs, pie charts, and pictographs.

Although executives resemble lay readers in many ways, there is also a significant difference between them. Much of the time lay people are reading primarily for interest. What they read obviously influences their lives and their decisions, but they seldom have to act directly upon it. Executives, however, must often make decisions based upon what they read. People and profits figure largely in executive decisions. Executives must also consider the social, economic, and environmental effects of their decisions upon the community at large. Aesthetics, public health and safety, and conservation are key decision-affecting factors today, and few executives would consider a report complete if it did not deal with them. All this means that executives are usually more interested in the implications of the data rather than the data themselves. Never fail to give executives the *so-whats* of the information you provide.

What questions do executives want you to answer in a report written for them? They want to know how a new process or piece of equipment can be used. What new markets will it open up? What will it cost, and why is the cost justified? What are the alternatives?

Why did you choose the new equipment over the other alternatives? Give some information about the also-rans. Convince the executive that you have explored the problem thoroughly. For all the alternatives include comments on cost, size of the project, time to completion, future costs in upkeep and replacement, and the effects on productivity, efficiency, and profits. Consider such aspects as new staffing, competition, experimental results, problems likely to arise. What are the risks involved? What environmental impact will this new development have? Figure 3-3, "What Managers Want to Know," taken from "What to Report" printed in the *Westinghouse Engineer*,[17] illustrates what information Westinghouse executives feel they need in a report.

Experts, in particular, often find writing for executives a difficult task. Experts are often most interested in methodology and theory. Executives are more interested in function. Excerpts from a salmon study done for the Alaskan Fish and Game Department illustrate the frame of mind the expert researcher should have while writing for the executive. In the introduction, the researcher poses the questions

Problems
What is it?
Why undertaken?
Magnitude and importance?
What is being done? By whom?
Approaches used?
Thorough and complete?
Suggested solution? Best? Consider others?
What now?
Who does it?
Time factors?

New Projects and Products
Potential?
Risks?
Scope of application?
Commercial implications?
Competition?
Importance to Company?
More work to be done? Any problems?
Required manpower, facilities and equipment?
Relative importance to other projects or products?
Life of project or product line?
Effect on Westinghouse technical position?
Priorities required?
Proposed schedule?
Target date?

Tests and Experiments
What tested or investigated?
Why? How?
What did it show?
Better ways?
Conclusions? Recommendations?
Implications to Company?

Materials and Processes
Properties, characteristics, capabilities? Limitations?
Use requirements and environment?
Areas and scope of application?
Cost factors?
Availablity and sources?
What else will do it?
Problems in using?
Significance of application to Company?

Field Troubles and Special Design Problems
Specific equipment involved?
What trouble developed? Any trouble history?
How much involved?
Responsibility? Others? Westinghouse?
What is needed?
Special requirements and environment?
Who does it? Time factors?
Most practical solution? Recommended action?
Suggested product design changes?

FIGURE 3-3 What Managers Want to Know

that will be answered in the report. They are the questions an executive would ask:

> Why have they gone? Can the runs be restored to any significant degree? Is it reasonable to base a large industry on the harvest cycle of a wild resource? What should be done? What should be done now?[18]

The stated purpose of the report further reassures the executive that the researcher is on the right track:

> Our approach has been first to gather and understand as much relevant information as could reasonably be found; and then to organize, interpret, and project toward the goal of defining a conceptual framework for successful actions by the State of Alaska through its Department of Fish and Game.[19]

Here it is obvious that scientific findings are going to be wedded to executive needs. Function—"successful action"—lies at the heart of the report.

Be honest. Remember that if your ideas are bought, they are *your* ideas. Your reputation will stand or fall on their success. Therefore, don't overstate your case. Qualify your statements where necessary.

Place yourself properly on the continuum that stretches between the sales representative and the scientist. (See pages 13–14.)

Give your conclusions and recommendations clearly. In writing any report for the executive, remember that you must interpret your material and present its implications, not merely give the facts. Souther points out that "the manager seldom uses the detail, though he often wants it available. It is the *professional judgment* of the writer that the manager wants to tap."[20] The researcher who amasses huge amounts of detail but neglects to state the implications, conclusions, and recommendations that follow from the facts has failed to do the complete job. Many technical people who have aspired to executive rank have not made it because they failed to grasp this simple fact.

Organize your report around executive reading habits. The author of "What to Report," Richard W. Dodge, writes that "Every [Westinghouse] manager interviewed said he read the *summary* or abstract; a bare majority said they read the *introduction* and *background* sections as well as the *conclusions* and *recommendations;* only a few managers read the *body* of the report or the *appendix* material." Figure 3-4 from "What to Report" illustrates how managers read reports. (See also pages 161–164.)

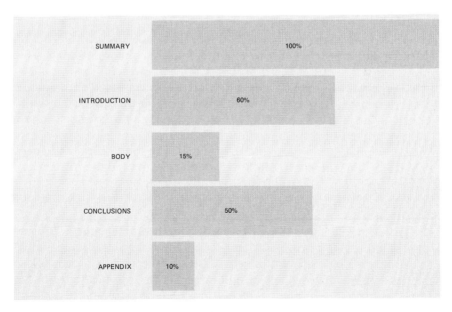

FIGURE 3-4   How Managers Read Reports

This research done at Westinghouse seems to indicate that if you want executives to read your conclusions and recommendations you should repeat them in the abstract and again in a separate section at the *front* of the report. And in modern practice that is where they are usually found.

If you feel you must report a large amount of technical data, put the data in an appendix where executives can read them if they want to (or assign experts to read them).

By way of a conclusion, here is a checklist from "What to Report" that describes an effective approach to writing reports for executives:

- State the purpose of this exercise.
- Give some background to set the stage.
- Explain the alternatives considered.
- Isolate and support the alternatives selected.
- Describe the next action to be taken or being recommended.

Good articles specifically written for executives are easy to find and worth examining; two excellent sources are *Fortune* and *Dun's Review and Modern Industry,* magazines aimed at executives in top and middle management. Our chapters on proposals, progress reports, and feasibility reports deal with specialized executive reports.

EXPERTS

To many lay people, writing meant for experts seems incredibly dull. But to the experts themselves, nothing could be more exciting. The well-written report in their specialty offers them solid facts to chew on—and experts are fascinated by facts. The report also draws inferences from those facts, inferences that experts can hail as new and valid or dispute as examples of faulty reasoning. An expert is akin to the symphony conductor who, while reading a musical score, can *hear it* and judge its potential. Lay people, like the musical audience, must wait for the interpretation by the conductor and the full orchestra before they can appreciate the same score.

Who are experts? For our purposes we will define them as people with either a master's degree or a doctorate in their fields or a bachelor's degree and years of experience, for example, college professors, industrial researchers, and engineers who design and build. They know their fields intimately. When they read in their own fields, they seldom look for background information. You may offer background information that you feel is particularly pertinent to the narrow subject at hand—such as a review of the experiments leading up to the one you have conducted. But this background will not be presented in simple terms, as it would be in an article for nonexperts. In some cases, rather than give background, you can refer experts to other sources—books or articles—where they can pursue background if they need to. (See also pages 405–408.)

Experts are very concerned with how and why things work. They want to see the theoretical calculations and the results of basic research. They want your observations, your facts—what you have seen, what you have measured. In reporting such things be as complete as time, space, and human patience allow. Facts that do not seem immediately important, that may even seem trivial, may assume great importance at a later time. Experts will suspect any conclusions drawn if they feel the facts are not complete. The technological explosion that boggles the mind with its immensity is built upon a foundation of many people's cooperatively working upon an accretion of facts, most of which seem trivial standing alone.

When writing for experts you may use any shorthand methods such as abbreviations, mathematical equations, chemical formulas, and scientific terms that you are sure your audience can comprehend. Complicated formulas and equations needed to support the conclusions, but not essential for understanding them, are often not placed in the body of the report. Modern practice usually places them in an appendix.

Another shorthand device found in expert reports is the use of tables and graphs. Tables provide an excellent way to lift classifications and groups of closely related facts out of the text and display them clearly. The graphs used most often are line graphs. They best portray the relationship between variables. Scientific concepts can be expressed in drawings such as the one in Figure 3-5. Maps and photographs or a line drawing, such as Figure 3-6, of unusual equipment also aid the expert reader significantly.

You do not normally have to define terms unless you have used them in some new or unusual way. However, a word of caution here. Abbreviations and symbols used, where possible, should be the standard ones for a given field, as defined by the authorities in that field. When they are not standard symbols they should be defined as in this example:

From Einstein's mass-energy equation, one can write the relations:

$E(\text{Btu}) = m \text{ (lb)} \times 3.9 \times 10$

m being the loss of mass.[12]

The writer does not define *E*, *Btu*, and *lb* because these are standard symbols and abbreviations for *energy*, *British thermal units*, and *pounds*. He does define *m* because it is not standard. Writers who do

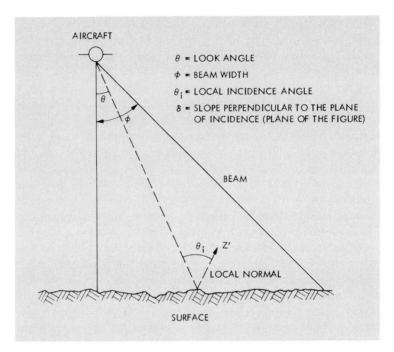

FIGURE 3-5 Imaging Radar Geometry. [From M. Daily et al., *Application of Multispectral Radar and LANDSAT Imagery to Geologic Mapping in Death Valley* (Pasadena, Calif.: Jet Propulsion Laboratory, California Institute of Technology, 1978), 13.]

not define nonstandard symbols cannot expect readers to know what they are. In fact, in a few years, returning to their own reports, *they* may not know what the symbols mean. Symbols and abbreviations may be defined as they are used, or if there are many of them, they can be defined in a glossary at the front of the report. (See pages 210–214.)

The fact that most professional and scientific fields have grown increasingly specialized leads to another caution. Experts in the same field but who have specialized in different aspects of the field may find they have difficulty in understanding one another. Not too long ago, we sat in on a meeting of agronomists being addressed by a fellow agronomist—one who specializes in plant genetics. The question-and-answer period that followed the talk made it painfully obvious that few of the agronomists had understood the speaker. He had not taken into account how far removed they were from his specialization. In this and similar cases, some of the same techniques recommended for the lay audience, such as giving background and defining terms, would be appropriate.

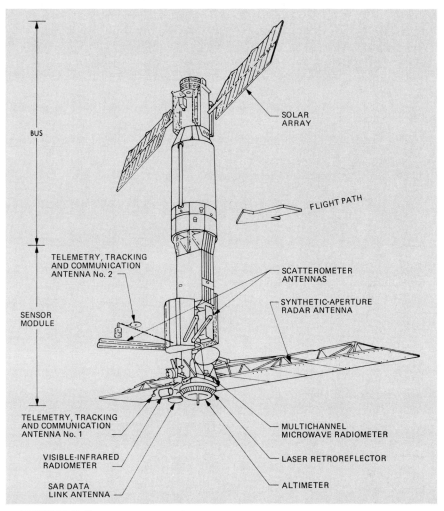

FIGURE 3-6 Seasat Bus and Payload Configuration. [Lee-Lueng Fu and Benjamin Holt, *Seasat Views Oceans and Sea Ice with Synthetic-Aperture Radar* (Pasadena, Calif.: Jet Propulsion Laboratory, California Institute of Technology), 6.]

Expert reports are, of course, not merely calculations and facts. Experts as well as executives want your inferences and conclusions. When you do draw inferences from your facts and observations be sure to make no unwarranted leaps. Stay within the bounds of the scientific method. In presenting your conclusions be careful that your language shows where you are certain and where you are in doubt. Because of this need for scientific honesty and caution, most expert discussions and conclusions do contain qualifications as well as positive statements. The following excerpt is an example of the balance between fairly positive statements and qualified statements that is typical of scientific discussion. (We have italicized the qualifiers.)

> Use of a LANDSAT intensity image provided additional information on varnish distribution on the alluvial fans, thus permitting discrimination of two fan gravel units that were not distinguishable in the radar data. The use of all four bands of LANDSAT *should* add the capability to distinguish compositions. We thus have seven dimensions of imagery available for computer processing techniques.
>
> Surely, particulate alluvial and colluvial regoliths, evaporites, and eolian deposits represent widespread geologic units on terrestrial planetary surfaces. On cloud covered planets (e.g., Venus), multiconfiguration radar data *appear* capable of discriminating many of the surficial geologic units likely to exist. On other planets, additional use of optical sensors will permit a more thorough compositional mapping. LANDSAT/radar combinations offer considerable *promise* for making high-quality geological maps of lithologic-textural units.[22]

Experts know that certainty in science is a hard-won achievement. They are content with probabilities until thorough experimentation and observation remove all reasonable doubt. Their style reflects their basic caution and honesty.

Our Chapter 18, "Empirical Research Reports," describes the most predominant type of expert report. You can find good samples of expert reports in the journals that cover your particular discipline. For examples of expert reports that aim at a somewhat more generalized but still expert scientific audience, we recommend to you *Science*, the journal of The American Association for the Advancement of Science.

TECHNICIANS

Technicians are at the heart of any operation. They are the people who finally bring the scientist's imaginative research and the engineer's calculations and drawings to life. They build equipment, and after it is built they maintain and use it. They are intensely practical people, perhaps with years of experience in a special area. They are the people who can say, "You know, if we used a wing nut here, instead of a hexagonal, the operator would have a much easier job getting that plate on and off," and they'll be right. Technicians are well worth listening to, and certainly you should write well for them.

Technicians' educational levels vary. Most typically they will range anywhere from a high school graduate to a junior engineer with a B.S. degree. They may have been trained in one of the many vocational schools. The high school graduate may have a great deal of on-the-job-training and experience. The junior engineer may be better trained in theory but have less practical experience. Technicians have limitations. They may not be able to follow complicated mathematics, and they'll grow restive with too much theory. Be careful of excessive sentence complexity. (See pages 173–182.)

You can assume a good deal of knowledge on the part of this audience but not as much as with the expert audience. Usually, with this audience, you will need to supply some background information and some definitions. An article from *Bell Laboratories Record* begins this way:

> Waveguide systems are often loosely called "plumbing." The name implies a network of empty pipes where electrical energy flows unimpeded. Actually a waveguide is a precisely designed, electrically tuned structure for propagating electromagnetic waves. It transmits certain determinable frequencies well, does not transmit some frequencies at all, and transmits others only with large losses.[23]

The author of this article has assumed general knowledge in the field on the part of his audience. For example, he does not define "electrically tuned structure" or "electromagnetic waves." His audience will recognize these terms. But he begins his article with background information on the subject: waveguide systems. He uses analogy—"plumbing"—and later describes in great detail what waveguides look like. Generally speaking, the technical audience cares more about the practical application of a theory than about the theory itself.

The author of the *Bell Laboratories Record* article on waveguides uses no equations whatsoever. Rather, he presents mathematical information in visual form or in simple tables. We took the two examples in Figure 3-7 from his article.

But the technician audience is educated and will want some theory, perhaps in a background section. Figure 3-8 shows two pages of a John Deere manual concerning the operation of the equipment used in hay and forage harvesting. Most of the manual is *how-to* information—practical advice about such things as mower preparation—for example, setting and adjusting springs and blades. But as the reproduced pages show, the writers took the time to explain some theory also. The technician reading these pages now knows, for example, why a smooth cutting section on a mower cutterbar is superior in some applications to a serrated one and vice versa. Note also the excellent use of graphics.

Keep theory explanations simple and fairly nonmathematical.

FREQ (Gc)	TYPE	ISO (db)	INS LOSS (db)	RETURN LOSS (db)	BAND-WIDTH (%)	FIGURE OF MERIT
				TYPICAL VALUES		
4	H-Plane	32	0.15	33	13	210
5	H-Plane	25	0.14	30	10	180
6	E-Plane	60	0.6	32	8	100
6	E-Plane	30	0.3	32	8	100
6	E-Plane	60	0.6	32	8	100
11	E-Plane	70	0.6	27	10	120
11	E-Plane	30	0.3	32	10	100
11	E-Plane	70	0.6	27	10	120
12	H-Plane	15	0.3	25	10	50

Resonance isolators developed at Bell Laboratories for Bell System use include designs for the 4-, 6-, and 11-gigacycle common carrier bands and other designs for special applications.

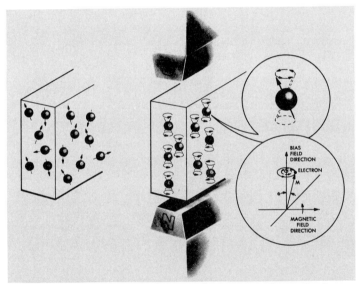

Operation of a resonance insulator hinges on the absorption of energy by electrons in a magnetic field. The left-hand drawing portrays spinning electrons in a piece of magnetic material. The arrows represent the spin axes and the magnetic moments that coincide with them, randomly oriented in the absence of any magnetizing influence. On the right, under the influence of an external magnetic field, the moments (spin axes) tend to align with the field. In lining up they precess about the magnetic lines of the external field as indicated in the upper detail. An rf field polarized in the same direction as the electrons precess can keep them precessing (lower detail).

FIGURE 3-7   Sample Table and Diagram. (From J. J. Degan, "Microwave Resonance Isolaters." *Bell Laboratories Record*, April 1966. Copyright © 1966 by Bell Telephone Laboratories, Incorporated. Reprinted by permission.)

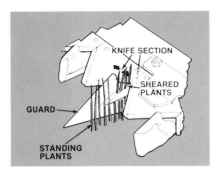

Fig. 8—Shearing Action of Mower Cutterbar

Fig. 9—Cutterbar Mower Cutting Hay

CUTTERBARS

The most commonly used hay-cutting device is a cutterbar with reciprocating knife that shears plant stems (Fig. 9). Cutterbars are used not only on mowers, but also on windrowers, forage harvesters, and grain combines. All cutterbars have the same basic components, but design variations are available to deal with specific problems. They are also called sicklebars, or sicklebar mowers.

Mower size is commonly designated by specifying the cutterbar length. Lengths commonly available are 5, 6, 7, 8, and 9 feet; however, most current mower cutterbars are 7 or 9 feet long.

CUTTERBAR COMPONENTS

The cutterbar is the heart of a mower and a complete understanding of its components (Fig. 10) and functions is essential. Basic cutterbar components are:

- Knife assembly
- Guards and ledger plates
- The bar
- Inner shoe
- Outer shoe
- Knife clips
- Wear plates
- Grass board and stick
- Yoke (cutterbar hinge)

Fig. 10—Basic Cutterbar Components

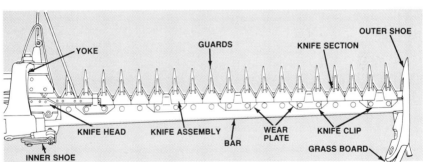

FIGURE 3-8   Theory Section for Technicians. [From *Fundamentals of Machine Operation—Hay and Forage Harvesting* (Moline, Ill. Deere and Company, 1976), 18–19. Copyright © 1976 Deere and Company, Moline, Illinois. Reprinted by permission.]

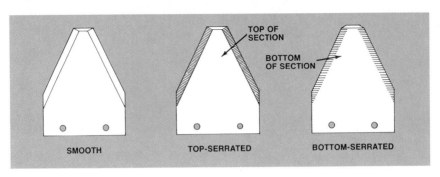

Fig. 11—Common Types of Knife Sections

KNIFE ASSEMBLY

Cutting is done between the knife assembly and the guards, so these components are the most important part of the cutterbar and will be discussed first. Proper knife and guard adjustment, operation, and maintenance are necessary for efficient mower operation. The knife is also known as a sickle.

The *knife head*, at the driven end of the knife, connects the knife and knife drive. The knife head is riveted securely to the knife back to permit easy replacement if needed.

The *knife back* is a flat steel bar to which all knife sections are securely riveted. Knife sections may be replaced if broken or badly worn.

A *knife section* is an individual cutting unit which must be matched to specific crop conditions (Fig. 11). Four types of sections are:

- Smooth
- Top-serrated
- Bottom-serrated
- Armored or hard-surfaced

Smooth sections are used in fine-stemmed crops, particularly where juices released during cutting tend to produce a buildup of residue. Chromed sections resist buildup of plant residue better than plain sections.

Top-serrated sections are used in coarse crops such as alfalfa, clover, timothy, straw, and other stiff-stemmed crops. The serrations have a crop-holding tendency which helps prevent stems from being pushed forward by the section movement instead of being cut. These sections retain their cutting ability without sharpening. Chrome surfacing provides a slick face; it resists buildup of plant-juice residue.

Bottom-serrated sections are used in somewhat the same conditions as are top-serrated sections. Be-

cause serrations are on the bottom, sections may be sharpened when edges become dull.

Armored sections (also called hard-surfaced) are coated on the underside with tungsten carbide. These specially-hardened sections may be used in all conditions suitable for serrated sections, and in most conditions suitable for smooth sections. Hard-surfaced sections are more susceptible to rock breakage than normal sections.

GUARDS

Mower guards are usually spaced on 3-inch centers. The most common knife stroke is also approximately 3 inches. However, the stroke varies on different mowers from 2½ to 3¾ inches—also with 3-inch guard spacing.

Guards have three functions. They protect the knife from solid objects (Fig. 12), hold the stationary *ledger plates* for efficient shearing of the crop, and divide the plants and guide them into the knife for easy cutting.

Fig. 12—Guards Protect Knife Sections

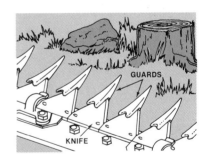

FIGURE 3-8 (con't)

While a junior engineer normally can handle simple calculus, the skilled technician may not be able to. For both groups, the wisest choice is to put most of the necessary math in graph form. If you feel equations must be given the technical audience, definitely put them in an appendix. The equations should supplement the graphical presentations. They should provide further depth for the reader who can use it. They should not be necessary for understanding the basic text.

Depend upon analogy when writing for the technical audience. Analogy bridges the gap between a reader's general information and the particular object or theory you are trying to explain. An excerpt from the *Bell Laboratories Record* article on waveguides illustrates the principle:

> Every electron orbiting about an atomic nucleus gives rise to magnetic fields. In some materials the field comes largely from the motion of the electron around the nucleus, but in ferromagnetic materials it depends more on the spin associated with the electron itself (see the drawing above). The spin creates a small magnetic movement that is precisely aligned with the axis of spin (this can be visualized as similar to the alignment of the earth's magnetic field between the north and south poles).[24]

In addition to the foregoing explanation, the article contains a simple illustration showing the spin axes and magnetic moments associated with them (the second of the two illustrations in Figure 3-7). An article designed for the expert audience doubtless would have explained the process in a series of equations. To understand the process in depth, the scientist needs the equations. Technicians do not need an in-depth explanation, nor would they be likely to understand it.

Technical manuals and operating instructions are the most common applications of writing for technicians. Both are described in Chapter 14, "Instructions for Performing a Process." The magazines *Popular Science* and *Popular Mechanics* offer fine examples of writing for the technician.

THE COMBINED AUDIENCE

Certainly the most difficult audience for a technical writer is the combined audience—perhaps composed of executives, experts, and technicians. Yet in industry such an audience is a common one. How do you avoid bombarding the manager with detail and at the same time satisfy the needs of the experts?

The best approach to the problem is to consider how your readers will use the report. It is probable that the executive will use the report to make a decision. The experts, perhaps, will use it to guide the executive in the decision-making process. Both experts and technicians may use information in the report to implement the decision.

Suppose, for example, you were writing a report for your company to television executives in another company proposing the adoption of a new microwave transmission system. Your report must satisfy the executives and their staff of experts and technicians. Obviously, you could not begin by talking about the transmission system in terms incomprehensible to the executives. Just as obviously you do not want to bore or irritate the experts with needless information. Also, there is a difference in point of view between the executives and the experts. The executives' major concern is whether the new system is a practical one. Will it represent enough of an improvement over previous systems to justify its cost? The executives may need a cost analysis and a cost comparison to previous systems. The experts need enough detail to understand the new design. The experts will be looking for holes in your theory. You must present it completely. The technicians will be worrying about construction and maintenance problems.

How do you satisfy all these varied needs? The answer is that you must compartmentalize your report. Many technical reports are simply not designed to be read straight through by one reader as is a novel or a short story. To allow your readers to select the part or parts they need, you divide your report into different segments designed to do different jobs.

You might begin with a general appreciation of the new system designed for the executives. Follow this with a section on cost analysis. Conclude with a summary and a recommendation in favor of the new system. These sections may provide all the information the executives need or want.

Following the material designed for the executives will be sections designed for the experts and technicians. Here you should use the techniques described in this chapter for these audiences. The experts and technicians will probably skim the parts designed for the executives and concentrate on the parts meant for them. The executives will base their decisions on the sections you wrote for them and on consultations with the experts and technicians.

When you put your report together, choose a format that supports rather than hinders the selective way in which your readers will go through your report. For example, use a table of contents and a system of headings. See our section on format, pages 161–165, and Chapter 11, "Formal Elements of Reports," for help in constructing a report for such selective reading.

Just as the decision will be a team decision, so the report may very well be a team effort. A technical editor might organize the report into the segments we have suggested. The team leader might write the general appreciation for the manager. Cost analysis would be farmed out to an accountant. Experts and technicians would write their appropriate sections. The technical editor would edit the sections, making sure that they can be read by their intended audiences, and put them into a proper format. Such team techniques are common in industry.

No matter how you approach the problem—as an individual or with a team—realizing that you can segment a report to suit the needs of a combined audience will simplify your task.

No matter what your audience, put its needs before your own. Be thinking of your intended audience at every stage of the writing process from research to final draft. Perhaps the advice that new staffers on *Life* magazine used to hear best sums up the attitude you should maintain toward all the audiences you may want to reach: "Never underestimate the intelligence of your readers and never overestimate their knowledge."

EXERCISES

1. Compare and contrast an article or report written for one kind of audience with another written for a different audience—perhaps an expert article with an executive one or an executive article with a lay one. Try to get articles on similar subjects and in your field. Ask and answer some of these questions about the articles.
    a. What stylistic similarities do you see between the two articles? What dissimilarities? (1) Which article has the greater sophistication? (2) What indicates this sophistication? (3) What is the average sentence length in each article? Paragraph length? (4) Which article defines the most terms?
    b. What is the author's major concern in each article? How do these concerns contrast?
    c. What similarities and dissimilarities of format and organization do you see?
    d. Are there differences in the kinds of graphic aids used in the two articles?
    e. Which article presents the most detail? Why? What kind of detail is presented in each article?
    f. How much background information are you given in each article? Does either article refer you to other books or articles?

2. Take an expert article from a journal in your field and rewrite it into a memorandum for an executive. Use the memorandum format shown on page 312.

3. Perhaps you are already planning a long paper for your writing course. Write an analysis of the anticipated audience.

Chapter 4

Finding Your Way in the Library

NO MATTER HOW well you write, no matter how good your style, organization, and mechanics, without adequate data you won't be convincing. Most of the papers you will write—proposals, process descriptions, feasibility reports—require library research. Even original empirical research reports require library research before you can begin. After all, you have to establish what is already known in your area of experimentation. You'll want to determine which previous experimental techniques have worked, which have failed. The library is the primary source for such information.

Because you will use the library for almost every writing chore you do, we have placed this chapter early in the book. Chapter 5, "Gathering and Checking Information," further describes the research process by telling you how to narrow your subject and how to take notes once you are in the library. It also explains nine ways to supplement library research. In addition, Appendix B, "Technical Reference Guides," provides an annotated list of the guides, reference works, and computer services particularly helpful to technical people.

This chapter explains how to work with the card catalog, reference works, periodicals, government publications, and computer services. Sometimes the task of library research can seem overwhelming, for good reason. Since Johannes Gutenberg first printed the Bible in 1456, over 30 million different *titles* have been printed.

Roughly 10 million books are contained in the main branch of the New York City Public Library. Widener Library at Harvard, the largest university library in the United States, holds over 9 million volumes. Even a small undergraduate library may hold as many as 50,000 volumes. The figures are staggering, but the task is not hopeless. Given a system, the curiosity of a good reporter, and a willingness to expend some shoe leather, you, too, can master the library.

THE CARD CATALOG

At the heart of every library is its catalog, listing most of the library's holdings—its books, reference works, and periodicals. Any listing needs a classification scheme, and the one used in most catalogs is subject, author, and title. If you know any one of the three, you can find out if the work you want is in the library. Currently, a card catalog consisting of drawers of three-by-five-inch cards is still the most common catalog to be found in libraries. Although this system is likely to remain the most used for some time, some libraries are beginning to replace their card catalogs with computer-generated catalogs, either in book or microfilm formats. However, despite this change in format, the purpose of the catalog remains the same, and entry to it is still primarily by subject, author, and title. In the years to come, library catalogs will almost certainly grow more computerized; when they do, a search for library materials may come to resemble the computer searches we describe on pages 63–71. But, for the moment, the system we describe here will stand you in good stead.

In most libraries, cards for author, title, and subject are filed together in one alphabetical listing; some libraries may have a separate subject file. Check also to see if your library alphabetizes letter by letter or word by word. It can make a significant difference in finding material, as you can see in the following example:

Letter by letter	*Word by word*
elk	elk
elkhorn	Elk Island National Park
Elkhorn City	Elk Mountains
Elkhound	Elk Point
Elk Island National Park	Elk River
Elkland	Elkhorn
Elk Mountains	Elkhorn City
Elko	elkhound
Elk Point	Elkland
Elk River	Elko

One method or the other is also used in alphabetical listings in reference works such as encyclopedias and indexes.

Finding a book by its author is the easiest and most obvious route if you are fortunate enough to have that information, so let's examine that entry to the card catalog first. Look carefully at Figure 4-1, and we'll explain the information that you can get from the card catalog without getting the book from the shelf.

In this case you knew the name of the author you were looking for, so you went directly to "Hitchcock," passed over "Hitchcock, Alfred" and "Hitchcock, Barton" and came to "Hitchcock, Christopher" printed at the top of the card. Directly under Hitchcock's name is the title of the book: *Plant Lipid Biochemistry*. A fuller explanation of the title tells you that the book is about "the biochemistry of fatty acids and acyl lipids with particular reference to higher plants and algae." You read on and discover that Hitchcock had a coauthor, B. W. Nichols, and that the book was published in London and New York by Academic Press in 1971.

The publisher's name can be important to you. As you come to know your field better, you may discover that some publishers have better reputations in the field than do others. The date may also be important. In some fields, medieval history for example, a 1925 book may be as good as or even better than a 1984 book. But in the fast-moving sciences, often the later the date, the more useful the book.

The next line tells you that the book has xiii (13) pages of prefatory

---

581.192
H631   **Hitchcock, Christopher.**
              Plant lipid biochemistry: the biochemistry of fatty
          acids and acyl lipids with particular reference to higher
          plants and algae [by] C. Hitchcock and B. W. Nichols.
          London, New York, Academic Press, 1971.

                xiii, 387 p.   illus.   24 cm.   (Experimental botany: an international series of
              monographs, v. 4)   £6.50

              Bibliography: p. [306]–332.

                1. Lipids.   2. Acids, Fatty.   3. Plants—Chemical analysis.   I. Nichols,
              Brian William, joint author.   II. Title.   III. Series: Experimental botany, v. 4.

          QK898.L56H5   581.1'9'247   77-170752
          ISBN 0-12-349650-0   MARC

          Library of Congress   72

FIGURE 4-1 Author Catalog Card

matter, which includes such items as the table of contents, illustration list, foreword, preface, and so on. The main body of the book has 387 pages and contains illustrations. It is 24 centimeters high (about 9½ inches, useful information for the librarian who must shelve it) and is the fourth volume in a series called *Experimental Botany*. The book costs £6.50, about $11.00.

The next line gives the useful information that the book contains a bibliography on pages 306–332. Research is much like throwing a rock into a pond, producing an ever-widening series of concentric circles. Each item researched usually leads to other items, in this case a 27-page bibliography.

The next listing on the card tells you the subject headings under which you can find this book in the card catalog: Lipids; Acids, Fatty; and Plants—Chemical analysis. But if you've already found the book, of what use are headings telling you where to find it? A good deal of use, actually. Under these subject headings you will find other books in the same field of knowledge—again, the widening concentric circles. The catalog also lists the book under the coauthor, Nichols; the title of the book; and the series, *Experimental Botany*—another place where you might find related books.

The numbers and letters at the bottom of the card provide additional information for the librarian that you need not concern yourself about. However, the numbers and letters at the upper-left corner—"581.192 H631"—are important to you. They represent the call number of the book. If you have access to your library's collection (open stacks), somewhere near the card catalog will be a map of the library, showing you where the various numbered sections are. If your library has closed stacks, you will need to give the call number to a librarian who will get the book for you.

Figure 4-2 shows a title card. Notice that it is identical to the author card except that the title of the book has been placed above the author's name. When looking for titles, ignore articles *(a, an, the)* and look for the entry under the first major word of the title.

Figure 4-3 illustrates a subject card. Again, it is the same as the author card, but in this case the subject has been placed above the author's name. Before conducting a subject search, you might want to look into the *Library of Congress Subject Headings*, a list of the subject headings used by the Library of Congress since 1897. You will find it a useful guide to the subject headings your library may use. Suppose, for example, you were interested in ice prevention. Looking in the *Library of Congress Subject Headings* under "Ice Prevention" reveals 14 other subject headings such as "Airplanes—Ice prevention," "Roads—Snow and ice control," "Telephone lines—Ice prevention," "Ice control," and "Icing."

581.192   PLANT lipid biochemistry.
H631
Hitchcock, Christopher.
   Plant lipid biochemistry: the biochemistry of fatty
acids and acyl lipids with particular reference to higher
plants and algae [by] C. Hitchcock and B. W. Nichols.
London, New York, Academic Press, 1971.

   xiii, 387 p.   illus.   24 cm.   (Experimental botany: an international series of
monographs, v. 4)   £6.50

   Bibliography: p. [306]–332.

   1. Lipids.   2. Acids, Fatty.   3. Plants—Chemical analysis.   I. Nichols, Brian
William, joint author.   II. Title.   III. Series: Experimental botany, v. 4.

   QK898.L56H5   581.1′9′247   77-170752
   ISBN 0-12-349650-0   MARC

   Library of Congress   72

FIGURE 4-2   Title Catalog Card

581.192   Acids, Fatty.
H631
Hitchcock, Christopher.
   Plant lipid biochemistry: the biochemistry of fatty
acids and acyl lipids with particular reference to higher
plants and algae [by] C. Hitchcock and B. W. Nichols.
London, New York, Academic Press, 1971.

   xiii, 387 p.   illus.   24 cm.   (Experimental botany: an international series of
monographs, v. 4)   £6.50

   Bibliography: p. [306]–332.

   1. Lipids.   2. Acids, Fatty.   3. Plants—Chemical analysis.   I. Nichols, Brian
William, joint author.   II. Title.   III. Series: Experimental botany, v. 4.

   QK898.L56H5   581.1′9′247   77-170752
   ISBN 0-12-349650-0   MARC

   Library of Congress   72

FIGURE 4-3   Subject Catalog Card

In any event, you must be imaginative and alert when you are searching in the catalog under subject headings. For example, looking for information about federal funding of interstate highways we found nothing of use under "Federal Funding" or "Funding, Federal." So we went at it from the other end and tried "Interstate Highways." We did not find any subject cards but did pick up two title cards: *Interstate Highway Maintenance Requirements* and *The Interstate Highway System.* The second card looked promising, with a cross-reference to "Road—Economic Aspects." This avenue led to five other books that dealt in part with interstate highway funding. On a catalog card for one of these five books, we found a cross-reference to "Express highways—Economic aspects" that led to three more books. So it goes. Often the problem is not in finding enough books but in knowing when to cut off the search.

You can sometimes look up periodicals (see Figure 4-4) in the card catalog. Some libraries list them there; some do not. Reference works (see Figure 4-5) are found in the card catalog. Figure 4-5 illustrates a few other points as well. Notice that you can look up reference works (and periodicals also) under subject headings as well as titles. For the *Demographic Yearbook,* "Population—Yearbook" is the subject heading. And notice that in this case, as in the case of many reference books, there is no author as such to be listed. So the corporate author (here, the United Nations) is often used for that purpose. Corporate authors also include names of professional organizations (The American Standards Association), names of institutions, and names

---

The National geographic magazine.   v. 1–
1888–
Washington, National Geographic Society.
   v.   illus. (part col.) ports, maps (part fold., part col.)   26 cm.

   Frequency varies.
   Editors: Jan. 1896–Oct. 1901, J. Hyde.—Nov. 1901–Feb. 1903, H. Gannett.—Mar. 1903–June 1954, G. H. Grosvenor.—July 1954–Jan. 1957, J. D. LaGorce.—Feb. 1957–
   Indexes:
   Vols. 1–42, Jan. 1899–Dec. 1922.   1 v.
   Vols. 1–49, Jan. 1899–Dec. 1925.   1 v.
   Vols. 1–66, Jan. 1899–Dec. 1934.   1 v.
   Vols. 1–70, Jan. 1899–Dec. 1936.   1 v.
   Vols. 1–78, Jan. 1899–Dec. 1940.   1 v.
   Vols. 1–100, Jan. 1899–Dec. 1951.   2 v.

(Continued on next card)
   14—7038*

FIGURE 4-4   Periodical Catalog Card

Population—Yearbook

Demographic yearbook. 1948–
Lake Success.

v.   30 cm.

Yearbooks for 1948–   issued with the United Nations publication sales
no.: 1949.xiii.1:
    Prepared 1948–   by the Statistical Office of the United Nations in
collaboration with the Dept. of Social affairs.
    In English and French.

1. Population—Yearbooks.   i. United Nations.   Statistical Office.

HA17.D45   312.058   50—641

Library of Congress   [69z³2]

FIGURE 4-5   Reference Book Catalog Card

of governmental agencies. Here, too, you can apply some imagination to your card catalog search. If you are interested in world demography—that is, vital statistics on births, deaths, and population change—it would be a good bet that the United Nations would have published information on the subject; as we see, that is the case.

The card catalog is your key to much that is in the library, and you should learn as much as you can about its use. If you need help, ask a librarian. Almost without exception, librarians are friendly and eager to help you, particularly if they see some sign that you are trying to help yourself.

REFERENCE WORKS

You are reading an article in the business and finance section of *Newsweek*, and the author tells you that $317 million worth of hair dryers were sold in the United States in 1980. An article in *Fortune* about the U.S. fishing industry reports that the value of the 1980 lobster catch in the United States was $75 million. An article in *Scientific American* states that in 1978, 123,000 psychologists were at work in the United States. Do you know how to locate such statistics for yourself when you need them? Or must you always rely on secondary sources such as magazines and newspapers? Do you know how to use reference books, books meant to be consulted for a specific piece of information? Beyond a doubt, you are familiar with a few of them, such as encyclopedias and dictionaries; but you may not be fully aware of the vast store of fundamental knowledge available in the reference section of even a small library. Or you may know the knowledge is there without knowing how to get at it, short of always going hat in hand to the reference librarian. Let us help you a bit.

The best introduction to the world of reference books is itself a reference book—Eugene P. Sheehy's *Guide to Reference Books.* Even a small library will have this book, and the librarian will know it by the familiar name of "Sheehy's." Ask your reference librarian for "Sheehy's" and spend a little time reading the introduction. You'll learn that Sheehy's covers about 10,000 reference works. You'll receive some information on how reference books are arranged— alphabetical order, chronological order, tabular order, regional order, and so forth. You'll be advised always to begin by reading the introduction to any reference book to see how it is arranged, what abbreviations it uses, and what little tricks you should know to extract information from it. Looking through Sheehy's table of contents, you'll see that it provides a comprehensive listing of both general reference works and reference works in the humanities, social sciences, history and area studies, and the pure and applied sciences.

Check out some of the listings. The general reference works section is a good place to begin. There you'll find many books that will help you find other books, such as *Books in Print* and the *Cumulative Book Index* (CBI). Both of these are found in every library and provide complete listings of books printed in America or, in the case of the CBI, printed in English anywhere in the world. Such bibliographical books supplement your library's card catalog. If a book you find in the CBI is not available in your library, don't be discouraged. Chances are good that you can get it through an interlibrary loan. Ask your librarian for help in this. But start early. Interlibrary loans do take time.

Now turn to the section of Sheehy's that covers your discipline. In the section on the pure and applied sciences, for example, there is a subsection on the biological sciences. Figure 4-6 is a reprint of one page of that section. It lists the major indexes and abstract journals that cover the field and lists biological dictionaries and periodicals. Other pages in the same section list guides to biological information, bibliographies, encyclopedias, handbooks, and style manuals. In short, Sheehy's lists all the essential reference works a student in the field would need. In every subject area it covers, Sheehy's lists whatever is available and appropriate.

The fact that Sheehy's *Guide to Reference Books* includes some 10,000 works suggests the rich diversity that is available. Encyclopedias are often good starting places for information in a field with

---

and science information; a selected bibliography, 1957–1961 . . . prep. in cooperation with the American University Center of Technology & Administration, School of Government & Public Administration. Paul C. Janaske, ed. Wash., 1962. unpaged.   **EC4**

Pt.2 of the report of the Seminar on Biological Science Communication, 1961. Pt.1 of the report was entitled *Information and communication in biological science.*

A selected, annotated list of 1,121 titles on data handling, information storage and retrieval, documentation, mechanical systems of information handling, etc. Arranged alphabetically by author with keyword title index.   Z699.2.A6

International catalogue of scientific literature: L, General biology. 1st–14th annual issues, 1901–14. London, 1901–19. 14v.   **EC5**

For full description *see* EA20.

Murray, Margaret Ransome and **Kopech, Gertrude.** A bibliography of the research in tissue culture, 1884–1950; an index to the literature of the living cell cultivated in vitro. N.Y., Academic Pr., 1953. 2v. (1741p.)   **EC6**

"Tissue culture has been defined as the maintenance of isolated portions of multicellular organisms in artificial containers outside the individual for considerable periods of time."—*Introd.* Represents 15,000 original articles from serials and books, expanded to 86,000 entries by cross-indexing, using a comprehensive subject classification scheme. Authors and subjects arranged in one alphabet. Entries include references to abstracts in *Biological abstracts.*

——— ——— Supplementary author list, 1950. (Incomplete and univerified, Oct. 1953) N.Y., Academic Pr., 1953. 11p.   Z6663.T5M7

Periodicals

U.S. Library of Congress. Science and Technology Division. Biological sciences serial publications; a world list, 1950–1954. Prep. under the sponsorship of the National Science Foundation. Philadelphia, Biological Abstracts, 1955. 269p.   **EC7**

A classified listing with geographical index, comp. by John Henry Richter and Charles P. Daly.   Z5321.U52

Abbreviations

Biological Council. Abbreviated titles of biological journals: a list culled with permission from the world list of scientific periodicals . . . 3d ed., comp. by P. C. Williams. London, The Council, 1968. 47p.   **EC8**

2d ed. (1954) had title *A list of abbreviations of the titles of biological journals.*

Based on standard abbreviations as approved by the International Organization for Standardization. Listing is alphabetical by full titles. More than 1,400 entries. Reprinted with minor corrections in 1969.   Z6945.A2B5

Indexes

Biological & agricultural index, a cumulative subject index to periodicals in the fields of biology, agriculture, and related sciences, 1964– . N.Y., Wilson, 1964– . v.50– . Monthly (except Aug.) with annual cumulation.   **EC9**

Continues the *Agricultural index* (EL31).

An alphabetic subject index to approximately 190 English-language periodicals in the agricultural and biological sciences. Title inclusion is decided by subscriber vote. Publications of U.S. and state government agencies and of university service and research facilities are not included. The list of periodicals indexed gives subscription information. No author entries.   Z5073.A46

Abstract journals

Biological abstracts from the world's biological research literature. v.1– , Dec. 1926– . Philadelphia, Biological Abstracts, 1926– . Semimonthly.   **EC11**

Subtitle varies. Frequency varies; now published semimonthly, with semiannual (formerly annual) cumulative indexes.

An abstracting journal of theoretical and applied biology, covering more than 5,000 periodicals published in 90 different countries. Preceded by *Abstracts of bacteriology,* v.1–9, 1917–25 (Baltimore, Williams & Wilkins), and *Botanical abstracts,* v.1–15, 1918–26 (Baltimore, Williams & Wilkins), which merged to form *Biological abstracts.*

Titles are given in the original language (except that the Oriental and Russian titles are transliterated), with English translation. Abstracts are in English and usually are signed. Arrangement and indexing vary. Each issue contains abstracts arranged by section and subsection, with subject (BASIC), author, systematic, generic, and cross indexes.

BASIC (Biological Abstracts Subjects in Context) is an indexing technique using computer methods. Each significant word is indexed and alphabetically positioned to the center of a line which includes several words preceding and following the keyword.

A list of new books and periodicals appears in each issue.

An annual "List of serials with word abbreviations" formerly appeared in *Biological abstracts.* This feature has been superseded by: QH301.B37

BIOSIS . . . List of serials with coden, title abbreviations, new, changed and ceased titles. [Philadelphia, BioSciences Information Service], 1971– . Annual.   **EC12**

International abstracts of biological sciences, v.1– , 1954– . London, Pergamon, 1954– . Monthly.   **EC13**

Title varies: v.1–3, *British abstracts of medical sciences.*
For complete record *see* EK64.   QH301.I475

Dictionaries

Abercrombie, Michael, Hickman, C. J. and **Johnson, M. L.** A dictionary of biology. 5th ed. Harmondsworth, Eng., Penguin, 1966. 284p. il.   **EC14**

1st ed. 1951.
For the student and layman.   QH13.A25

Cowan, Samuel Tertius. A dictionary of microbial taxonomic usage. Edinburgh, Oliver & Boyd, 1968. 118p.   **EC15**

Intended for the microbiologist, but includes selected terms from botanical and zoological systematics. Modeled on Fowler's *Dictionary of modern English usage,* so that comparisons of terms and discussions of broader subjects generally accompany the definitions.   QR9.C65

Dumbleton, C. W. Russian-English biological dictionary. Edinburgh, Oliver & Boyd, 1964. 512p.   **EC16**

Gives equivalent English terms for Russian terms in the biological sciences, excluding pathology. Includes the scientific names of species. Assumes a knowledge of Russian grammar and syntax.   QH13.D78

Gray, Peter. The dictionary of the biological sciences. N.Y., Reinhold, [1967]. 602p. il.   **EC17**

A new work by the author of the *Encyclopedia of the biological sciences* (EC26), made up largely of brief definitions for terms which did not warrant entries in the *Encyclopedia.* Users should note that all words from a given root are listed under that root, usually without cross reference.   QH13.G68

Henderson, Isabella Ferguson and **Henderson, William Dawson.** A dictionary of biological terms: pronunciation, derivation, and definition of terms in biology, botany, zoology, anatomy, cytology, genetics, embryology, physiology. 8th ed.

FIGURE 4-6   Sheehy's Guide to Reference Books. (From Eugene P. Sheehy, ed., *Guide to Reference Books,* 9th ed. Copyright © 1976 by the American Library Association. Reprinted by permission of ALA.)

which you are not thoroughly familiar. The general encyclopedias such as *Americana, Britannica,* and *Colliers* provide a variety of articles written for nonexperts. In addition to the general encyclopedias, there are numerous specialized ones: *The Encyclopedia of the Biological Sciences,* the *Encyclopaedic Dictionary of Physics,* the *Universal Encyclopedia of Mathematics,* the *International Encyclopedia of the Social Sciences,* the *Encyclopedia of Textiles,* and so forth. Indeed, there is an encyclopedia for almost every subject you might be interested in.

For unadorned facts, the various almanacs, yearbooks, and handbooks are indispensable aids. You are probably familiar with general almanacs such as *The World Almanac and Book of Facts.* It includes such diverse items as the altitudes of the world's cities; how to address an ambassador; the table of atomic weights; gazetteer information about every nation in the world, including population, area, and principal cities; the depth of the oceans; the casualties suffered in World War II (or the Boer War, for that matter); the public debt of the United States from 1870 ($1.5 billion) to the present (over $1 trillion); and the odds against a royal flush in poker (649,739 to 1).

Yearbooks cover important events in any given year. There are general yearbooks, such as those that encyclopedias produce annually to supplement their basic volumes. And there are specialized yearbooks that cover a political unit or a specialized area of knowledge. One of the most useful yearbooks is the *Statistical Abstract of the United States,* published by the U.S. Bureau of the Census. This book is the source for the information on hair dryers, lobsters, and psychologists that began our section on reference books. It presents tables and figures that abstract almost all the information gathered each year about the United States. It contains sections on population, education, geography and environment, energy, science, forests and forest products, and many more. Figure 4-7 presents a typical page of figures from the *Statistical Abstract.* In addition, the original sources of all statistics are supplied, so the reader can pursue the data further if needed. The United Nations publishes a similar work, the *Statistical Yearbook,* which offers data on population, finance, agriculture, trade, education, and much more for more than 150 countries.

Handbooks are books containing information needed in specialized fields. A very partial listing suggests the possibilities: *Handbook of Mathematical Economics, Handbook of Chemistry and Physics, Handbook of Latin American Studies, Handbook of Private Schools,* and *Handbook of American Popular Culture.* Students might be wise to look at the biennial *Occupational Outlook Handbook* published by the U.S. Bureau of Labor Statistics. It gives the latest information on the major occupations in the United States, including training

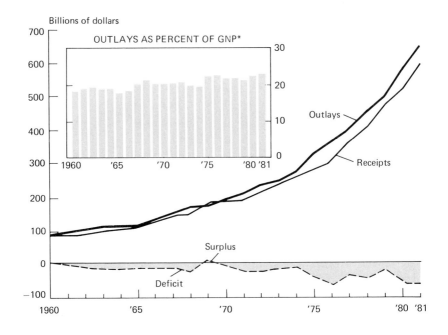

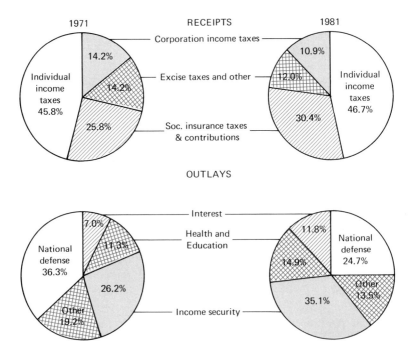

required, earning potential, locations, and much more. You can find bibliographic information in *Sources of Business Information* and economic information in the *Editor and Publisher Market Guide*. The latter covers the United States. It is arranged by state and city and gives information for each city listed on location, transportation, population, industries, banks, retail outlets, newspapers—even the drinking water, as in this entry for Liberal, Kansas: "Alkaline, medium hard, not fluoridated."

Atlases are another rich source of information. To learn how to use them, begin with the general atlases that are a collection of maps and information about population, climate, resources, and so forth and read the introduction to any good atlas such as *Goode's World Atlas*. As with other types of reference works, there are many specialized atlases as well as general ones. *The Rand McNally Commercial Atlas and Marketing Guide* gives economic information of use to business people. Another example is the *Hydrological Atlas of Canada*, which provides information concerning Canada's surface and underground water resources.

We hope we have given you the idea. Many people in both the governmental and private sectors gather data about our world, its people, and their activities day by day, year by year. Most of these data find their way into a reference book. When properly used, such books are tools that can take the guesswork out of many of the decisions that all of us have to make.

PERIODICALS

For our purposes here, we'll define periodicals as commerical magazines, newspapers, and professional journals. Each is discussed below.

Commercial Magazines

The numerous commercial magazines—*Fortune, Scientific American, Sports Illustrated*, and so forth—provide a constant stream of information on almost every subject imaginable. They often serve as useful sources for initial research into a topic because they are written in easy-to-understand prose for the nonexpert. Also, they often provide, within the text of an article, numerous names and titles that can serve as sources for additional information if needed.

FIGURE 4-7 Statistical Abstract of the United States. [From U.S., Bureau of the Census, *Statistical Abstract of the United States*, 102nd ed. (Washington, D.C.: U.S. Government Printing Office, 1981), 244.]

When you consider extracting information from any periodical, you should refer to periodical indexes, that is, published guides to the subjects, authors, and titles that you wish to find. One guide—the *Readers' Guide to Periodical Literature*—covers more than one hundred American commercial periodicals of a broad, general nature in all subjects. The *Readers' Guide*, published every two weeks, provides a constant up-to-date source of information. The biweekly publications (like many such guides and indexes) are gathered in cumulative volumes quarterly and yearly. (Some such cumulative system is standard for most guides and indexes of this sort.) In one alphabetical list, the *Readers' Guide* provides a subject and author listing of all the articles printed in the magazines it indexes during the time period covered by that particular *Guide*. The subject index is carefully cross-referenced. The entry for an article includes the title of the article, the author's name, the magazine the article appeared in, the date and volume number of the magazine, and the page number—in short, all the information you need to find the article you are after in the library's periodical section.

Chances are that you are already familiar with the *Reader's Guide*. If not, seek it out in the reference section of your library. Read its introduction and then use it to find a few articles on some subject that interests you. The *Readers' Guide* is a most useful index in its own right, and it also will familiarize you with the correct procedure for using most such indexes.

Newspapers

Finding information in newspapers is not nearly so easy as finding it in magazines. Fortunately, there is one reliable source to one reliable newspaper—the *New York Times Index* for the *New York Times*. The *New York Times* dates back to 1851. For most of that time, it has been indexed, and all back copies of the paper are currently available on microfilm. Most libraries have a copy of the *Index;* most university and large city libraries also have a set of the microfilms. The *New York Times Index* is organized like the *Readers' Guide* and provides some additional features that make it useful even if you don't have access to the microfilms. The *Index* summarizes major news stories and includes a chronological list of the major events in any story. The *Index* also contains the pictures, maps, and graphs that originally accompanied the stories. (See Figure 4-8.)

Once the *Index* has revealed the date of a story on the subject you are researching, you can turn to your local newspaper for national and international stories on the same topic or event. Most newspapers maintain an indexed file of their back copies in a library familiarly called the "morgue." The morgue is usually accessible to the

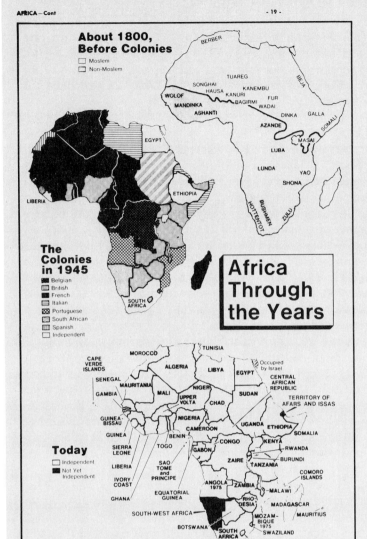

About 1800, Before Colonies

☐ Moslem
☐ Non-Moslem

BERBER
TUAREG
SONGHAI
HAUSA KANEMBU
WOLOF KANURI FUR
MANDINKA BAGIRMI WADAI
ASHANTI DINKA GALLA
AZANDE
EGYPT
MASAI
LUBA
LUNDA
YAO
SHONA
ETHIOPIA
LIBERIA
BUSHMEN HOTTENTOT ZULU
BEJA
SOMALI

The Colonies in 1945

▨ Belgian
■ British
■ French
▤ Italian
▩ Portuguese
▥ South African
▥ Spanish
☐ Independent

SOUTH AFRICA

Africa Through the Years

Today

☐ Independent
■ Not Yet Independent

TUNISIA
MOROCCO
CAPE VERDE ISLANDS
ALGERIA LIBYA EGYPT
SENEGAL
MAURITANIA MALI NIGER CHAD SUDAN
GAMBIA UPPER VOLTA
GUINEA-BISSAU NIGERIA CENTRAL AFRICAN REPUBLIC
GUINEA CAMEROON UGANDA ETHIOPIA
SIERRA LEONE BENIN CONGO KENYA SOMALIA
LIBERIA TOGO GABON RWANDA
IVORY COAST SAO TOME and PRINCIPE ZAIRE TANZANIA BURUNDI
GHANA EQUATORIAL GUINEA ANGOLA 1975 ZAMBIA MALAWI COMORO ISLANDS
SOUTH-WEST AFRICA RHO-DESIA* MADAGASCAR MAURITIUS
BOTSWANA MOZAM-BIQUE 1975
SOUTH AFRICA SWAZILAND LESOTHO

TERRITORY OF AFARS AND ISSAS
Occupied by Israel

*(Declared itself independent in 1965)

US Sen Foreign Relations Com declaring that because of Cuba's possible further involvement in Africa 'Ford Adm is planning blockage of Cuba' (S), Mr 26.5:2; US Repr Thomas P O'Neill (Mass) says that Pres Ford should repudiate publicly or explain statements by US Sec of State Kissinger and White House that 'decisive action' would be taken against any further Cuban mil ventures in Africa and elsewhere (S), Mr 28,51:4

News analysis of US Sec of State Kissinger's recent warnings to Cuba and USSR that US will not accept further Communist mil intervention in Africa contends that if warnings are meant to strengthen US st ... ... they

in N Amer and 'create most extreme difficulty for US,' HR Internatl Rel Com testimony (S), Mr 30,1:5

Phelps-Stokes Fund pres Franklin H Williams Ir urges US Sec of State Kissinger to formulate consistent African policy based firmly upon respect for human freedom; says most important cornerstone of African policy must be unqualified support of African Unity Orgn, Ap 1,30:3

US officials rept that US Sec of State Henry A Kissinger has decided against visiting S Africa during tour of black Africa; trip, scheduled to begin Apr 23 or 24, will include Kenya, Tanzania, Zambia, Zaire, Nigeria, Ivory Coast, Lib ... ... Senegal (S), Ap 3.14:5

Evelyn Jones Rich and Apr 4 cartoon in Travel and presumably preparir with criticisms, Ap 18,X Ed, commenting on Sc maintains that Kissinger is not in position to say leaders most want to he. Lewis series on US Sec holds that Rhodesia is n Kissinger will have on m Ap 22,33:1

African-Arab min conf of 8-point charter of coop Orgn and Arab League de for approval at African-Ar proposals for closer links b provisions noted (S), Ap 23

Kissinger says that he is c black majority rule in Rhode pol unity and econ developr DC; says that he will stress equally determined to prev involvement in African aff remarks is warning that A pol or econ support if the Kissinger says efforts to l 'suspended' so long as C agitates for Puerto Ricar James Reston article cor Kissinger has not made recommend if USSR's 'ir intervention in Africa co to ask Kissinger about d nationalist movements ir **expected to become insi of 7 African countries; T and Zambian Pres Kenn hard for US assurances c primarily for guerrilla fo. minority Govt of Rhodes** illus;(L), L), Ap **24,1:3;** K (S), Ap 24,2:4; sr US of US Sec of State Kissinge African trip will be diffic mil aid against Rhodesia African visit by declarin aspirations with US aspi scheduled to visit with F 25 (M), Ap 25,5:1; **Kissi Kenyatta and Tanzanian** seeks to establish tone of airport statement, Tanzar has come to learn views formulate comprehensive mentioned USSR and Cu simply that US is commit should be solved by Africa that Kissinger is 'seeking t he is talking to Kenyatta a Kenya and neighbors; Tan Ibrahim Kafuma comment paints gloomy picture of pr several 3d-world diplomats Kissinger's trip for any sigr new opportunities for USS Ap 26,5:1; Anthony Lewis State Kissinger's African to may be taking Africa serio Angola as example; discuss future US policy, especially Kissinger arrives in Tanzar for concept of majority ru welcomed with slightly q Pres Julius K Nyerere; il major policy speech at s' Kenneth D Kaunda (M) speech; illus, Ap 28,16:1 delighted by Zambian P Kissinger's proposals (S) Zaire; Zambian Pres Ker Kissinger's speech on ne African policy; Kissinge: has canceled his visit; G been under pressure from (M), Ap 28,17:4; ed says traditional Amer commit rule, racial equality says Se Sulzberger article says US restructuring of US Africa 'remiss' for 20 yrs, Ap 28 'certain foreign countries' to cancel visit there (M), Kissinger's major policy cancels Apr 29 activities stomach ache; is schedu originally was scheduled canceled visit on ground Kutu Acheampong was cancellation to pressure Kissinger leaves Zaire ar 7:3; leaves Liberia and a billion rescue operation Africa and appeals to wo Kissinger is scheduled to

FIGURE 4-8   The New York Times Index. (From *The New York Times Index 1976*. Copyright © 1976 by the New York Times Company. Reprinted by permission.)

public, often even by phone and mail. Many newspapers will provide photocopies of requested stories and articles for a small fee. Knowing the dates of a major story through your research in the *New York Times Index* will simplify your search for the same story in your local newspaper's morgue.

Professional Journals

As you advance in your field, the most useful periodicals for you will be the professional journals relating to that field. These journals report the latest research over an extended period of time. Journals contain much information that may never get into book form or that may appear in a book too late to do you any good. Therefore, any serious member of a profession has to master research in its journals. This task may seem complex on the surface but is really not that difficult. Three reference books will simplify it considerably.

The *Applied Science and Technology Index* constitutes a subject index to over 200 journals in numerous scientific fields. It's a good starting point for more advanced research. Figure 4-9 shows you a typical page from this index. Notice that good cross-referencing is provided, as under the entries for "Physical Geography" and "Physics." The entries provide a fair amount of information about each item. Look at the first entry under "History" in the right-hand column. After consulting the instructions and abbreviations in the front of the index, you would be able to translate this entry as follows: The title of the article is "Publications in the History of Physics during 1973." The author is S. G. Brush. The article contains a five-page bibliography. It appeared in the *American Journal of Physics*, volume 43, pages 850 to 860, in October 1975.

But most scholarly journals are not covered by the *Applied Science and Technology Index*, so you'll have to broaden your search. We've already mentioned Sheehy's *Guide to Reference Books*. Remember that this guide is arranged by disciplines. The indexing and abstracting services for a particular discipline are listed under that discipline. Look again at Figure 4-6, a reproduction of a page covering the biological sciences. Under the heading "Indexes," we find listed the *Biological and Agricultural Index*. If you read the entry, you'll see that this index covers 190 English-language periodicals in the field.

Below the indexes are listed the abstract journals. We might take a moment here to explain the term *abstract journal.* An abstract journal provides the same service that an index does, plus one valuable additional feature. As the name implies, each article listed is also abstracted, that is, summarized. Sometimes the abstract itself will provide all the information you need. At the very least you will know after reading it whether you want to read the original article. Abstracts are great time-savers, and whenever you can use them

292 APPLIED SCIENCE & TECHNOLOGY INDEX

PHOTOMULTIPLIER tubes
Double beam, single photomultiplier spectrofluorimeter. W. H. Melhuish. diags Sci Instr 8:815-16 O '75
Inexpensive device for the determination of photomultiplier current gains. N. W. Bower and J. D. Ingle, jr. bibl diags Anal Chem 47:2069-72 O '75
Photomultiplier anode pulse integration circuit. G. S. Beddard and P. Williams. il diag Sci Instr 8:720-1 S '75
Photomultiplier window materials under electron irradiation; fluorescence and phosphorescence. W. Viehmann and others. bibl App Optics 14:2104-15 S '75

PHOTOMULTIPLIERS. See Photomultiplier tubes

PHOTONS
Simple approach to the multiphoton states of a thermal radiation field. S. A. Afsenius. bibl diag Am J Phys 43:882-4 O '75
Spin and photon echoes. D. C. Haueisen and T. D. Bonifield. bibl Am J Phys 43:824-9 S '75
See also
Mössbauer effect

Measurement

Digital light scattering photometer. S. P. Lee and others. bibl diags R Sci Instr 46:1278-82 S '75
High-sensitivity appearance potential spectrometer for the detection of photons, electrons, and ions. A. Kluge. bibl diag R Sci Instr 46:1179-81 S '75
Simple digital clipped correlator for photon correlation spectroscopy. S. H. Chen and others. bibl il diags R Sci Instr 46:1356-67 O '75

PHOTOOXIDATION. See Oxidation

PHOTOSENSITIVE glass. See Glass, Photosensitive

PHOTOSYNTHESIS
Alternative sources for power generation. B. Francis. il diags Electronic Eng 47:37-9 O '75
Optically detected zero-field magnetic resonance studies of the photoexcited triplet state of the photosynthetic bacterium rhodospirillum rubrum. R. H. Clarke and others. bibl Am Chem Soc J 97:7178-9 N 26 '75
Source of photosynthetic oxygen in bicarbonate-stimulated Hill reaction. A. Stemler and R. Radmer. bibl Science 190:457-8 O 31 '75

PHOTOTYPESETTING
Putting words on paper. il Comp Air Mag 80:6-7 O '75

PHOTOVOLTAIC cell. See Photoelectric cells

PHOTOVOLTAIC effect. See Photoelectric effect

PHTHALATES
Properties of materials; diallyl phthalates; tables. Materials Eng 82:140 Mid-S '75

PHTHALIC anhydride
Control phthalic anhydride emissions. C. W. Moores. il diags Hydrocarbon Process 54:100-3 O '75

PHTHALIMIDE
Kinetics of intramolecular electron transfer between phthalimide groups linked by $-(CH_2)_n-$ or $-(CH_2CH_2O)_mCH_2-$ chains; studies of chain flexibility. K. Shimada and others. bibl Am Chem Soc J 97:5834-41 O 1 '75
Solvent effects in the kinetics of electron transfer between *N-n*-butyl-phthalimide radical ion and its parent molecule. Y. Shimozato and others. bibl Am Chem Soc J 97:5831-3 O 1 '75

PHTHALOYL group
Azetidinone antibiotics; removal of phthaloyl protective group from acid and base sensitive compounds. S. Kukolja and S. R. Lammert. bibl Am Chem Soc J 97:5582-3 S 17 '75

PHYSICAL apparatus and instruments
1975 AAPT apparatus competition. F. E. Christensen. Am J Phys 43:935 N '75
See also
Astronomical instruments
Electric instruments
Geophysical instruments
Goniometers
Optical instruments
Scientific apparatus and instruments
Sound—Apparatus

PHYSICAL geography
See also
Climatology
Drainage (physical geography)
Lagoons
Paleogeography
Topography
Valleys

PHYSICAL measurements
See also
Dimensional analysis
Doppler effect
Electric measurements
Temperature—Measurement
Vibration—Measurement
Viscosity—Measurement

PHYSICAL metallurgy. See Metallography

PHYSICIANS
Physician migration reexamined. R. Stevens and others. bibl Science 190:439-42 O 31 '75

PHYSICISTS
Motivations of physicists. M. J. Moravcsik. Phys Today 28:9 O '75

PHYSICS
See also
Atoms
Ballistics
Compressibility
Diffusion
Dimensional analysis
Dynamics
Elasticity
Electricity
Electromagnetic theory
Electrons
Evaporation
Expansion (heat)
Fluids
Fluorescence
Force and energy
Free energy
Friction
Geophysics
Gravitation
Gravity
Heat transmission
Inertia (mechanics)
Ionization, Gaseous
Ions
Kinematics
Light
Liquids
Magnetism
Matter
Mechanics
Meteorology
Molecules
Music—Acoustics and physics
Particles, Elementary
Pauli exclusion principle
Photoelasticity
Plasma (Physics)
Plasticity
Porosity
Pressure
Radiation
Relativity (physics)
Relaxation time (physics)
Reflection (optics)
Sound
Space and time
Statics
Steam
Temperature
Thermodynamics
Turbulence
Velocity
Vibration
Viscosity
Vortex motion
Wave mechanics
Work function

Bibliography

Publications in the history of physics during 1973. S. G. Brush. Am J Phys 43:850-60 O '75

Examinations

Repechage, a second chance at learning. M. S. Weinhous and W. R. Webber. Am J Phys 43:831-2 S '75

Experiments

Best measuring time for a Millikan oil drop experiment. J. I. Kapusta. Am J Phys 43:799-800 S '75
Conservation of angular and linear momentum on an air table. R. R. Rockefeller. diags Am J Phys 43:981-3 N '75
Kerr effect in nitrobenzene; a student experiment. A. W. Knudsen. bibl diags Am J Phys 43:888-94 O '75
Optical activity demonstration. G. Freier and B. G. Eaton. diag Am J Phys 43:939 N '75
See also
Nuclear physics—Experiments

History

Publications in the history of physics during 1973. S. G. Brush. Am J Phys 43:850-60 bibl (p856-60) O '75
Role of music in Galileo's experiments. S. Drake. il Sci Am 232:98-104 Je '75; Discussion. E. Carterette. 233:8-9+ N '75

Study and teaching

Experience with a course in physics for visual arts majors. R. C. Bearse and D. W. McKay. Am J Phys 43:907-9 O '75
Exploratorium and other ways of teaching physics. F. Oppenheimer. Phys Today 28:9+ S '75
Karate strikes. J. D. Walker. bibl il diags Am J Phys 43:845-9 O '75

PHYSICS, Mathematical. See Mathematical physics

PHYSICS laboratories
Life science-related physics laboratory on geometrical optics. T. H. Edwards and others. il Am J Phys 43:764-5 S '75

FIGURE 4-9 Applied Science and Technology Index. (From *Applied Science & Technology Index*. Copyright © 1976, 1977 by The H. W. Wilson Company. Material reproduced by permission of the publisher.)

instead of indexes, it makes sense to do so. Under the heading "Abstract Journals," Sheehy's lists two such services. The first, *Biological Abstracts from the World's Biological Research Literature*, "covers more than 5,000 periodicals published in 90 different countries."

The third major source in conducting a literature search in professional journals is *Ulrich's International Periodical Directory*. This reference, published every two years with supplements in the intervening years, is found in even the smallest libraries and is known to librarians simply as "Ulrich's" (pronounced *ol-ricks*). It covers approximately 60,000 in-print periodicals. Ulrich's classifies by subject and contains a title index as well. Its subject listing covers all fields of knowledge and is well divided and subdivided. For example, there are 12 subdivisions under biology ranging from "Biological Chemistry" to "Microscopy," "Ornithology," "Physiology," and "Zoology."

Figure 4-10 reproduces a page from Ulrich's. Look at the entry for the periodical *Economic Botany*. Again, after reading the instructions in the front of Ulrich's, we are able to translate this entry easily: *Economic Botany* is devoted to applied botany and plant utilization. It was first published in 1947 and is published quarterly. A subscription costs $22 a year. The editor's name and address are provided. Among its special features are abstracts, advertisements, bibliographies, book reviews, and illustrations. Its circulation is 1,800.

The next bit of information given about *Economic Botany* is your key to easy access to the journal. Ulrich's tells which indexing and abstracting services (if any) cover the journal, in this case, *Biological Abstracts*, *Biological and Agricultural Index*, *Chemical Abstracts*, and *Science Citation Index*. You now know where to look up articles printed in *Economic Botany*. This indexing information is given for all periodicals listed in Ulrich's when applicable; for a researcher, it is one of the most valuable features of the book.

Finding your way in the scholarly journals, then, is not that difficult. The *Applied Science and Technology Index* will locate many journal articles for you. Between them, Sheehy's and Ulrich's will help you locate any journals you're likely to need and tell you what indexing and abstracting services cover them. With that information, you're on your way.

GOVERNMENT PUBLICATIONS

Government publications are a motherlode of information waiting to be mined. The United States government publishes pamphlets, periodicals, research reports, and books. It addresses every audience, from the expert to the elementary school child. Many libraries have at least a limited collection of government publications. But 50

206 BIOLOGY — BOTANY

CARIBBEAN RESEARCH INSTITUTE
QUARTERLY REPORT. see *BIOLOGY*

581 US ISSN 0008-7475
CASTANEA. 1936. q. membership. (Southern
Appalachian Botanical Club) West Virginia Univ.,
Morgantown, W.V. 26506. Ed. Jesse F. Clovis. adv.
bk. rev. bibl. index. circ. 800. Indexed: Biol.Abstr.

581
CATALOG OF FOSSIL SPORES AND POLLEN.
(Text in various languages) 1957. irreg.(3/yr) $13.50
to individuals; institutions $35. Pennsylvania State
Univ., Coal Research Section, Deike 517, College of
Earth & Mineral Sciences, Univ. Park, PA 16802.
Ed. A. Traverse. index. circ. 600. (looseleaf format;
also avail. in cards)

589.2 CS ISSN 0009-0476
CESKA MYKOLOGIE. (Text and summaries in
various language) 1947. 4/yr. fl.44. (Ceskoslovenska
Akademie Ved, Czechoslovak Scientific Society of
Mycology) Academia, Publishing House of the
Czechoslovak Academy of Sciences, Vodickova 40,
112 29 Prague 1, Czechoslovakia (Distributor in
Western countries: John Benjamins B.V., Amsteldijk
44, Amsterdam (Z.), The Netherlands) Ed.Bd. bk.
rev. abstr. charts. index. circ. 850. Indexed:
Biol.Abstr. Chem.Abstr.

CHEEKWOOD MIRROR. see *ART*

CORNELL PLANTATIONS. see *GARDENING
AND HORTICULTURE*

CORRIERE FITOPATOLOGICO. see *BIOLOGY —
Entomology*

COTON ET FIBRES TROPICALES. see
AGRICULTURE

581 SP ISSN 0011-2372
CUADERNOS DE BOTANICA CANARIA:
comunicaciones sobre flora y vegetacion del
Archipielago Canario. (Text and summaries in
English and Spanish) 1967. 3/yr. 300 ptas.($5.)
(Excmo. Cabildo Insular de Gran Canaria, las
Palmas) Finca Llano de la Piedra, Santa Lucia de
Tirajana, Gran Canaria, Spain. Ed. Guenther
Kunkel. bk. rev. index. circ. 500. Indexed:
Excerp.Bot.

CURRENT ADVANCES IN PLANT SCIENCE. see
BIBLIOGRAPHIES

581 UK ISSN 0011-4073
CURTIS'S BOTANICAL MAGAZINE. 1787. 2 yr.
$9.25 per mo. Bentham- Moxon Trust, Royal
Botanic Gardens, Kew, Richmond, Surrey, TW9
3AB. Eng. Ed. D. Hunt. illus. index. per vol. of 4
parts. circ. 600. Indexed: Biol.Abstr.

500.9 AG ISSN 0011-6793
DARWINIANA. (Summaries in English, French,
German and Latin) 1922. s-a. Arg.$180.($23.) ‡
(Academia Nacional de Ciencias Exactas, Fisicas y
Naturales de Buenos Aires) Instituto de Botanica
Darwinion, Lavarden y del Campo, San Isidro,
Buenos Aires, Argentina. Ed. Arturo Burkart. bk.
rev. bibl. illus. circ. 900. Indexed: Biol.Abstr.
Chem.Abstr. Excerpt.Bot.

581, 635 CN ISSN 0045-9739
DAVIDSONIA. 1970. q. $4. University of British
Columbia, Botanical Garden, Vancouver, B.C. V6T
1W5, Vancouver 8, B.C. Canada. Ed. Dr. C.J.
Marchant. bibl. illus. circ. 300. (tabloid format)
Indexed: Biol.Abstr.

DEUTSCHE BAUMSCHULE. see *GARDENING
AND HORTICULTURE*

581 GW ISSN 0011-9970
DEUTSCHE BOTANISCHE GESELLSCHAFT
BERICHTE. 1882. 12/yr. DM.156. Gustav Fischer
Verlag, Wollgrasweg 49, Postfach 720143, 7000
Stuttgart 72, W. Germany. Ed. Dr. A. Pirson. bibl.
charts. illus. index.

581 SX ISSN 0012-3013
DINTERIA: contributions to the flora and vegetation
of south west africa. (Text and summaries in
Afrikaans, English & German) 1968. price varies.
South West Africa Scientific Society, Box 67,
Windhoek, South Africa. Ed.W.Giess. illus.

581 UK ISSN 0012-396X
DISTRIBUTION MAPS OF PLANT DISEASES.
1942. m.(3 maps per mo.) £3($7.80) Commonwealth
Agricultural Bureaux, Commonwealth Mycological
Institute, Ferry Lane, Kew, Surrey, Eng. Ed. A.
Johnston. charts. maps.
Mycology

581 US ISSN 0012-4982
DOKLADY BOTANICAL SCIENCES. English
translation of: Doklady Akademii Nauk S.S.S.R.
vols.208-213,1973. s-a. $60. (Academy of Sciences
of the U. S. S. R. Botanical Sciences Section)
Consultants Bureau, 227 W. 17th St., New York,
NY 10011. Ed. A. I. Oparin. bibl. charts. illus.
index. circ. 292. (also avail. in microfilm) Indexed:
Biol. & Agri.Ind. Biol.Abstr. Chem.Abstr.

581 US ISSN 0013-0001
ECONOMIC BOTANY. devoted to applied botany
and plant utilization. 1947. q. $22. New York
Botanical Garden, Bronx, NY 10458. Ed. Dr.
Richard E. Schultes. adv. bk. rev. abstr. bibl. illus.
circ. 1,800. Indexed: Biol.Abstr. Biol. & Agri.Ind.
Chem.Abstr. Sci. Cit. Ind.

581
EGYPTIAN JOURNAL OF BOTANY. (Text in
English; summaries in English and Arabic) 1958. 3/
yr. $9.50. (Botanical Society, U.A.R.) National
Information and Documentation Centre, (NIDOC),
Cairo, U.A.R. Ed. A. I. Naguib. charts. illus.
Indexed: Biol.Abstr. Chem.Abstr.
Formerly: Journal of Botany of the United Arab
Republic (ISSN 0021-9363)

EUPHYTICA, Netherlands journal of plant breeding.
see *AGRICULTURE*

581 GW ISSN 0014-4037
EXCERPTA BOTANICA. SECTIO A.
TAXONOMICA ET CHOROLOGICA. (Text in
English, French and German) 1959. 1-2 vols per
year(7 nos. per vol.) DM.134. Gustav Fischer
Verlag, Postfach 720 143, Wollgrasweg 49, 7000
Stuttgart 72, W. Germany. Ed.Bd. adv. bk. rev.
cum.index.

581 GW ISSN 0014-4045
EXCERPTA BOTANICA. SECTIO B.
SOCIOLOGICA. (Text in English, French and
German) 1959. 1 vol. per yr.(4 nos. per vol.)
DM.70. Gustav Fischer Verlag, Wollgrasweg 49,
Postfach 720143, 7000 Stuttgart 72, W. Germany.
Ed. Dr. R. Tuexen. adv. bibl. index.

632, 581 UN ISSN 0014-5637
F A O PLANT PROTECTION BULLETIN: a
publication of the world reporting service on plant
diseases, pests, and their control. (Editions in
English, French and Spanish) 1952. bi-m. $4 ‡
Food and Agricultural Organization, Plant
Production and Protection Division, Viale delle
Terme di Caracalla, Rome, Italy. Ed. Dr. A. V.
Adam. bibl. charts. illus. index. circ. 6,000.
Indexed: Biol.Abstr. Biol. & Agri.Ind. Chem.Abstr.

581 GE ISSN 0014-8962
FEDDES REPERTORIUM. Zeitschrift fuer
botanische Taxonomie und Geobotanik. (Text in
English, French and German) 1906. 10/yr. M.15.
per no. Akademie-Verlag GmbH., Leipziger Str. 3-4,
108 Berlin, E. Germany. Ed.Bd. bibl. charts. illus.
index. Indexed: Biol.Abstr.
Formerly: Feddes Repertorium Specierum
Novarum Regni Vegetabilis.

581 US ISSN 0015-0746
FIELDIANA BOTANY. 1895. irreg. a change basis
or sale (price avail. on inquiry) Field Museum of
Natural History, Roosevelt Rd. & Lakeshore Drive,
Chicago, IL 60605. Ed. Patricia M. Williams. bibl.
charts. illus. circ. 500. Indexed: Biol.Abstr.
Chem.Abstr.

580 UR ISSN 0015-3303
FIZIOLOGIIA RASTENII. (Summaries in English)
1954. bi-m. $36. Izdatel'stvo "Nauka", Podsosensky
21, Moscow, U.S.S.R. Ed. A. L. Kursanov. bk. rev.
charts. illus. index. circ. 2,800. Indexed:
Biol.Abstr. Chem.Abstr.

500.9 EC ISSN 0015-380X
FLORA. (Text in English and Spanish) 1937.
exchange basis. Instituto Ecuatoriano de Ciencias
Naturales, Apartado 408, Quito, Ecuador. Ed. Dr.
M. Acosta-Solis. bk. rev. abstr. bibl. charts. illus.
Indexed: Biol.Abstr.

581 DK ISSN 0015-3818
FLORA OG FAUNA. (Text in Danish; summaries in
English) 1894. q. Kr.40.($7.) Naturhistorisk
Forening for Jylland, Skjaersoevej 5, Risskov,
Denmark. Ed. Edwin Norgaard. bk. rev. illus.
index. circ. 1,000. Indexed: Biol.Abstr. Zoo.Rec.

581 SA ISSN 0015-4504
FLOWERING PLANTS OF AFRICA. (Text in
Afrikaans and English) 1945. 2/yr. R.1.50 per no. ‡
Dept. of Agricultural Technical Services, Private
Bag X144, Pretoria, South Africa. Ed. Dr. D. J. B.
Killick. illus. cum.index. circ. 750. Indexed:
Biol.Abstr.

581 CS ISSN 0015-5551
FOLIA GEOBOTANICA ET
PHYTOTAXONOMICA. (Text and summaries
inEnglish, French, German and Russian) 1966. q.
fl.75. (Ceskoslovenska Akademie Ved, Botanical
Institute) Academia, Publishing House of the
Czechoslovak Academy of Sciences, Vodickova 40,
112 29 Prague 1, Czechoslovakia (Distributor in
Western countries: Dr. W. Junk B.V., 13, van
Stolkweg, The Hague, Netherlands) Ed. S. Hejny.
bk. rev. charts. index. circ. 1,100. Indexed:
Biol.Abstr.
Formerly: Folia Geobotanica et Phytotaxonomica
Bohemoslovaca.

581 SA ISSN 0015-847X
FORUM BOTANICUM. (Text in Afrikaans and
English) 1962. m. R.5 for non-members. South
African Association of Botanists, Private Bag X101,
Pretoria, 0001, South Africa. Ed. E.W. Oliver.
bibl. index. cum.index. circ. 350. (processed)
Formerly: South African Forum Botanicum.

581 US ISSN 0092-1793
FREMONTIA. 1965. q. $8. California Native Plant
Society, 2490 Channing Way, Berkeley, CA 94704.
Ed. Margedant Hayakawa. adv. bk. rev. circ. 2,600.
(back issues avail)
Formerly: California Native Plant Society.
Newsletter.

589.2 DK ISSN 0016-1403
FRIESIA. Nordisk mykologisk tidsskrift. (Text in
Danish, English, French, German or Swedish) 1932.
1-2 nos per year. Kr.60($10) Svampekundskabens
Fremme - Danish Mycological Society,
Thorvaldsens Vej 40, 1871 Copenhagen V,
Denmark. Ed. N. F. Buchwald. bk. rev. charts.
illus. cum.index every 5 years. Indexed: Biol.Abstr.
Mycology

581 US ISSN 0016-2167
FRONTIERS OF PLANT SCIENCE. 1948. s-a.
contr. free circ. ‡ Connecticut Agricultural
Experiment Station, New Haven, CT 06504. Ed.
Paul Gough. illus. circ. 7,500. Indexed: Biol.Abstr.
Chem.Abstr.

581 US ISSN 0016-4585
GARDEN JOURNAL. 1951. bi-m. membership(non-
members $5.) New York Botanical Garden, Bronx,
NY 10458. Ed. Mary E. O'Brien. adv. bk. rev. illus.
index. circ. 6,500. Indexed: Biol.Abstr. Biol. &
Agri.Ind. Sci. Cit. Ind.
Formerly: Garden Journal of the New York
Botanical Garden.

GARDENS ON PARADE. see *GARDENING AND
HORTICULTURE*

581 CL ISSN 0016-5301
GAYANA: BOTANICA. 1961. s-a. Universidad de
Concepcion, Instituto Central de Biologia,
Concepcion, Chile (Subscr. To: Comision Editora,
Casilla 301, Concepcion, Chile) Ed. Bd. Indexed:
Biol.Abstr.

GENETICA POLONICA, Polish journal of theoretical
and applied genetics. see *BIOLOGY — Genetics*

581 IT ISSN 0017-0070
GIORNALE BOTANICO ITALIANO. (Text in
English, French, German and Italian; summaries in
English and Italian) 1844. bi-m. L.12.000($20.)
Societa Botanica Italiana, Via Lamarmora N. 4,
50121 Florence, Italy. Ed. Prof. Pier Virgilio
Arrigoni. charts. illus. index. circ. 600. Indexed:
Biol.Abstr. Chem.Abstr. Excerp.Bot.

FIGURE 4-10 Ulrich's International Periodical Directory. (From
Ulrich's International Periodicals Directory, 16th ed. Copyright © 1976 by
the R. R. Bowker Company. Reprinted by permission of the publisher.)

regional libraries nationwide are designated depositories for government documents, and that is where you would find the most complete coverage. For the depository library nearest you, see *Government Publications and Their Use* by Laurence F. Schmeckebier and Roy B. Eastin, or ask your local librarian.

Most libraries with a substantial number of government documents catalog and shelve them separately from the rest of the collection. Before beginning any research in government publications, ask the librarian in charge of them to familiarize you with what is available and how to get at it. The brief coverage that follows should provide enough information about some of the basic tools to help you survive in what is sometimes a confusing research area.

Good starting points for any novice in the use of government publications are *Government Publications and Their Use* and Joe Morehead's *Introduction to United States Public Documents*. Browsing through both books is a good introduction to the world of government documents.

The best general guide to government documents is the *Monthly Catalog of United States Government Publications*, published by the U.S. Superintendent of Documents. Each monthly catalog lists the documents published that month. The list is organized by government agency—"Environmental Research Laboratories," "Federal Aviation Administration," "Federal Energy Office," and so forth. The agency-by-agency listing is followed by subject, author, and title indexes. The December catalog each year contains author, subject, and title indexes covering the entire year. The *Monthly Catalog* indexes most, although not all, of the documents printed by government agencies. It also tells the reader how to order government publications.

The Superintendent of Documents frequently issues separate, updated *Price Lists* covering the approximately 80 fields in which the government publishes, such as home economics, geology, maps, scientific tests, standards, and consumer information. These lists are particularly helpful in locating recent publications in a given subject area.

When looking for less recent government documents, the best guide is the 15-volume *Cumulative Subject Index to the Monthly Catalog of United States Government Publications 1900–1971*, compiled by William W. Buchanan and Adna M. Kanely. This collection indexes 72 years of government documents.

The federal government conducts and commissions research on an enormous number of subjects, such as "Severe Local Storms Research" (conducted by Purdue University) or a "Survey of Sonic Boom Phenomena for the Non-Specialist" (conducted by a private research company). For the report literature on this research, the best guide is *Government Reports Announcements and Index*, published every two weeks by the National Technical Information Service of the Department of Commerce. It indexes reports from 22 subject areas, including agriculture, behavioral and social sciences, biology, chemistry, physics, and space technology. The introduction to each issue gives complete instructions on how to use the index. Figure 4-11 shows a typical page.

For keeping track of the work done by congressional committees, the *CIS/Index to Publications of the United States Congress*, published by the Congressional Information Service of Washington, D.C., is the best guide. It provides indexes and abstracts to House and Senate hearings and reports. It's a valuable guide to the many documents that grow out of congressional committee work.

A good many states publish yearbooks, manuals, data books, statistical abstracts, and so forth that contain a wealth of information on various subjects. The table of contents from *The State of Hawaii Data Book 1981* reprinted in Figure 4-12 illustrates the range of material available. States also publish reports, brochures, monographs, and similar items in a manner similar to the federal government. For a listing of the major publications of the 50 states, see the *Monthly Checklist of State Publications* published by the Library of Congress.

Many libraries also carry some of the documents published by the United Nations. A good many of these documents pertain to science and technology. A sensible procedure here is to read through *The Yearbook of the United Nations* to survey the activities of the UN's major scientific agencies during that year—the International Atomic Energy Agency (IAEA), Food and Agriculture Organization (FAO), United Nations Educational, Scientific, and Cultural Organization (UNESCO), World Health Organization (WHO), and World Meteorological Organization (WMO).

Armed with information gathered from *The Yearbook*, you can find your way around in the *United Nations Documents Index*, which keeps track of UN publications. Also check your library's card catalog under "United Nations" or the names of the agencies listed above. Check also to see if your library carries any of the more specialized indexing and abstracting services published by the UN, such as *FAO Documentation* or *Aquatic Sciences and Fisheries Abstracts*. Both of these are published by the FAO and exemplify the services that the UN provides.

COMPUTERIZED INFORMATION RETRIEVAL

Many libraries now offer access to computerized sources of reference material. That is, instead of your doing all the work normally involved in tracing useful sources of information, you can hire a

Prompt Copper Recovery from Mine Strip Waste.
Rept. of investigations,
B. W. Madsen, R. D. Groves, L. G. Evans, and G. M. Potter. Mar 75, 25p BuMines-RI-8012

Descriptors: *Copper ores, *Strip mining, *Leaching, Wastes, Materials recovery, Flotation, Leaching, Molybdenum, Silver, Gold, Copper, Size separation, Reclamation, Extractive metallurgy.
Identifiers: *Mine wastes.

In conventional dump leaching of sulfide waste, generally less than one-quarter of the copper is extracted. Gold, silver, and molybdenum that may be present in the dump are not recovered. Much of the sulfides in mine strip waste occurs along natural fracture planes. Some of these sulfides are liberated with enrichment of the fines fraction when the waste is broken during mining or in subsequent crushing. In research to take advantage of such enrichment, the treatment of three sulfide copper mine stripping waste samples was investigated by sizing or by crushing and sizing, followed by floatation of the enriched fines and leaching the remaining coarse rock. In the treatment of a chalcocite-bearing monazite mine waste sample, as much as 35 percent of the copper, 38 percent of the molybdenum, and 19 percent of the silver originally in the waste were recovered by floatation of the fines. An additional 33 percent of the copper was recovered by leaching the coarse fraction for 500 days.

PB-241 218/7GA PC$3.25/MF$2.25
Bureau of Mines, Albany, Oregon. Albany Metallurgy Research Center.
Recovery of Nickel and Cobalt from Low-Grade Domestic Laterites.
Rept. of investigations,
R. E. Siemens, P. C. Good, and W. A. Stickney. Mar 75, 19p BuMines-RI-8027

Descriptors: *Cobalt, *Nickel, *Laterites, Reduction(Chemical), Carbon monoxide, Leaching, Extraction, Electrowinning, Ammonium hydroxide, Ammonium sulfate, Extractive metallurgy, Magnesium.
Identifiers: Low grade deposits.

A process is being developed by the Bureau of Mines to selectively recover nickel and cobalt from low-grade domestic laterites. In laboratory evaluation of the process, the oxides in the laterite were selectively reduced with carbon monoxide at temperatures from 350 to 600C. For material containing more than about 5 percent magnesia, pyrite additions or post reduction heat treatments were necessary to achieve satisfactory nickel and cobalt extraction for this range of reduction temperatures. Multistage leaching of the reduced material at ambient temperature and pressure in the presence of oxygen, ammonium sulfate, and ammonium hydroxide extracted up to 92 and 87 percent of the contained nickel and cobalt, respectively. The nickel was selectively recovered from the leach solution by solvent extraction and was then stripped from the loaded organic with dilute sulfuric acid to provide a nickel-rich electrolyte. Treatment of the raffinate with hydrogen sulfide, resulted in the recovery of cobalt as a sulfide. The only contaminant in solution was magnesium which was removed by ion exchange or precipitation.

PB-241 343/3GA PC$4.75/MF$2.25
Missouri Univ., Rolla. Dept. of Geological Engineering.
Effect of Mining Operations on Ground Water Levels in the New Lead Belt, Missouri.
Completion rept.,
Don L. Warner. Dec 74, 92p W75-06986, OWRT-A-060-MO(2)
Contract DI-14-31-0001-3825

Descriptors: *Ground water, *Mining, Water supply, Aquifers, Lead ore deposits, Mine waters, Dewatering, Pumping, Missouri.
Identifiers: *Water levels.

In order to work the underground mines of the New Lead Belt, it is necessary to dewater the Bonneterre Formation of southeast Missouri. This requires pumpage of several hundred to several thousand gpm, depending on the mine location. The effect of this pumping on groundwater levels was evaluated. The study showed that it is probable that mine pumping does not effect groundwater levels in the deep aquifer beyond a distance of about five miles from any of the mines. Major areas of influence are still more restricted. As mining continues, the area influenced will extend further from north to south, but will probably not expand much eastwest.

PB-241 504/0GA PC$3.75/MF$2.25
Continental Oil Co., Ponca City, Okla. Research and Development Dept.
Seismic Mine Monitor System. Phase II.
Research rept. Aug 73-Jun 74,
James C. Fowler. Oct 74, 46p BuMines-OFR-24-75
Contract H0133112

Descriptors: *Seismic detection, *Monitors, *Search and rescue, Mining, Safety, Mines(Excavations).
Identifiers: Mine safety.

The research studied the possibility of using a seismic listening system to detect and locate miners trapped in areas where other communications were not available. This report covers the installation and testing of a permanent seismic monitor system at the Loveridge Mine in West Virginia. The testing of the system showed the following: Getting the seismic data back to the central computer in a usable form is possible; Detecting small explosions is possible; Using this system as an auxiliary system to detect hammer blows from miners trapped in the area of one of the geophones is possible.

PB-241 629/5GA PC$4.75/MF$2.25
Geological Survey, Denver, Colo. Geologic Div.
Selected Bibliography Pertaining to Uranium Occurrence in Eastern New Mexico and West Texas and Nearby Parts of Colorado, Oklahoma, and Kansas.
Interim rept.,
Warren I. Finch, James C. Wright, and Michael W. Sullivan. 1975, 100p USGS/GD-75/003

Descriptors: *Uranium ores, *Bibliographies, Geology, Groundwater, Stratigraphy, Geophysical prospecting, New Mexico, Texas, Colorado, Kansas, Oklahoma.

Nearly 500 selected references to uranium and to stratigraphy, structure, and groundwater geology related to uranium-bearing formations in eastern New Mexico and West Texas and nearby parts of Colorado, Kansas, and Oklahoma are indexed topically and geographically. The list is nearly complete through 1972 and contains some references with later dates.

8J. Physical Oceanography

AD-A009 442/5GA PC$3.25/MF$2.25
Woods Hole Oceanographic Institution Mass
Measurements of Vertical Fine Structure in the Sargasso Sea.
Technical rept.,
S. P. Hayes, T. M. Joyce, and R. C. Millard, Jr. 13 Jul 74, 10p Rept nos. WHOI-75-24, WHOI-Contrib-3453
Contract N00014-66-C-0241
Office of Naval Research, Arlington, Va.
Availability: Pub. in Jnl. of Geophysical Research, v80 n3 p314-319, 20 Jan 75.

Descriptors: *Sargasso Sea, *Internal waves, Microstructure, Temperature, Depth, Sea water, Bathythermograph data, Electrical conductivity, Salinity, Spectrum analysis, Thermoclines, North Atlantic Ocean, Reprints.

The vertical fine structure statistics in the northwest Atlantic (midocean experiment area) have been studied by using the WHOI/Brown CTD (conductivity, temperature, and depth). Five depth intervals including water masses representative of the entire water column have been subjected to spectral analysis of the temperature fine structure. All intervals show a similar power law dependence upon vertical wave number of -2.5, although spectral levels vary by more than a factor of 1000, whereas differences of only a factor of 2 exist on estimated vertical displacement spectra. Results are not inconsistent with the supposition that internal waves are responsible for much of the variability. On scales smaller than 10 m in the main thermocline, features with the characteristic signature of sheets and layers begin to appear.

AD-A009 444/1GA PC$3.25/MF$2.25
Woods Hole Oceanographic Institution Mass
The Temperature and Salinity Fine Structure of the Mediterranean Water in the Western Atlantic.
Technical rept.,
S. P. Hayes. 14 Mar 74, 14p Rept nos. WHOI-75-27, WHOI-Contrib-3273
Contract N00014-66-C-0241
Availability: Pub in Deep-Sea Research, v22 p1-11 1975.

Descriptors: *Sea water, *Sargasso Sea, North Atlantic Ocean, Mediterranean Sea, Microstructure, Mixing, Temperature, Salinity, Advection, Density, Depth, Gradients, Temperature inversion, Reprints.
Identifiers: Pycnoclines, West Atlantic Ocean.

The temperature and salinity fine structure of the Mediterranean Water in the Sargasso Sea (28 deg N, 70 deg W) have been studied with the Woods Hole Oceanographic Institution/Brown CTD microprofiler. The features observed are shown to be due to the advection of water with different temperature and salinity characteristics along isopycnal surfaces. The vertical density gradient of the water column is unaffected by the presence of the temperature inversion. (Author)

AD-A009 526/5GA PC$3.25/MF$2.25
Scripps Institution of Oceanography La Jolla Calif
NORPAX Highlights. Volume 3, Number 3, April 1975. Large-Scale Thermal Structure during POLE. Pacific Sea Level Stations,
T. P. Barnett, S. D. Rearwin, and Klaus Wyrtki.
Apr 75, 19p
Contract N00014-69-A-0200-6043
See also report dated Feb 73, AD-755 755.

Descriptors: *Ocean tides, *Bathythermograph data, *Pacific Ocean, Temperature, Bathythermographs, Heat, Flowmeters, North Pacific Ocean, Ocean currents, Sea level.

Contents:
Large-scale thermal structure during POLE; Pacific Sea level stations.

AD-A009 602/4GA PC$3.25/MF$2.25
Oregon State Univ Corvallis School of Oceanography
Temporal Variability of Suspended Matter in Astoria Canyon,
William S. Plank, J. Ronald V. Zaneveld, and Hasong Pak. 3 Apr 74, 7p Rept no. Ref-74-21
Contract N00014-67-A-369-0007
Availability: Pub. in Jnl. of Geophysical Research, v79 n30 p4536-4541, 20 Oct 74.

FIGURE 4-11 Government Reports Announcements and Index. (From U.S., Department of Commerce, National Technical Information Service, *Government Reports Announcements and Index* 75, no. 14 (1975), 70.)

68

CONTENTS

SECTIONS

FIGURE 4-12   The State of Hawaii Data Book, 1981

computer to do much of it for you. Computerized information retrieval systems are now available in such areas as agriculture, business, chemistry, environmental studies, and education. For a listing of major systems, see Appendix B, pages 519–520.

The United States government, through its National Technical Information Service (NTIS), now offers computer searches covering the thousands of documents published by the federal government or under its sponsorship. The way in which NTIS operates is typical of many such services.

NTIS offers both published searches and custom searches. For a published search, you look in the NTIS catalog, *NTISearch*, for published searches in your field. For example, under the subject heading "Air Pollution" you'll find the catalog entry concerning automobile air pollution (reproduced in Figure 4-13). If you order this search, using the appropriate number listed, you'll receive 190 bibliographical citations and abstracts for research reports concerning automobile air pollution. Figure 4-14 shows one such citation and abstract. The cost for such a search is currently $30.

NTIS, and most other computerized retrieval systems, can go beyond merely giving you citations and abstracts. Most will offer you, in either paper or microfiche, copies of the cited reports. Microfiche is a form of microfilm that saves both money and storage space. Look at the bottom line of Figure 4-14. If the cited report appears useful to you, a copy of it can be obtained from NTIS by using the appropriate order number. In paper it will cost you $5; in microfiche, $3. Obviously, for a price, you can save yourself a lot of legwork.

NTIS also offers customized searches. Using this system, you (usually after consultation with an information retrieval specialist) provide NTIS with a list of key words called descriptors. (See Figure 4-14 for a list of typical descriptors.) NTIS then searches its entire data base for any articles stored under the descriptors you have furnished. Such searches can cost $100 and up.

Computerized information retrieval systems have multiplied significantly since the 1970s. Their success indicates that such services will play an increasing research role in the 1980s and beyond.

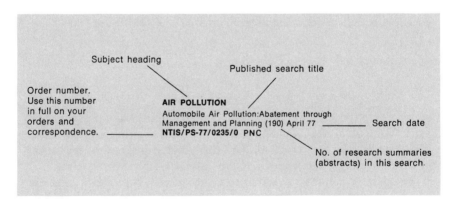

FIGURE 4-13   NTIS Catalog Entry. [From U.S., Department of Commerce, National Technical Information Service, *NTISearch* (Springfield, Va., 1978), 3.]

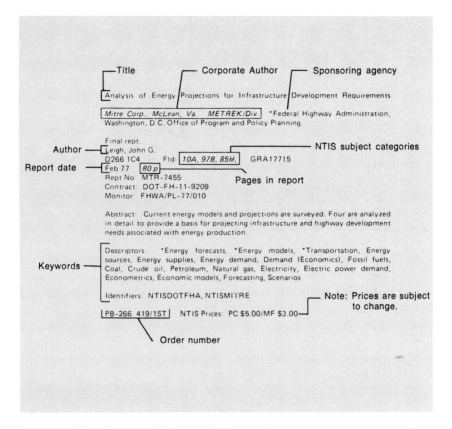

FIGURE 4-14 NTIS Citation and Abstract. [From U.S., Department of Commerce, National Technical Information Service, *NTISearch* (Springfield, Va., 1978), 4.]

CONCLUSION

Our aim in this chapter has been to help you find your way in any library, regardless of its size. Only when you are armed with the facts can you really be a good writer. A knowledge of the facts makes you a more valuable member of your profession and, indeed, a better citizen. Knowledge is power, and libraries contain most of our knowledge. If you want to go beyond the basic information provided here, check out from your library some of the books listed under *Library Research* in Appendix C, "A Selected Bibliography." Also look through Appendix B, "Technical Reference Guides."

EXERCISES

1. See if your library prints a guide to its holdings and operation. If it does, obtain a copy for future use.

2. Go to your library's card catalog. Locate a book of interest to yourself. Find the book in the stacks, take it to the circulation desk, and check it out. Write out full bibliographical information for the book:

author(s) or editor(s)	publisher
title	city of publication
edition	date of publication

3. Go to your library's reference division. Obtain the following information:
a. From Sheehy's, list the indexes and abstract journals in the discipline closest to your intended major or in a discipline in which you think you may be interested.
b. In Ulrich's, find a journal in your field of major interest. Give the following information for the journal:

name of journal
editor's name
circulation
full names (not abbreviations) of indexes and/or abstracts that cover the journal.

c. From the *Oxford English Dictionary*, copy the first three lines of a definition of a word that begins with the same first two letters as your last name (example, Smith—smack).
d. In each of the following categories, furnish the names and call numbers of at least one reference work that you found in the reference division:

encyclopedias	bibliographies
dictionaries	yearbooks
atlases	indexes
biographical works	abstracts

4. See if your library has the *New York Times* on microfilm. If so, find the following information:
a. Two major headlines from the front page of the *New York Times* for the day that you were born.
b. The name and price of an article of clothing listed in an advertisement in the *New York Times* on the day that you were born. Name also the firm that placed the ad.

5. Go to your library's periodical division and find the following information:
a. The cover story of a news magazine such as *Time* or *Newsweek* for the month and year in which you were born.
b. The name of the first periodical shelved in the reference division whose title begins with the same letter as your last initial. In the event that no periodical begins with the same letter, go forward or backward in the alphabet until you find one.

6. See if your library has any holdings in government publications. If so, find the following information:
a. From *Government Reports Announcements and Index* find the descriptors from an article in the discipline closest to your major.
b. In a recent *Monthly Catalog of United States Government Publications* find the name of a report published by a government agency that relates to your major field.

7. See if your library has access to any of the computerized information retrieval services listed in Appendix B. If so, list the services available.

Chapter **5**

Gathering and
Checking Information

IN CHAPTER 2, we discuss how to get started in a writing project, how to determine topic, audience, purpose, and your role as a writer. When you have arrived at some clear statement of all these variables (preferably in writing), it's time to begin gathering the information you'll need. In addition to information gathering methods, you'll need some techniques for checking and reviewing your information as you gather it, both to stay on course and to correct your course as necessary. The methods and techniques you may use include but are not limited to the following:

Calling upon Your Memory
Searching the Literature
Generalizing from Particulars and Particularizing from Generalizations
Inspecting Local Sites and Facilities
Administering a Questionnaire
Checking Reader's Attitudes and Requirements
Interviews
Letters of Inquiry
Performing Calculations and Analyses
Reviewing the Information Already Gathered

CALLING UPON YOUR MEMORY

In gathering information begin with yourself. What do you already know about your subject? You may already possess all the information you need; then again, you may not. Let's continue, as an example, the research problem we set in Chapter 2, concerning the effect of the aging of America on a small restaurant chain. We ask you to imagine once again that you are a young restaurant executive planning a report for your new boss. (See pages 17–18.) Inventory what you already know. Again, it pays to do this in writing. Here are some assorted facts you may have in your head:

- America's birthrate is declining and people are living longer. Therefore, the average age of the American population is increasing.
- The birthrate during the ten years or so after the end of World War II in1945 was higher than normal. The people born then are now something of a bulge moving through the population. An article in *Newsweek* last year had useful information about this.
- In a restaurant management course, the professor referred to a highly successful restaurant chain in Chicago. The chain has a tricky name, Lettuce Entertain You, and its customers are primarily what your professor, David Lockwood, laughingly called YUPS, for young urban professionals.
- The Cornell University School of Hotel and Restaurant Management publishes a journal that contains excellent articles on hotel and restaurant management. The journal should be worth browsing through.
- Somewhere recently an article appeared on how to evaluate the effect a restaurant's environment had on its customers. Could it have been in the Cornell journal? The article said that lights too dim had a negative effect on customers, and bright lights increased the speed with which people ate and left the restaurant.

If you are normally observant, you could extend this list several times over. Someone interested in restaurant management could no doubt extend the list to several dozens of items. Here, near the beginning of what is often called the prewriting stage, you can get more down in writing than you realize. Even some notions about organization may appear at this stage.

SEARCHING THE LITERATURE

After you have searched your own brain for information on your subject, the next logical step is to search the brains of others. The library is the best starting point to learn what others have found out and recorded about your subject. In a very real way, the library is our collective memory. (If you do not yet feel very expert in using a library, we recommend to you our Chapter 4, "Finding Your Way in the Library" and Appendix B, "Technical Reference Guides.")

If you are a college student, you may have a choice of three or more libraries; your town library, the main library on campus, the library maintained by each school of your college and, sometimes, by a department within a school.

Because of its nearness to your living quarters, or some equally practical consideration, you decide to try the main library on central campus. After an hour of searching through the subject card files and a few indexes and consulting with the reference librarian, you are pleased to have located these six items as a start on your "working bibliography":

Working Bibliography

Backus, Harry. 1977. Designing restaurant interiors. New York: Lebhar-Friedman Books.

Lawson, Fred. 1973. Restaurant planning and design. New York: Van Nostrand Reinhold.

Davern, Jeanne M., ed. 1976. Places for people. New York: McGraw-Hill.

U.S. Bureau of the Census. 1981. Statistical abstract of the United States. Washington, D.C.: U.S. Government Printing Office.

Lambert, Carol U. 1981. Environmental design: The food-service managers's role. The Cornell Hotel and Restaurant Administration Quarterly 22, no. 1:62–68.

1980. Counting the house: The 1980 U.S. census. The Cornell Hotel and Restaurant Administration Quarterly 20, no. 4:14–16.

Eventually, you will be able to extend this initial working bibliography to several times its present length. But at the start, you have evidence that a sizeable quantity of reading matter will be available to you.

Now you have an important decision to make—which information source to read first. In general, it is best to select a published source that is large, recent, authoritative. On the basis of these criteria, the fourth source probably should be your first choice. You find the book and start to open it. But hold on a moment! How are you going to extract and use any pertinent information it contains? Having a method of doing this will prove critically important to you later as you turn to organizing and writing your report.

Many researchers and report writers find that converting recorded data into prose reports is unbearbly torturous and time-consuming. It is true, of course, that there are no magical shortcuts or miracle methods that will produce "instant reports." However, there is no sense whatsoever in plunging blindly into the information-recording-and-compiling process or in continuing with methods that have proved cumbersome in the past. Has it been, perhaps, your practice to copy research information onto the sheets of a notebook?

If so, you know the battle that ensues when your individual notes have to be sorted and somehow placed in usable order for compilation into a report. You have to leaf back and forth through the notebook to identify and locate notes bearing upon the topic of present interest. Some notes may escape your attention: others may accidentally be used twice. Still worse, the resulting text is likely to be disjointed and badly organized.

Abandon the notebook practice for taking notes. Instead, take your notes on cards of four-by-six-inch size, or even larger. On one set of cards you will keep the bibliographic data you need, one card per reference. By having each reference on a separate card, you'll be able, when it comes time to do your final bibliography or references, to sort your cards into any sequence, alphabetical or otherwise, that you need.

On a second set of cards, keep the notes that you extract from your sources. Limit the information on each card to a single narrow topic—one that you may label with a few identifying words such as *Installation Cost, Service Procedures*, or *Maintenance*. Place the topic identifier prominently on each card, perhaps at the upper-left corner. You thus make it possible to sort through the entire collection of cards and organize into subpacks all cards bearing the same topic identifier. Now you can, at any stage of the process, copy onto a sheet of paper all of the topic identifers and thus be able to inspect the extent of your information in topical array.

When you reach the final organization stage, you will be able to draft one or more tentative organizational plans to hold and arrange the information for the body of your report—and at the same time be able to detect omissions, duplications, overlaps, and irrelevancies. (For further information on organizing your material, see Chapter 8.)

Once you have created the subpacks, you can make a still finer sorting of cards in each of the subpacks until you have them arranged in order for actual composition. It is a relatively straightforward process to pick up card 1 of subpack 1 and consume its information in order to write the first paragraph of your first body chapter. Prior sorting, permitted by the card note-taking process and the use of topic identifiers, will have freed your mind of the necessity to hunt and sort while composition is in process. Hence you can give your full attention to the digestion of information and to its clear and coherent expression in prose.

If you do not have a note-taking and note-using method that works for you, try the method we have outlined. The method works for thousands of researchers and report writers, both student and professional.

To return to our example problem: at this point, you have selected your first reference, *Statistical Abstract of the United States.* Make a bibliographic identification card for it, as in Figure 5-1. Notice the brief identifier placed in the upper-right corner. Such bibliographical identifiers will save you much time and writing later, as you will shortly see.

Now let your reading and note taking begin. In the *Statistical Abstract* you are looking for demographic data to confirm that the average age of the American population is increasing. Looking through the table of contents you note a table labeled "Resident Population, by Age, Sex, and Race 1960–1980." Turning to it, you find that you don't need the entire table but that you can extract from it exactly the information you are seeking. You copy the information onto a note card, as illustrated in Figure 5-2. The label, *Stat ab*, in the upper-right corner of the card in Figure 5-2 identifies the source as that fully described by the bibliographic identification card you made out earlier. The label in the upper-left corner, *Age groups, 1960, '70, '80*, is the topic identifier. The information at the bottom of the card is the number and name of the table and the page it is found on. You will need this information to document your source when you write your report or to find the source again should you need to.

You search on through the *Statistical Abstract* extracting further useful information. You place each piece of information on a separate card with appropriate labels and identifiers. The more careful you are at this stage, the easier your job will be when it comes time to write the report.

When you have extracted the information you need from the *Statistical Abstract*, you turn your attention to the article "Counting the

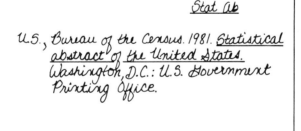

FIGURE 5-1 Bibliographic Identification Card

age Groups   Stat ab
year   age Groups

    5-9 10-14 15-19 20-24 25-29 30-34 35-39
60 18,692 16,773 13,219 10,801 10,869 11,949 12,481
70 19,969 20,804 19,084 16,383 13,486 11,437 11,113
80 16,697 18,241 21,162 21,313 19,518 17,558 13,963
(Numbers in thousands)

Table 29, "Resident Population, by age, sex
and Race: 1960 to 1980," p. 26.

FIGURE 5-2 Notecard

House: The 1980 U.S. Census." You read through it and decide it contains information you need. As a first step before extracting information, you prepare a bibliographical card for it, as illustrated in Figure 5-3. Notice that for an article, the complete information needed includes the volume number (20), the issue number (4), and the inclusive page numbers (14–16).

In this article you find a passage that you think may lend authority to your paper. Because the language of the passage is economical and informative, you decide to copy it verbatim onto a card as shown in Figure 5-4. Note again the use of labels and identifiers on the note card. The label "p. 14" in the lower-right corner indicates the page from which the extract was taken. Because this is a verbatim quote, you have put quotation marks around it.

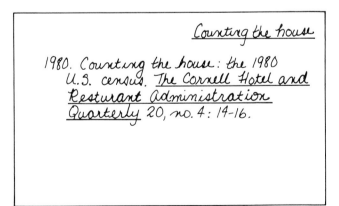

Counting the house

1980. Counting the house: the 1980
U.S. census. The Cornell Hotel and
Resturant Administration
Quarterly 20, no. 4: 14-16.

FIGURE 5-3 Bibliographic Identification Card

Census data are invaluable tools for the savvy marketer. They tip him off to the profitable business prospect by telling him where the money is, who's got it, and where they live.

For example, knowing that members of the baby-boom population bulge have reached maturity is critical to those whose products are aimed at the dwindling youth market. Older Americans, who have historically enjoyed the dubious distinction of being the target of *no one's* marketing effort, are growing not only in numbers but also in affluence. Census data are the primary source of information on such trends.

> Baby boom   Counting the house
>
> ". . . knowing that members of the baby-boom population bulge have reached maturity is critical to those whose products are aimed at the dwindling youth market."
>
> p. 14

FIGURE 5-4 A Verbatim Abstract

Further on in the article, you find a paragraph that contains valuable information. In this case, you don't need the entire passage, so you extract only the information you do need, as shown in Figure 5-5. When you condense or parapharase material taken from a source, avoid using the sentence structures of the original. In that way, you eliminate any hint of plagiarism, always a troublesome thing in research papers.

Your reading and note taking continue. After finishing "Counting the House: The 1980 U.S. Census," you read other books and articles. You return to the card catalog and indexes to add to your original list of readings. You uncover additional sources referred to in your readings. Things go well. You read, take notes, and make plans.

GENERALIZING FROM PARTICULARS AND PARTICULARIZING FROM GENERALIZATIONS

Now that you have done some reading in the literature, you may want to lean back in your chair and take a long hard look at what you learned so far. You have learned, for example, that research data on how different restaurant environments affect people are distressingly scarce. What makes one environment work and another fail is somewhat unclear. You have a few facts that seem solid. People do not like restaurants to be too dimly lit. They want to be able to read the menu without the help of a match. If the restauranteur's objective is a rapid turnover of customers, bright lights help. People seem to eat more quickly and leave under such conditions. You know that

Business observers who optimistically brandish terms like "discretionary income" and "four-day work week"—and their name is legion in the hospitality industry—may be lulled into reflective silence when they peruse the '80 census results. In comparison to the 34-percent increase registered in median family income between 1960 and 1970, the four-percent rise observed between '70 and '78 looks anemic indeed. And the factors that have slowed income gains to a crawl—economic sluggishness, a rise in the number of single-parent families, the changing age makeup of the population—have, if anything, been more salient during the last two years than in the earlier part of the decade.

```
Family Income                    Counting the house

Median family income increase
down:
        1960-70     34% increase
        1970-78      4% increase

                                    p. 15
```

FIGURE 5-5   A Condensed Extract

spacing is important to people and that Americans' reactions to space have been fairly well mapped out into four zones:

Intimate distance: 0–18 inches
Personal distance: 18 inches–4 feet
Social distance: 4–12 feet
Public distance: 12 feet and beyond

From that set of data, you generalize that most tables in a restaurant should be spaced so that diners are at least 18 inches away from diners at other tables. But how much over 18 inches? Would it have to be as much as 4 feet? Space is expensive. Furthermore, you have had some cheerful experiences in crowded restaurants. You remember one in particular. On your eighteenth birthday, your mother and father had given you your present while at dinner in a crowded restaurant where the tables were extremely close together. The people at the next table had asked to see the present and had passed it around commenting on how nice it was. Rather than resenting the intrusion, you and your parents had enjoyed it. Somehow it made the occasion more festive.

You combine this personal experience with a researcher's report that in crowded situations people are more likely to talk to one another. Would conversations carried on between tables be desirable or undesirable for the young urban professionals (YUPS) you are beginning to see as your target clientele? You remember that in the *Statistical Abstract* you learned that the numbers of single people and people living alone have greatly increased since 1960. In 1960, 7 million people lived alone. By 1980, that number had increased to almost 18 million. Could some of the people living alone be lonely and therefore enjoy the chance for a more intimate restaurant atmosphere that encourages conversation?

You think and you ponder about these particular facts and draw generalizations from them. Some of your generalizations seem pretty firm; others you are doubtful about. They will need more checking.

The kinds of reasoning you have been following here have their formal names in logic. Reasoning from one or more particular facts to reach a covering generalization is called *induction*. Notice the flow of particulars in this excerpt from an article on how people use space in cities:

> When people stop to have a conversation, we wondered how far away do they move from the main pedestrian flow? People didn't move out of it. They stayed in or moved *into* it, and the great bulk of the conversations were smack in the center of the flow. The same gravitation characterized "traveling conversations"—the kind in which two people move about, alternating the roles of "straight man" and principal talker. Although there is a lot of apparent motion, if you plot the orbits, they turn out to be quite restricted.
>
> People also sit in the mainstream. At the Seagram plaza, the main pedestrian paths are on diagonals from the building entrance to the corners of the steps. These are natural junction and transfer points and the site of lots of activity. They are also a favored place for sitting and picknicking. Sometimes there will be so many people that pedestrians have to step carefully to negotiate the steps. The pedestrians rarely complain. While some will detour around the blockage, most will thread their way through it.[1]

The generalization to be drawn from these similar particulars is stated in the article: "What attracts people most, it would appear, is other people."[2]

The other kind of reasoning called *deduction* leads from the general to the particular. In deduction you apply a generalization to deduce a conclusion. For example, as the young restaurant executive you have been musing about the spacing required between tables. The particulars of the research just cited as well as other research and your own personal experience have given you the generalization that people really don't mind being crowded a bit. In fact, urban people may even enjoy it. Using this generalization, you deduce that spacing between tables in a restaurant designed for YUPS could be minimal, perhaps right at 18 inches. Your deduction has given you a conclusion. Rational thinking is based upon this constant flow of induction and deduction—generalizing from particulars and particularizing from generalizations.

Our experience would also suggest another generalization: to continue your research you must delve into an additional source of information, namely the local scene.

INSPECTING LOCAL SITES AND FACILITIES

Library research is vital to most investigations. A few hours in the library can turn up data that represent months and even years of work. Library research can keep you from reinventing the wheel. Nonetheless, in most investigations there comes a time when you want to go and look and see, listen and hear, touch and feel for yourself.

On your tour of inspection, take along a notebook to jot down whatever you observe. At this time, do not attempt to be particularly logical: your on-the-spot notes can always be organized later. In this case, you decide to observe people in a local restaurant, Scarpelli's, that you know is popular with people in their twenties and thirties. You call the owner, Sandy Dolan, get his permission, and spend several hours people watching during the evening meal hour. Your jottings read as follows:

- Dress varies, perhaps in accordance with whether people came directly from work or not. Some people are in obvious business attire, others in jeans, open-necked shirts, and casual jackets. Footwear varies from well-shined black boots to scuffed jogging shoes. Despite the apparent casualness, people are not sloppy. They seem to care about how they look.
- The room is clean and well lighted, though not enough to produce a sense of glare. The walls are white. Tablecloths are white and heavy feeling. There are large mirrors in dark frames around all the walls. There are coat hooks between the mirrors, and along the walls tables are closely lined up between the coat hooks. Looks a bit like a restaurant in a French movie.
- People seem to know each other. There is some conversation between tables, and some people get up and speak to people at other tables. Atmosphere is relaxed and friendly.
- Waiters, male and female, are dressed in black slacks, white shirts, and short white aprons—kind of androgynous.
- Tables throughout the room are quite close together. Barely room for waiters to get through. People are cheerful about it. They hitch up their chairs almost without thought to make room when necessary. The noise level is fairly high. People are not shouting, but there is a steady sound of voices. People are talking as much as eating. They are enjoying themselves.
- Tables seem to fill and clear out at about 50-minute intervals. People have to wait for a table but not too long.

Depending upon your powers of observation and your knowledge of the facts of restaurant life, you might fill several notebook pages with such comments in an hour or two of observing. In any case, you will have familiarized yourself with a restaurant of the type you are beginning to envision yourself recommending. You still have a lot of questions, maybe more than answers, but that is entirely normal and healthy in the early stages of research.

To continue your research, several courses are now open to you. In this state of open-mindedness, it occurs to you that you would like to get some reactions from people about restaurants, perhaps your restaurant and the one you have just observed, Scarpelli's. A questionnaire administered to patrons might yield some interesting results. You decide to try it.

PREPARING AND ADMINISTERING A QUESTIONNAIRE

It is not easy to construct and administer questionnaires and tabulate their replies without years of study and experience. Even the "experts" do not pretend to have solved all the problems. We do not intend to make an instant expert of you, but we can offer some basic advice that should enable you to handle questionnaires of limited complexity and scope.[3]

It is always an imposition to ask someone to fill out a questionnaire. You should keep this fact in mind when the time comes to design and administer a questionnaire yourself. Fortunately, you can usually approach your subjects in such a way that the questionnaire does not seem an imposition.

First, be sure that a questionnaire is really necessary. Questionnaires should be used only when the desired information cannot be obtained from another source—interviews, personal observation, searches through daily newspapers, courthouse and library files, and so on. Time, money, and effort can be saved by avoiding the questionnaire process whenever possible.

Whether or not you decide a questionnaire is indicated, the research you do before making that decision is not wasted. If you find no questionnaire is needed, that means your research yielded a lot of the answers you thought might require a questionnaire, and you're farther along than you had thought. If you *do* need to use the questionnaire approach, the information you've obtained about your topic will be invaluable in designing that questionnaire. Among other things, your research should help you to limit the number of questions. As a general rule of thumb, the fewer questions you ask, the more likely you are to get cooperation from your respondents. Ask only for pertinent information and only for information that you can't get through other research avenues.

The extent of your preliminary research will vary, of course, depending on the amount of time involved. If the questionnaire is part of a project for a one-semester or one-term course, you will have to tailor your topic to suit the available time and plan your research accordingly. With a relatively narrow, straightforward topic (like our restaurant example), a week or so of study should suffice to establish your credentials as an authority and enable you to write an authoritative questionnaire. But if the questionnaire is part of a graduate-level thesis, your topic will undoubtedly be much broader—and you may need to spend six months or more simply acquiring the information that will serve as a basis for the questionnaire.

Drafting the Questions

When you've thoroughly studied your topic and determined that a questionnaire is definitely the best way to obtain at least some of the data you need, it's time to make up a list of questions. For several reasons, this part of the project is somewhat more difficult that it sounds.

First, your questions have to take into account the factor of *reliability*. That is, your questions should be worded so that they measure what you want to measure. In our restaurant problem, if you wanted to measure patrons' reactions to the environment of a restaurant, you would word your questions so that their reactions, good or bad, to the food do not color their answers. For instance, a general question such as "Did you enjoy your experience here tonight?" might measure overall customer satisfaction very well. But you could not use it by itself to interpret what part environment, food, or service played in the level of satisfaction. You have to word your questions specifically to measure what you hope to measure. As we'll see, our restaurant researcher managed to word the questionnaire to obtain fairly reliable results.

Second, your questions must be worded to ensure a *valid* interpretation of each question by the respondents—that is, an interpretation that matches your own. This requirement overlaps the first to some extent but also goes beyond it. In trying to write questions that express your meaning so that respondents cannot possibly misinterpret it, you must try to read each question through their eyes. Remember that although you have studied the topic in depth, the questionnaire may be their first introduction to it. Some people will know more than others, of course, but the idea is to formulate questions that will be clear to *everyone*. Returning to the restaurant problem, you would not learn much from the question, "Did you enjoy the ambiance of the restaurant?" Some of your repondents might not really know what is meant by *ambiance*. Even if they do know the meaning of the word (the atmosphere or environment surrounding one), they may not know what to include about ambiance in their answers. For instance, should they include the lighting, the waiters' attitudes, the presentation of the food, and so forth? Too many decisions are thus left to the respondents. With such a question both validity and reliability fly out the window. As in ensuring *reliability*, to ensure *validity* you have to word your questions very specifically, and you must use words that you know your respondents will interpret in the same way you do.

Third, you have several different *types* of questions to choose among in drafting your questionnaire. Once you have selected the type that best suits your purpose, in general, stick to it throughout the questionnare. Changing from one type of question to another in midstream may confuse your respondents. The six types of questions most commonly used on questionnaires are as follows:

1. *Dichotomous.* A *dichotomy* is a division into two separate parts, and a dichotomous question offers the respondent a choice between two answers—*yes* or *no, before* or *after, manual* or *electric*, and so on. Thus, this type of question provides a relatively limited span of responses and is primarily useful when the questionnaire topic can be analyzed in black-and-white terms.
2. *Multiple-choice.* A multiple-choice question allows the respondent to choose among several possible answers:
   In preparing copy for a printer, if you must attach printed pages to a backing sheet, which of these methods do you prefer?

   stapling __ taping __ gluing __ using paper clips __
   If you think that there may be other possible answers, you can add a blanket category, followed by a line for the respondent's answer: other _____. This approach is probably the most common for questionnaires; it permits a fairly wide range of responses, but the choices are sufficiently controlled by the administrator to permit easy interpretation of results.
3. *Ranking.* In ranking, the respondent is asked to rank several possibilities in order of personal preference:
   If your company required you to transfer to one of the following cities, what would be your order of preference? Place the appropriate number on each line, 1 for first choice, 2 for second choice, and so on:

   Boston __ Chicago __ Denver __ Miami __
4. *Rating.* In rating, the respondent is asked to rate the importance of an item to herself or himself:
   By placing a check at the appropriate place, indicate the importance of the following items in measuring your satisfaction with a hotel:

	Very Important	Somewhat Important	Not Important
Front desk service			
Room service			
Lounge service			
Restaurant service			

5. *Fill-in-the-blank.* This short-answer approach may be used to elicit either factual answers or opinions:

How many children do you have? (If none, write "none.") _____

How long have you held your present job? years _____ months
_____

What is your gross annual income (to the nearest thousand)? _____

6. *Essay.* An essay questionnaire allows the respondent maximum freedom in composing an answer. Questions are somewhat general and followed by an inch or more of space for the answers:

Do you approve of the president's fiscal policies? Why or why not?

_____

Despite the obvious advantage of complete and detailed answers, the essay method is the least efficient of the six described here, and it is therefore the least suitable for short-term projects. (By the same token, it could be the best for a long-term project in which thorough individual responses are important.) Essay questionnaires are time consuming for both administrator and respondent, handwriting may be illegible and difficult to decipher, and replies may be very difficult to tabulate.

Administering the Questionnaire

Once the questions are drafted to measure what you want to measure with no chance of reader misinterpretation and are formatted consistently, you have nothing more to worry about, nothing that is except the following: How many dozens or thousands of questionnaires should you administer? Should the administration procedure involve mailing, telephoning, or personal interviews? Will the local telephone directory supply enough respondents or should you use another source? Is there any guarantee that the respondents to your questionnaire will be representative of the population you hope to study?

There is no set formula for solving these problems. The answers will vary with each project, depending on the issue being investigated, the time period involved, any special circumstances that must be dealt with, and so on. For the sake of this discussion, however, let us assume that you as our restaurant researcher, with the help of an article in the *Cornell Hotel and Restaurant Administration Quarterly*, have drafted a satisfactory group of questions. You have decided to administer the questionnaire to the patrons of three restaurants: your own, Scarpelli's, and an expensive hotel dining room. You call the managers of the two restaurants you are not connected with and get their cooperation by agreeing to share the results of your questionniare with them. The questionnaires and sufficient pencils will be given to everyone sitting at a table by the waiter when he or she presents the check. The questionnaire is shown in Figure 5-6.

```
                    CUSTOMER ATTITUDE SURVEY

      Completing this survey will take only a few moments of
  your time.  Your answers will help us measure your reaction
  to this restaurant's physical environment (color, lighting,
  decoration, spacing, furnishings, and so forth.)  Please do
  not let your answers be influenced by the quality of food
  or service you have received tonight.  Please complete the
  survey and return it to your waiter or cashier when he or
  she asks for it.  Thank you very much for your help and
  cooperation.

  1.  Please tell us your age and sex.  Age____ Sex____

  2.  In each pair of statements below, please circle
  the statement that most accurately describes your
  response to the physical environment of this restaurant.

                  appealing / unappealing
                 attractive / unattractive
                     bright / dull
                    cheerful / gloomy
                comfortable / uncomfortable
                distinctive / ordinary
                  expensive / inexpensive
                     formal / informal
             good lighting / poor lighting
                   inviting / repelling
                     ornate / plain
                   pleasant / unpleasant
                   tasteful / tasteless
```

FIGURE 5-6 Questionnaire. [*Source:* Adapted from Carolyn U. Lambert, "Environmental Design: The Food Service Manager's Role," *The Cornell Hotel and Restaurant Administration Quarterly* **22**, no. 1 (1981): 62–68.]

Even the most intelligently devised questionnaires sometimes miss their mark, and the responses prove unrewarding or hard to tabulate. Therefore we suggest that you administer the first version of any questionnaire you devise on a trial basis. Then make whatever revisions seem to be indicated.

Do not expect questionnaires to produce miracles. If you mail them (stamped return-address envelope included, of course), do not be shocked if only 35 percent respond. House-to-house canvassing may or may not produce a higher percentage of responses, depending upon the time of day and the mood of the community among other things. Administering a questionnaire by telephone is a tedious procedure, but it permits you to cover a large geographical territory without pounding the streets. A combination of these means of administration may prove to be the best overall answer. Whatever method you choose, try to select your subjects on a random basis. By doing so, you can be surer of obtaining a representative sample.

In our restaurant situation, having licked the questionnaire problem, you stop to take stock again. It begins to seem likely that you will recommend some drastic changes in restaurant environment to your boss, Jane Lewis. You wonder if it would not be a good idea to check out her attitude toward such changes. You decide you had better talk to her.

CHECKING READERS' ATTITUDES AND REQUIREMENTS

In writing on the job, you usually have some notion of who your readers are, and you may even have close and daily contact with them. They may have a client relationship with you in that they may have hired you to do the research for them. Or, as in our restaurant example, there may be a boss-subordinate relationship. Whatever, if you do have such contact, you should check with your readers from time to time. It's another way of keeping your project on track.

In a large formal undertaking, you may stay in contact through progress reports. (See Chapter 16, "Progress Reports.") In a smaller, more informal situation, a few words of conversation or a short memo may do the job. Suppose, for example, your project is to write a set of instructions for calibrating a television set. You have reached the point in gathering information where you are investigating needed test equipment.The thought occurs to you that the company technicians who will be your readers may not know how to operate the test equipment. Do not continue in the dark. Go see or call the technicians and ask them what they know. If they are unfamiliar with the test equipment, you will have to include operating instructions for it. If they are familiar with it, you can refer to it and move on. In such ways, you can shape your report to your readers.

In our restaurant example, you may now be realizing that switching a restaurant from a teenaged clientele might be an expensive proposition. It might call for extensive remodeling. For various reasons, your boss may not want to undergo such expense. Call her or send her a memo. (See pages 311–313.) Summarize what you know to date, and ask if you should continue. If so, are there restrictions upon what you should be planning or thinking about? You do, and she gives you a green light. "Go ahead," she says. "Tell me what you find out and what you think I ought to do about it. We need to plan for the future."

Good writing is an act of constant refining of your material to suit your purpose and audience. Some of this refining goes on during the writing and rewriting stages. But much of it, and by no means the least important, goes on during this prewriting stage when you are gathering and checking your material.

INTERVIEWS

You have begun to sift through the answers to your questionnaire concerning the physical environment in your three selected restaurants. Some preliminary answers are becoming obvious. The young adults who answered the questionnaires definitely favor the environment of Scarpelli's. They do not care very much for the environment in either your restaurant or the expensive hotel dining room. There are interesting mixes of adjectives for each restaurant, but the pointers are becoming clear. For instance, young adults find your restaurant "bright" but "unappealing" and "ordinary." They find the hotel dining room "distinctive" but "gloomy" and "unappealing." They find Scarpelli's "bright," "cheerful," and "appealing." You decide that it's time to flesh out what you have learned through the literature, on-site inspections, and questionnaires with some interviews.

You list some people who may be able to help you:

Carl Strickland, Associate Professor of Food Management
Virginia Book, Professor of Interior Design
Sandy Dolan, owner-manager of Scarpelli's

Your instincts suggest that Sandy Dolan, who has run a series of successful restaurants for 20 years, may well have the broadest view of what it takes to please customers. You call him, and he agrees to talk to you on Thursday, three days hence.

If you have watched interviews on TV you probably have noticed that most experienced interviewers keep a tablet or clipboard in front of them and from time to time glance at their prepared questions. This tactic is helpful for the expert—it will be essential for you.

Prepare a list of questions in advance. In general, stay clear of "dead end" questions that can be answered with a simple yes or no. (The experts call such questions bipolar.) Instead, phrase your questions in a way that invites the person being interviewed to expand upon his or her first answers. Pass the ball to the interviewee and let that person do practically all the carrying. Phrase your questions clearly and economically; avoid ten-line "essay" questions that will eat up valuable time when your interviewee should be doing the talking. On the other hand, do not insist upon asking every question on your prepared list. If the interviewee discourses productively, let the discourse continue until he or she runs out of things to say. Also, do not insist upon posing the questions in their prepared order; instead, use the cues provided by what has just been said to guide you in selecting a reasonably relevant next question. If your interviewee permits it, tape-record the interview for later reference.

When you arrive for the interview, introduce or reintroduce yourself, giving your name, business or professional connections, subject matter for discussion, reasons for the interview, and the use to be made of the information gained. You should both sense the mood of the occasion and try to color the mood. For a half-hour interview, the opening five minutes may be given to generalities and personalities apparently far afield from the real reasons for your interview. Once rapport has been established, take some key remark to lead into one of your prepared questions. What follows may read, in part, like this:

INTERVIEW

Q. Mr. Dolan, I notice that your color scheme in Scarpelli's is simple, basically mirrors with dark frames and white walls. Why have you chosen this scheme?

A. Well, yes, I guess that's right. We try to keep it simple. I dislike busy decoration if you know what I mean—all candles and swords and capes and phony ancestral portraits. Keep it simple, bright, and cheerful is what we try to do.

Q. Why the mirrors?

A. Well, that's part of it, really. In a restaurant, in a very real sense, your customers are your decoration. They come in here dressed casually but pleasingly. They're informal but not grubby. They add the needed color to the room. The mirrors reflect all that back and intensify it. Think of all the different scenes played out in these mirrors—much better than static paintings.

Q. I sense that the room is French in feeling. Is that accidental?

A. No, not really accidental, probably more subliminal.

Q. I'm not sure I understand what you mean by that.

A. OK. I mean I probably didn't consciously design the room to look like a French café. But I've traveled in France, and I like French restaurants. I associate them with good food. I was just following my own instincts in here, and the design, probably in a subliminal way, reflected my admiration for things French.

Q. Do you think your customers react favorably to the French look?

A. I think so. They keep coming back anyway. Actually, if you were to analyze it, an ethnic look that's not overdone, that's kept simple, is probably a successful look today. People expect ethnic food—French, Italian, Mexican, Middle Eastern, what have you—to be good. Yes, maybe an ethnic feel in the decoration helps. But keep it simple. For instance, you could catch an Arabian look with stucco in a sand color and maybe a discreet use of brass and leather. But don't, for heaven's sake, fill up the place with burnooses on the wall and camel saddles for seats. See what I mean?

Q. Yes, I do. Thank you.

And so it goes. During an interview like this, you are looking for cues and clues. Most well-informed people, like Sandy Dolan, will talk at length if you let them. But do not expect blueprints and detailed estimates. Those will come later. You plan other interviews, perhaps with Professor Strickland and Professor Book. You also realize that it would be a good idea to follow up your questionnaires with some in-depth interviews of patrons of Scarpelli's. In the meantime, you drop a note of thanks to Sandy Dolan for his candid and helpful replies.

LETTERS OF INQUIRY

Like questionnaires and interviews, letters of inquiry can provide you with useful information that may not be available in a library. When writing a letter of inquiry, keep in mind that you are imposing upon the other person, asking for time and attention. Therefore, brevity without curtness is in order. Sometimes a list of specific questions is appropriate—plus a general solicitation for advice and information. But be careful. If you ask for too much, you may get nothing at all. If there is any favor you can offer in return, introduce this thought toward the end of the letter. For additional information on how to write letters of inquiry, see pages 289–293.

As the young restaurateur of our example, you have used an article by a Professor Carolyn U. Lambert in setting up your questionnaire. You have found the article quite useful. In one place, the article describes in a general way some research that might measure the effect of various table spacings and sizes in a restaurant. You wonder if a letter to Professor Lambert might not obtain more specific advice on how to go about such research. You write her as follows:

```
Dear Professor Lambert:

As a manager in the Jane E. Lewis, Inc., restaurant
chain, I am studying the possibility of remodeling
one of the restaurants in the chain. Specifically,
we are exploring the possibility of shifting the
restaurant's environment and menu to suit a young-
adult clientele rather than the present teenaged
clientele.
```

```
Your 1981 article, "Environmental Design: The Food
Manager's Role," has been most useful.  A questionnaire
based upon your sample adjective checklist elicited
precisely the information we were looking for.

One of the problems still facing us is deciding what
size tables to use and how to space them.  In your
article, you state that "Initial research in food-
service settings could identify and compare the space
required to eat and the space actually occupied by
patrons, as well as determine the impact of making
changes in a given space on conversation levels,
turnover, and other patterns of behavior."

The questions you identify in this passage are the
very questions we seek to answer.  Since 1981, has
any additional research been done in this area?
Could you, perhaps, specify more exactly how such
research could be accomplished?

I'll be happy to share the results with you of any
research I do in this area.  I enclose for you now
a summary of the preliminary results obtained using
your adjective checklist.

Thank you for your attention and for your very useful
article.

Sincerely,
```

You mail your letter of inquiry and turn to a problem that has been much on your mind. Given the cost of remodeling for a different clientele, will the payback be sufficient to make it worthwhile? You sit down with calculator, pencil, and paper to see.

PERFORMING CALCULATIONS AND ANALYSES

A month ago when you began your research into a new clientele for a restaurant, you did not think very much about what the cost could be to carry out any recommendation you might make. With the passing days, however, your ideas have begun to take shape. You're excited about the possibility of converting one of the restaurants in the chain from serving a teenaged clientele to what you have come to think of as YUPS. It now occurs to you to consider the financial aspects of your project. What will be the cost? How much will it cost to borrow needed money? Will the return on investment be worthwhile? You decide to make some rough calculations.

Several articles about restaurant renovation in one of the leading professional journals in the field, *Restaurant Hospitality*, have convinced you that $200,000 is about the minimum you can expect to spend for the kind of changes you have in mind. You decide to use a figure of $250,000 to allow for inflation and the extra costs that always seem to come up. A call to your boss, Jane Lewis, reveals that the figure doesn't frighten her off. She could put up $50,000 for a worthwhile project. The firm would have to borrow the remaining $200,000.

A call to a local bank reveals that a loan of $200,000 for business purposes would have to be for no more than five years at an interest rate of 14 percent. Because Lewis is a good customer of the bank, it will allow a "balloon" (that is, a *lump sum*) repayment of the principal at the end of five years. You calculate the cost of the $200,000 loan. There would be ten semiannual interest payments of $14,000 for a total of $140,000, a tidy sum of money.

The remodeled restaurant should seat 150 people. With good turnover, you should be able to count on two lunch sittings each business day of the week, that is, 300 people a day. On Saturdays, you plan on only 200 for lunch. For dinner, evidence from similar restaurants suggests you can serve approximately three sittings a night and thus serve 450 for dinner, Monday through Friday. The experience of the Lettuce Entertain You group that you have read about in *Restaurant Hospitality* indicates that you could plan on serving 900 people on Saturday night.

Let's see:

$$5 \times 300 \text{ lunches} = 1{,}500$$
$$\text{Saturday lunch} = \underline{200}$$
$$1{,}700 \text{ lunches per week}$$

$$5 \times 450 \text{ dinners} = 2{,}250$$
$$\text{Saturday dinner} = \underline{900}$$
$$3{,}150 \text{ dinners per week}$$

Sunday brunch business is booming with YUPS, and experience in restaurants such as R. J. Grunts in the Lettuce Entertain You group indicates that 900 for Sunday brunch would be a reasonable estimate. You plan to close the restaurant on Sunday nights.

Now let's figure possible gross revenue and net profits for all these meals. Drawing upon information gleaned from *Restaurant Hospitality* and your interview with Sandy Dolan of Scarpelli's, you have some numbers to work with:

Lunch checks average $7.50 per person: $1{,}700 \times \$\,7.50 = \$12{,}750$
Dinner checks average $12.50 per person: $3{,}150 \times \$12.50 = \$39{,}375$
Brunch checks average $ 9.75 per person: $900 \times \$\,9.75 = \underline{\$\,8{,}775}$
                                                    Total $60{,}900

Your calculations show that a weekly gross revenue of $60,900 would be reasonable. You know that with good management the net should be about 7 percent of the gross or about $4,300 a week.

Next you'll compare this projected net profit against the current net of the teenaged operation. After figuring in other expenses that you can project over the five-year period, you'll be able to get a grasp of whether the planned new restaurant will repay the $250,000 invested in it and, if so, how long it will take.

Because our example is an executive report, your entire research, calculations, and analysis are heading toward a single point—a decision. Will remodeling for a new clientele be worthwhile—yes or no? If yes, what are the risks involved, and what are the first steps to be taken to get the project under way? In essence, your calculations and analysis are creating new knowledge based upon what you have been able to discover in your research.

REVIEWING THE INFORMATION ALREADY GATHERED

At many stages in your research, perhaps weekly, you should take stock of your situation. What have you learned? Where are you now? What step should you take next? Without an occasional review you will most assuredly waste much of your time and go off on tangents. Keep track of your purpose and your audience's purpose. They may quite properly shift a bit as you gain more insight into the problem, but don't let the shift be accidental. Know what you are doing and where you are heading.

In our restaurant problem, your review might reveal that you have established the following points:

- The arguments for shifting from a teenaged clientele to YUPS seem real and convincing. Your preliminary calculations seem to indicate that the investment will pay off. You know, though, that you must also compare the investment in the restaurant with an alternative investment. That is, how well will the $250,000 invested in the restaurant do against the same sum invested in, say, a mutual fund or a money market fund. You make a note to do the comparison.
- Somewhat to your surprise, you find you already possess substantial information on American age groups and on restaurant management. Much of it needs to be checked and verified, but you are gaining in knowledge every day.
- Most if not all the information you need is available from individuals and libraries.

- You now have an extensive set of note cards filled with information on population growth, age groups, the post-World War II baby-boom bulge, and restaurant environments. You have a much firmer grip on the financial side of things than you had a few weeks ago. You wonder if you can find some information on income by age groups. You decide to have a look at the *Statistical Abstract of the United States* again.
- You have moved back and forth between particulars and generalizations. You have concluded that a restaurant of the type you envision can seat people more densely than you first expected.
- Your on-site inspection of Scarpelli's confirms your generalization about spacing and gives you many ideas about environment.
- Your questionnaire has given you many useful insights into how YUPS react to different restaurant environments. Simplicity seems to be a key factor. Decor that is overelaborate or too formal seems to be out. Some whimsy without excessive cuteness seems to be OK. That needs more checking.
- You learned a lot in your interview with Sandy Dolan. He confirms many of your notions about taste, particularly about "keeping it simple."
- You still await a reply from Professor Lambert. You hope it will aid you in further research into table size and spacing.
- You realize that your purpose has changed as you have gained additional information. Originally, you had planned, somewhat vaguely, simply to report on demographic changes and to indicate what these changes might mean to the restaurant chain. Now you see quite clearly that you have enough evidence to recommend a change. Furthermore, you are beginning to shape what the changes should be and are gathering information to demonstrate how the changes will pay off. You have kept in touch with your boss, and she has agreed to this shifting and tightening of your purpose.

The larger the project and the longer it takes, the more such stock-taking as we have demonstrated here is essential. Notice, too, that research has a snowballing effect. Information begets information; the more you learn, the more adroit you will be in devising new means of adding to your knowledge. It's a comforting thought.

Understand that the material under each heading of this chapter is simply illustrative. We have shown you one questionnaire; you may need to prepare and administer several of them, or none at all. You may need to arrange for a dozen interviews, or none at all. In other words, you will have to cycle back and forth through the information gathering and checking methods we have shown you. In some projects you will use many of the methods extensively; in others you will need only a few.

How do you know when enough is enough, when you have gathered all the information you need? There is no easy answer to that question. Sometimes you'll find you are taking notes and suddenly you will realize that you already have the information in notes taken earlier. If that happens often enough, it's a good sign that your research may be complete. Perhaps a better sign is when a weekly review reveals that you have a firm purpose for what you are doing and that you possess the information you need to fulfill that purpose. When you reach that happy point it will be time to move into the next stage, organizing the report.

We discuss organizing the report in Chapter 8, but a word before we leave this chapter. You will, of course, really be organizing and even writing your report all through the information gathering and checking process. As you gather information from whatever source, you will take notes under headings that will suggest organizational units to you. Gradually, as your purpose, audience, and information take shape, the overall organizational pattern may become clear. In this way, you may skip from gathering and checking your information directly into organizing it without even realizing it. In the same way, you will be writing bits and pieces of your report as you go along. Many of your notes, whether based upon your research or your own thinking, with a little scrubbing and polishing will go right into your report. The results of your mathematical calculations may need only to be formalized into tables or graphs to be presentable. Writing is truly difficult if you sit down empty headed except for the thought you must fill a certain number of pages with writing. If you build toward that writing moment with careful planning, intelligent information gathering, and even some early writing and graphing, you will find that many of the terrors you may have concerning the task will disappear.

EXERCISES

In the exercise at the close of Chapter 2, "Getting Started," you began planning a future writing project. Now, using the same project, complete the following exercises.

1. Calling upon your memory, construct a list of things that you already know about your subject.

2. Go to the library and, using the reference tools available to you, construct a working bibliography. We suggest that you write your bibliography on cards.

3. Consult one of the sources listed in the bibliography you constructed for Exercise 2. On a note card, write a direct quotation from the source. On another card, condense or paraphrase the quote on the first card.

4. Using the subject area of your projected report, complete one or more of the following assignments.
   a. Inspect a local site or facility. Write up what you learn in a report of about 500 words.

b. Construct a questionnaire that you could use with people who can provide you information in the subject area.

c. Interview some expert on the subject. Write up the interview in a report of about 500 words.

d. Write a letter of inquiry to some expert in the subject area. Ask only for information that you cannot obtain in any other way.

5. Using the various review techniques—such as induction, deduction, calculations, and analyses—discussed in this chapter, conduct regular reviews of your projected writing project. When requested, write up these reviews in a manner given by your instructor.

Chapter **6**

Technical Exposition

IN THIS CHAPTER we discuss exposition—in the next, narration, description, and argumentation. These are the basic modes of discourse from which all technical reports stem. The modes are strategies that enable you to present your material in a persuasive way. All writing, all purposeful communication, has a persuasive element. You must, at least, persuade your audience that you have mastered your material. For instance, even a simple set of instructions has to persuade the reader that the process described is a reasonable way to go about the task. The literature review of a well-written physical research report does far more than narrate the history of past experiments. It must persuade the reader that the present experiment is one more step in a historical process, a step taken by a careful experimenter with a regard for the past and an eye on the future. No tricks are involved. If the experimenter merely piles up historical facts to fill up space, no rhetorical skill will gloss over the emptiness of the endeavor. But if the writer's historical reseach does relate to the present experimentation, a rhetorically sound literature review will demonstrate the relationship in a persuasive way. If you realize from the very beginning of any writing task that you must write persuasively, your research and your thinking may be better than it otherwise would be.[1]

In exposition you inform, clarify, explain, and instruct. As the Latin roots of *exposition* suggest, you *put forth* or *expose* your material. In this expository paragraph from *Scientific American,* the writer explains the difficulty in understanding how language is learned:

> In order to understand how language is learned it is necessary to understand what language is. The issue is confused by two factors. First, language is learned in early childhood, and adults have few memories of the intense effort that went into the learning process, just as they do not remember the process of learning to walk. Second, adults do have conscious memories of being taught the few grammatical rules that are prescribed as "correct" usage, or the norms of "standard" language. It is difficult for adults to dissociate their memories of school lessons from those of true language learning, but the rules learned in school are only the conventions of an educated society. They are arbitrary finishing touches of embroidery on a thick fabric of language that each child weaves for herself before arriving in the English teacher's classroom. The fabric is grammar: the set of rules that describe how to structure language.[2]

Most of the writing in this book, like that you see every day in newspapers and magazines, is expository. Most of the writing you have done in high school and college has been expository. You can use various rhetorical devices in your exposition: topical arrangement, exemplification, definition, classification and division, comparison, and causal analysis.

You can use any of these devices as an overall organizing principle for an entire paper. For example, an entire paper could be written as a causal analysis or a definition. In order to justify some choice, you might write a report based upon a comparison organizational plan. But you also use these devices as subordinate methods of development within a larger framework. For example, within a paper organized topically you might have small sections based upon exemplification, definition, causal analysis, and so forth.

This interlocking nature of rhetorical devices should become clear to you as we explain each device and give you examples. But we point it out to you now so you'll understand that the two uses of the rhetorical devices are mutually supportive and not in conflict with one another.

TOPICAL ARRANGEMENT

Writing projects sometimes begin with a topic. Your topic might be "a choice between electric and oil heat." Or it might be something like "Christmas tree farming." As you'll see when you get further into this chapter (pages 112–113), the heating choice topic probably lends itself to a definite organization plan based upon comparison. The Christmas tree topic, however, does not immediately suggest a definite plan. When such is the case, a topical plan is often the answer. That is, you look for subtopics under the major topic. These should serve as umbrella statements beneath which you can gather yet smaller sub-subtopics and related facts. In the case of the Christmas tree topic, the subtopics might very well be "production" and "marketing." "Production" could be broken down further into "planting," "maintaining," and "harvesting." "Marketing" could be broken down into "retail," "wholesale," and "cut-your-own." With some thought, you can break most topics down into umbrella-sized subtopics.

While you are settling upon the topic and subtopics for your paper, you should also be aware of the need for topic limitation. Students, in particular, often hesitate to limit their topic sufficiently. They fear, perhaps, that if they limit their material too severely they will experience difficulty in writing essays of a sufficient length. The truth of the matter is really the reverse. You will find it easier to write a coherent, full essay of any length if you *limit the scope of your topic*. With you scope limited and your purpose clearly defined, you can fill your paper with concrete, specific facts and examples. When your scope remains broad and your purpose vague, you must deal in abstract generalizations.

Suppose you want to deal with the topic of lasers in about a thousand words. If you keep the purpose simply as "explaining lasers," what can you say in a thousand words? Probably just a few simple-minded generalizations about laser theory, history, and applications. Suppose, however, you limit the subject by asking yourself what smaller topics you can break it into. We have already indicated a few: theory, history, and applications. Any one of these is probably limited enough for a fairly decent thousand-word paper.

For the best paper, you would probably limit even more. Let's take "laser applications." What topics can this be broken into? To name just a few, communication, industrial uses, weaponry, medicine, and radar. You could go still further. Medicine could be broken down into eye surgery and cancer treatment. You could choose one of these. With a few minutes' thought, *before writing or even organizing*, you can limit your topic to manageable size.

Keep in mind that once your organizational plan is set, you're not restricted from using the other rhetorical devices in carrying out your plan. You can support a topical arrangement with exemplification, definition, classification and division, comparison, and causal analysis.

EXEMPLIFICATION

Expository writing sometimes consists largely of a series of general-izations supported by examples. Many writers deal too often in gen-eralizations, failing to support them with the concrete examples that would make them understandable and believable. We have just given you a generalization about writers. We now have two options. We can trust that you will believe us and move on, or we can provide you with an example. One example will not prove our case, but it will make it more understandable and persuasive. We feel that, normally, a writer faced with this choice should stop to give an example. Look at the following faulty piece of student writing. The student is giving some background information for a paper on "Lasers in the Field of Communications."

> Since its recent development, the laser has been experimented with by hundreds of different groups. Many different materials have been used to make lasers, and efforts are continuing to find more new substances that can be used to make lasers. Besides lasers made of different mate-rials, different types of lasers have been built.

The student continues in this vein, generalization upon generaliza-tion without ever giving any solid examples of the point in question. With nothing to hold onto, the reader is left asking: "What is recent?" "What different groups?" "What materials?" "What types of lasers?" In the following example, the student gives background information for a paper on "Gyroscopes in Modern Aircraft Instruments."

> Gyroscopes have been in practical use since the eighteenth century. In 1774 the gyroscope was used as an artifical horizon so that navigators could take accurate sextant readings on hazy days. In 1896 Obry installed the gyro in the guidance system of a self-propelled torpedo. Not long after, Elmer Sperry developed the gyrocompass, an instrument that aligns itself at a right angle to the earth's axis of rotation. During World War II the Germans used an inertial guidance system with a platform stabilized by gyros and gimbals in their V-2 rockets. The United States acquired some of the German scientists, and as a result our early guid-ance systems were much like those of the Germans. The gyro-stablilized platform is currently used in radar antennas, attitude control of planes and ships, and gunlaying equipment.

The second example is greatly superior to the first. It contains only one generalization, "Gyroscopes have been in practical use since the eighteenth century." The writer then follows the generalization with a host of examples, providing a good historical introduction to the uses of gyroscopes.

There are two common ways to use examples. One way is to give one or more extended, well-developed examples. We have used this method in showing you two large samples of student writing. The other way is to give a series of short examples that you do not develop in detail. The writer on gyroscopes has used this approach, carrying it to an acceptable extreme in the last sentence of the paragraph.

Examples give your writing life and believability. They give your audience something concrete to hold onto. Readers will remember and understand your generalizations when you have amply illustrated them.

DEFINITION

As a writer about technical subjects, you will constantly be defining terms. Of course, it is true that you can sometimes substitute simpler, everyday words for specialized vocabulary. But, often, to do so would cause the reader more trouble than to use the specialized term and define it. Mark Twain illustrated the problem nicely by describing, without a single technical term, how to hitch up a team of horses:

> It may interest the reader to know how they "put horses to" on the continent. The man stands up the horses on each side of the thing that projects from the front end of the wagon, and then throws the tangled mess of gear on top of the horses, and passes the thing that goes forward, through a ring, and hauls it aft, and passes the other thing through the other ring and hauls it aft on the other side of the other horse opposite to the first one, after crossing them and bringing the loose end back, and then buckles the other thing underneath the horse, and takes the other thing and wraps it around the thing I spoke of before, and puts another thing over each horse's head, with broad flappers to keep the dust out of his eyes, and puts the iron thing in his mouth for him to grit his teeth on, up hill, and brings the ends of these things aft over his back, after buckling another one around under his neck to hold his head up, and hitching another thing that goes over his shoulders to keep his head up when he is climbing a hill, and then takes the slack of the thing which I mentioned a while ago, and fetches it aft and makes it fast to the thing that pulls the wagon, and hands the other things up to the driver to steer with. I never have buckled up a horse myself, but I do not think we do it that way.[3]

You should define any term that you feel is not in your reader's normal vocabulary. The less expert your audience, the more you will need to define. Definitions range in length from a single word to long essays or even books. Sometimes, but not usually, a synonym inserted into your sentence will do. You might write, "The oil sump, that is, the oil reservoir, is located in the lower portion of the engine crankcase." Synonym definition serves only when a common interchangeable word exists for some bit of technical vocabulary you wish to use.

Most often you will want to use at least a one-sentence definition containing the elements of a logical definition: *term = genus or class + differentia*. Do not let the technical terms frighten you. You have been giving and hearing definitions cast in the logical pattern most of your life. In the logical definition you state that something is a member of some genus or class and then specify the differences that distinguish this thing from other members of the class:

Term	=	*Genus or Class*	+	*Differentia*
An ohmmeter	is	an indicating instrument		that directly measures the resistance of an electrical circuit.
A legume	is	a fruit		formed from a single carpel, splitting along the dorsal and the ventral sutures, and usually containing a row of seeds borne on the inner side of the ventral suture.

The second of these two definitions, particularly, points out a pitfall you must avoid. This definition of a legume would satisfy only someone who was already fairly expert in botany. Real lay people would be no further ahead than before, because such terms as *carpel*, *ventral sutures*, and so forth are not familiar to them. When writing for nonexperts you may wish to settle for a definition less precise but more understandable, such as "A legume is a fruit formed of an easily split pod that contains a row of seeds, such as a pea pod." Here you have stayed with plain language and given an easily recognized example. Both of these definitions of a legume are good. The one you would choose depends on your audience.

To make sure you are understood, you will often want to extend a definition beyond a single sentence. The most common devices for extending a definition are description, example, and comparison. The following definition taken from *Chamber's Technical Dictionary* goes beyond the logical definition to give a description:

> *anemometer:* An instrument for measuring the velocity of the wind. A common type consists of four hemispherical cups carried at the ends of four radial arms pivoted so as to be capable of rotation in a horizontal plane, the speed of rotation being indicated on a dial calibrated to read wind velocity directly.

In this example a drawing might also prove useful.

In our lay definition of *legume*, an example was given: "such as a pea pod." Often comparison is valuable: "A voltmeter is an instrument for measuring electrical potential. It may be compared to a pressure gauge used in a pipe to measure water pressure." The following definition of the Richter magnitude scale is a good example of an extended definition intended for an intelligent lay audience. Notice how the writer begins by defining terms that will be needed in defining the Richter scale:

> Seismic waves are the vibrations from earthquakes that travel through the Earth; they are recorded on instruments called seismographs. Seismographs record a zig-zag trace that shows the varying amplitude of ground oscillations beneath the instrument. Sensitive seismographs, which greatly magnify these ground motions, can detect strong earthquakes from sources anywhere in the world. The time, location, and magnitude of an earthquake can be determined by seismograph stations.
>
> The Richter magnitude scale was developed in 1935 by Charles F. Richter of the California Institute of Technology as a mathematical device to compare the size of earthquakes. The magnitude of an earthquake is determined from the logarithm of the amplitude of waves recorded by seismographs. Adjustments are included in the magnitude formula to compensate for the variation in the distance between the various seismographs and the epicenter of the earthquakes. On the Richter scale, magnitude is expressed in whole numbers and decimal fractions. For example, a magnitude of 5.3 might be computed for a moderate earthquake, and a strong earthquake might be rated as magnitude 6.3. Because of the logarithmic basis of the scale, each whole number represents a tenfold increase in measured amplitude; as an estimate of energy, each whole number step in the magnitude scale corresponds to the release of about 31 times more energy than the amount associated with the preceding whole number value.
>
> At first, the Richter scale could be applied only to the records from instruments of identical manufacture. Now instruments are carefully calibrated with respect to each other. Thus magnitude can be computed from the record of any calibrated seismograph.
>
> Earthquakes with magnitude of about 2.0 or less are usually called microearthquakes; they are not commonly felt by people and are generally recorded only on local seismographs. Events with magnitudes of about 4.5 or greater—there are several thousand such shocks annually—are strong enough to be recorded by sensitive seismographs all over the world. Great earthquakes, such as the 1906 San Francisco earthquake and the 1964 Good Friday earthquake in Alaska, have magnitudes of 8.0 or higher. On the average, one earthquake of such size occurs somewhere in the world each year. Although the Richter scale has no upper limit, the largest known shocks have had amplitudes in the 8.8 to 8.9 range.
>
> The Richter scale is not used to express damage. An earthquake in a densely populated area which results in many deaths and considerable damage may have the same magnitude as a shock in a remote area that does nothing more than frighten the wildlife. Large-magnitude earthquakes that occur beneath the oceans may not even be felt by humans.[4]

As this writer has done, extend your definition as far as is needed to ensure the level of reader understanding desired.

Sometimes you may wish to begin a definition by telling what something is *not*, as in the following definition, again from *Chamber's:*

> *metaplasm:* Any substance within the body of a cell which is not proto-plasm; especially food material, as yolk or fat, within an ovum.

You should, of course, avoid circular definition. "A botanist is a student of botany" will not take the reader very far. However, some-times you may appropriately repeat on both sides of a definition com-mon words you are sure your reader understands. In the following *Chamber's* definition it would be pointless to drag in some synonym for a word as readily understandable as *test*.

> *Gmelin's test:* A test for the presence of bile pigments; based upon the formation of various colored oxidation products on treatment with con-centrated nitric acid.

You have several options for placement of definitions within your papers: (1) within the text itself, (2) in footnotes, (3) in a glossary at the beginning or end of the paper, and (4) in an appendix. Which method you use depends upon the audience and the length of the definition.

Within the Text. If the definition is short—a sentence or two—or if you feel most of your audience needs the definition, place it in the text with the word defined. Most often, the definition is placed after the word defined, as in this example:

> The word *Bantu* is an exclusively linguistic label and has no other primary implications, either of race or of culture. The family of dialects was named by a 19th-century German philologist, Wilhelm H. I. Bleek. The words *ki-ntu* and *bi-ntu* in these African dialects respectively mean "a thing" and "things"; the words *mu-ntu* and *ba-ntu*, "a man" and "men." Thus *bantu* can be construed as "people" or "the people," and this seemed to Bleek a name eminently suitable for a language held in common by so many.[5]

Sometimes, the definition is slipped in smoothly before the word is used—a technique that helps break down the reader's resistance to the unfamiliar term.[6] The following definition of *supernova* is a good example of the technique:

> The vast majority of stars in a typical galaxy such as our own are extremely stable, emitting a remarkably steady output of radiation for

millions of years. Occasionally, however, a star in an advanced stage of evolution will spontaneously explode, and for a few months it will be several hundred million times intrinsically more luminous than the sun. Such a star is a supernova, and at the time of its greatest brillance it may emit as much energy as all the other stars in its galaxy combined.[7]

When you are using key terms that must be understood before the reader can grasp your subject, define them in your introduction.

In Footnotes.   If your definition is longer than a sentence or two and your audience is a mixed one—part expert and part lay—you may want to put your definition in a footnote. A lengthy definition placed in the text could disturb the expert who does not need it. In a footnote it is easily accessible to the lay person and out of the expert's way.

In a Glossary.   If you have many short definitions to give and if you have reason to believe that all members of your audience will not read your report straight through, place your definitions in a glossary. (See pages 210–213.) Glossaries do have a disadvantage. Your readers will be disturbed by the need to flip around in your paper to find the definition they need. When you use a glossary be sure to draw your readers' attention to it, both in the table of contents and early in the discussion.

In an Appendix.   If you need one or more extended definitions (say, more than 200 words each) for some but not all members of your audience, place them in an appendix. (See pages 227–228.) At the point in your text where readers may need them, be sure to tell readers where they are.

CLASSIFICATION AND DIVISION

Classification and division are useful devices for bringing order to any complex body of material. You may understand classification and division more readily if we explain them in terms of the *abstraction ladder*. We borrow this device from the semanticists—people who make a scientific study of words. We will construct our ladder by beginning with a very abstract word on top and working down the ladder to end with a concrete term:

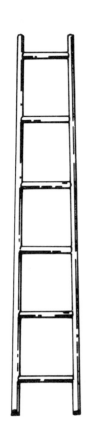

Factor: Almost anything can be a factor. You could be a factor in someone's decision. So could wealth, climate, and geography, to name but a few significant concepts. Without more specific references, we cannot determine what is specifically meant at this rung of the ladder.

Wealth: Now we have moved down a rung on the ladder. We have added specific information. Wealth could be money. It also could be stocks and bonds, land, furniture, or any of the other numerous material objects that people value highly.

Furniture: We now have become much more specific. We can mean beds, tables, chairs, desks, lamps, and so on.

Table: Now we are zeroing in on the object. However, *table* still refers to a huge class of objects: coffee tables, kitchen tables, dining room tables, library tables, end tables, and so on.

Kitchen table: Now we know a good deal more. We know the function of the object. People eat at kitchen tables. Cooks mix cakes on them. We can even generalize somewhat about their appearance. Kitchen tables are usually plain objects, compared to end tables, for example. Many of them are made of wood and have four legs.

John Smith's kitchen table: Now we know precisely what we are talking about. We can describe the size, weight, shape, and color of this table. We know the material of which it is made. We know that John Smith's family eats breakfast at this table but not dinner. In the evenings John Smith's daughter does her homework there.

We must keep one important distinction in mind: even "John Smith's kitchen table" is not the table itself. As soon as we have used a word for an object, the abstraction process has begun. Beneath the word is the table *we see* and beneath that is the table *itself,* consisting of paint, wood, and hardware that consist of molecules that consist of atoms and space.

In classification you move *up* the abstraction ladder, seeking higher abstractions under which to group many separate items. In division you move *down* the abstraction ladder, breaking down higher abstractions into the separate items contained within them. We will illustrate classification first.

Suppose for the moment that you are a dietician. You are given a long list of foods found in a typical American home and asked to comment on the value of each. You are to give such information as calorie count, carbohydrate count, mineral content, vitamin content, and so forth. The list is as follows: onion, apples, steak, string beans, oranges, cheese, lamb chops, milk, corn flakes, lemons, bread, butter, hamburger, cupcakes, and carrots.

If you try to comment on each item in turn as it appears on the list, you will write a chaotic essay. You will repeat yourself far too often. Many of the things you will say about milk will be the same things you say about cheese. To avoid this repetition and chaos you need to classify the list, to move up the abstraction ladder seeking groups that look like the following:

Food
I. Vegetables	IV. Cereal
A. Onions	A. Corn flakes
B. String Beans	B. Bread
C. Carrots	C. Cupcakes
II. Fruit	V. Dairy
A. Apples	A. Milk
B. Oranges	B. Cheese
C. Lemons	C. Butter
III. Meat	
A. Steak	
B. Lamb chops	
C. Hamburger	

By following this procedure, you can use the similarities and dissimilarities of the different foods to aid your organization rather than have them disrupt it.

In division you move down the abstraction ladder. Suppose your problem now to be the reverse of the foregoing. You are a dietician and someone asks you to list examples of foods that a healthy diet should contain. In this case you start with the abstraction, food. You decide to divide this abstraction into smaller divisions such as vegetables, fruit, meat, cereal, and dairy. You then subdivide these into typical examples such as cheese, milk, and butter for dairy. Obviously, the outline you could construct here might look precisely like the one already shown. But in classification we arrived at the outline from the bottom up; in division, from the top down.

Very definite rules apply in using classification and division.

Keep All Headings Equal. In the preceding example, you would not have headings of "Meat," "Dairy," "Fruit," "Cereal," and "Green vegetables." "Green vegetables" would not take in a whole class of food as the other headings do. Under the heading of "Vegetables," however, you could have subheadings of "Green vegetables" and "Yellow vegetables."

Apply One Rule of Classification or Division at a Time. In the preceding example, the classification is done by food types. You would not in the same classification include headings *equal* to the food types of such subjects as "Mineral content" or "Vitamin content." You could, however, include such subheadings under the food types.

Make Each Division or Classification Large Enough to Include a Significant Number of Items. In the preceding example, you could have many equal major headings such as "Green vegetables," "Yellow vegetables," "Beef products," "Lamb products," "Cheese products," and so forth. In doing so, however, you would have overclassified or overdivided your subject. Some of the classifications would have included only one item.

Avoid Overlapping Classifications and Divisions as Much as Possible. In the preceding example, if you had chosen a classification that included "Fruits" and "Desserts," you would have created a problem for yourself. The listed fruits would have to go in both categories. You cannot always avoid overlap. (No one, for example, has devised a classification for the rhetorical modes that completely avoids overlap). But keep it to a minimum.

As long as writers observe the rules, they are free to classify and divide their material in any ways that best suit their purposes. An accountant, for example, who wished to analyze the money flow for construction of a state's highways might choose to classify them by source of funding: federal, state, county, city. An engineer, on the other hand, concerned with construction techniques, might choose to classify the same group of roads by surface: concrete, macadam, asphalt, gravel.

In a fact sheet about controlling the insects that prey on trees, the author chose to classify the insects according to damage produced:

TYPES OF DAMAGE

Leaf Chewers

A number of insects, mostly caterpillars and beetles, damage foliage by consuming all or parts of leaves or needles. This may take the form of skeletonizing, leaf mining, or free feeding. The skeletonizers consume the soft leaf tissues leaving a lacy pattern of the veins and sometimes the upper or lower leaf surface. The miners work between the leaf surfaces causing brownish or papery blotches or winding narrow trails. The free feeders eat the complete leaf and sometimes fold or roll the leaves or web leaves together.

Broad-leafed trees in otherwise good condition can withstand a complete defoliation without serious injury. They will produce a second crop of leaves the same season if defoliation occurs before August. Extensive defoliation for several successive seasons can kill the branches (called die-back) and possibly the tree.

Most evergreens, however, will die if completely defoliated. Any branch that is stripped of all its needles should be pruned out.

Sap-Sucking Insects and Mites

Aphids, plant bugs, scales, thrips, and mites extract the sap from buds, leaves, twigs, or stems. Their attacks frequently cause curling, spots, galls, yellowing, mottling, or deformed leaves and flowers. A healthy, well-established tree will withstand most infestations without permanent damage. However, some die-back may follow heavy infestations and ornamentals are frequently disfigured. This detracts from their value in the landscape.

Some aphids and scale insects also secrete a sweet sticky material called honeydew. During periods of heavy infestations, honeydew coats the leaves and branches of the trees as well as sidewalks, lawn furniture, and other objects under the trees. A black sooty-mold fungus grows on this honeydew, giving the trees a sooty appearance.

Disease Carriers

Some insects are vectors of disease organisms. The most common of these is Dutch elm disease, which is spread by elm bark beetles. Chemicals are sometimes needed to control these insects, which carry diseases but which may not otherwise be harmful to trees.

Borers

Trees and shrubs may be attacked by several kinds of insects which bore or tunnel inside trunks, branches, or twigs. Borers generally infest weakened or dying trees rather than healthy, vigorous ones. Proper pruning, fertilizing, and watering will help prevent borer problems. There are few practical chemical controls for borers. Some species may be controlled in individual specimen trees by injecting the burrows or tunnels with carbon tetrachloride and then plugging the treated burrows with clay or putty.[8]

Using this classification scheme, the author teaches his readers how to recognize whatever pests are plaguing their gardens at any given moment. The rest of the article explains how to deal with each type of pest. Obviously, insects can be classified in many ways; the types of damage they produce suggested the best classification scheme for this author, in view of the article's purpose and the needs of the audience.

COMPARISON

In comparison you show how something is like or not like something else. Contrast is implied in comparison. You frequently use comparison. Someone asks you, "What is Charlie Jones really like?"

You answer, "Well, he's a pretty unusual guy; he's a fine athlete. He's as good a football player as Pete Smith. But he's different from Pete. Pete is strictly an athlete. He doesn't care whether school keeps or not. But Charlie is a good student. He really hits the books—Bs and As all the time."

You assume here that the listener is familiar with Pete Smith. By comparing the unfamiliar Charlie Jones to Pete, you aid the listener in understanding Charlie.

In technical writing you will use comparison in two different ways. First, you will sometimes structure a whole paper, or large parts of one, on comparison. Second, you will often use analogies— comparing the unknown to the known—in definitions and explanations. We examine the more complete use of comparison first.

Frequently you will have to compare two or more processes or mechanisms. Begin by explaining the things to be compared and the points on which you are comparing them. Following these explanations, you have your choice of two different organizational approaches. We can best illustrate these approaches in outline form.

Assume that you are comparing two desalination processes: freezing and flash evaporation. Your points of comparison are (1) cost, (2) ease of maintenance, and (3) purity of the water produced. After the necessary explanations of the processes and the points of comparison, you might organize your material this way:

  I. Freezing
  A. Cost
  B. Ease of maintenance
  C. Purity
 II. Flash evaporation
  A. Cost
  B. Ease of maintenance
  C. Purity

In this organizational approach, you take one process at a time and run through each of the points of comparison and contrast. This arrangement has the advantage of giving you the whole picture for each process as you discuss it. The emphasis is on the processes. If your purpose was argumentative, you would call the points of comparison and contrast *criteria* or *standards*.

In the second approach, you discuss each process point by point:

I. Cost
   A. Freezing
   B. Flash evaporation
II. Ease of maintenance
   A. Freezing
   B. Flash evaporation
III. Purity
   A. Freezing
   B. Flash evaporation

The second point-by-point arrangement has the advantage of sharper comparison. Often, in an argument, it is the better of the two arrangements. It also has an advantage for readers who read selectively. Not every reader will have equal interest in all parts of a report. For example, an executive reading this report might be most interested in cost, an engineer in ease of maintenance, a consumer in purity. (See pages 163–164 for more on selective reading of reports.)

In addition to organizing whole papers around comparison, you will frequently use comparison as an aid in defining and explaining. These comparisons, called analogies in this use, may be quite short or fairly extended. Here is a famous extended analogy by Sir James Jeans that explains why the sky is blue.

> Imagine that we stand on an ordinary seaside pier, and watch the waves rolling in and striking against the iron columns of the pier. Large waves pay very little attention to the columns—they divide right and left and reunite after passing each column, much as a regiment of soldiers would if a tree stood in their road: it is almost as though the columns had not been there. But the short waves and ripples find the columns of the pier a much more formidable obstacle. When the short waves impinge on the columns, they are reflected back and spread as new ripples in all directions. To use the technical term they are "scattered." The obstacle provided by the iron columns hardly affects the long waves at all, but scatters the short ripples.
> We have been watching a sort of working model of the way in which sunlight struggles through the earth's atmosphere. Between us on earth and outer space the atmosphere interposes innumerable obstacles in the form of molecules of air, tiny droplets of water, and small particles of dust. These are represented by the columns of the pier.
> The waves of the sea represent the sunlight. We know that sunlight is a blend of many colors—as we can prove for ourselves by passing it through a prism, or even through a jug of water, or as Nature demonstrates to us when she passes it through raindrops of a summer shower and produces a rainbow. We also know that light consists of waves, and that the different colors of light are produced by waves of different

lengths, red light by long waves, and blue light by short waves. The mixture of waves which constitutes sunlight has to struggle through the obstacles it meets in the atmosphere just as the mixture of waves at the seaside has to struggle past the columns of the pier. And these obstacles treat the light waves much as the columns of the pier treat the seawaves. The long waves which constitute red light are hardly affected, but the short waves which constitute blue light are scattered in all directions.

Thus the different constituents of sunlight are treated in different ways as they struggle through the earth's atmosphere. A wave of blue light may be scattered by a dust particle, and turned out of its course. After a time a second dust particle again turns it out of its course, and so on, until finally it enters our eyes by a path as zigzag as that of a flash of lighning. Consequently the blue waves of sunlight enter our eyes from all directions. And that is why the sky looks blue.[9]

Sir James has used analogy in a very imaginative way. By comparing light waves (an unfamiliar concept) to ocean waves (a concept familiar to his English readers), he has made it easy for his readers to grasp his meaning. Such extended analogies are most useful when you are writing for lay people.

You should frequently use short, simple analogies when you are writing. Many people have difficulty in understanding the immense power released by nuclear reactions. A completely technical explanation of $E = mc^2$ probably would not help them very much. But suppose you tell them that if one pound of matter—a package of butter, for instance—could be converted directly to energy in a nuclear reaction, it would produce enough electrical power to supply the entire United States for 41 hours (that is, over 11 billion kilowatt hours). Such a statement reduces $E = mc^2$ to an understandable idea.

Analogies serve particularly well in definitions or descriptions. If, after describing a transistor, you tell a lay audience that it is similar to a water faucet in that you can use it to control the flow of electrons or shut them off completely, you make the concept more understandable.

Throughout your writing, use analogy freely. It is one of your best bridges to the uninformed reader.

CAUSAL ANALYSIS

In causal analysis, you state and defend the proposition that x causes y; or the converse proposition: y is the effect of x. Of course, x can represent more than one cause and y more than one effect.

Inductive and deductive reasoning are necessary ingredients in any causal analysis. We introduce you to inductive and deductive reasoning in "Generalizing from Particulars and Particularizing from Generalizations" (pages 80–83). Remember that in inductive reasoning, you move from the particular to the general. You present a series of facts and from them draw some general inference. To illustrate: Fifty people out of a hundred fall ill after a church picnic. An investigation shows that the 50 people who became ill ate potato salad. The 50 who did not become ill did not eat potato salad. A reasonable inference would be that the potato salad caused the illness. Final confirmation of the inference would depend upon an analysis of the potato salad that proved it was contaminated.

In deductive reasoning you move from the general to the particular. You start with some general principle, compare it to a fact, and draw a conclusion concerning the fact. Although you will seldom use the form of a syllogism in writing, we can best illustrate deductive reasoning with a syllogism:

1. All men are mortal.
2. Aristotle is a man.
3. Therefore, Aristotle is mortal.

Most often the general principle itself has been arrived at through inductive reasoning. For example, from long observation of lead, scientists have concluded that it melts at 327.4°C. They have arrived at this principle inductively from many observations of lead. Once they have inductively established a principle, scientists can use it deductively as in the following syllogism:

1. Lead melts at 327.4°C.
2. The substance in container A is lead.
3. Therefore, the substance in container A will melt at 327.4°C.

In writing, the syllogism is rarely presented in its formal logical arrangement. It will be presented in a more informal way, as it is in this causal analysis:

> Methanogens, like other primitive bacteria, are anaerobic: They live only in areas protected from oxygen. This makes sense, since there was virtually no oxygen in the atmosphere when bacteria first evolved. But once bacteria developed chlorophyll *a*, the pigment of green plants, they began to use carbon dioxide and water for photosynthesis and produced oxygen as a waste product. When massive colonies of these photosynthetic bacteria developed, they pumped large amounts of oxygen into the atmosphere. Oxygen is a powerful reactive gas, and most early bacteria were not equipped to survive with it. Some bacterial species that were adapted to the new gas, including the oxygen producers themselves, continued to thrive. Others presumably evolved special metabolisms to protect them from oxygen, found anaerobic environments, or disappeared.[10]

Presented formally, the syllogism in this paragraph would go something like this:

1. Methanogens cannot live in oxygen.
2. Oxygen was introduced into the methanogens' environment.
3. Therefore, methanogens either evolved special metabolisms to protect them from oxygen, found anaerobic environments, or disappeared.

Usually you will argue both inductively and deductively within the same paper. First, you examine a body of facts and establish a generalization. Then, in turn, you will use that generalization to arrive at some conclusions about another body of facts.

Causal analysis may go forward or backward. When we examine an effect and want to determine its cause, we move backward. When we want to forecast, we examine existing conditions and try to predict what effects these conditions will cause. In discussing cause and effect we are normally dealing with time. Therefore, causal analysis often takes the form of a narrative.

In causal analysis we may be involved with more than one cause, that is, with complex causes. In this case, you will in your paper list and explain all the causes. Often you will attempt to show a necessary relationship among them. Sometimes you will attempt to prove that one or another is the main cause.

The following paragraphs illustrate both the narrative movements of causal analysis and the interrelationship among complex causes:

Most experts advise parents to deal with their children's fears—of strangers, loud noises, darkness, animals or imaginary monsters—by helping them to feel secure and by not demanding that these fears be overcome quickly. But parents also must learn more about these excessive and unnecessary fears: how they develop, and how they either can be cured or prevented.

Some of your baby's fears begin during infancy. If he remains afraid of new people, strange places or noises, he will not grow independent enough to explore, and will miss out on many of life's extremely important experiences.

There are basic fear reactions that are quite normal for a helpless infant. Certain unexpected events, such as loud noises, abrupt changes in surroundings or sudden pain, may startle an infant, causing him to cry, gasp, become rigid and thrash his arms. However, those objects or events that naturally cause fears to develop may lead to other unnecessary fears called *learned* fears. Three processes are involved in the development of a learned fear:

Association: Previously neutral events become associated with experiences that babies naturally fear. This can occur within the first few months of life. For example, the infant who is placed in overly hot bath water one time may learn to fear bath time in general.

Imitation: Many fears are acquired through watching and imitating the reactions of adults or other children. This process usually begins at around two or three years of age. Parents can easily teach their children to fear spiders, darkness or thunderstorms, if they themselves consistently show fear in response to these things.

Symbolic Representation: As children grow in their ability to imagine, frightening fairy tales or television programs provide new sources of fear. Children are particularly susceptible between the ages of three and five. They may see scary witches, bears or monsters in shadows on the wall or hear ghostly moans when a branch grazes the side of the house.

Although most children's fears are perfectly understandable, the origins of others are not as easy to pinpoint. However, knowing the original cause of a fear is not always necessary in order to diminish it.[11]

The following two paragraphs illustrate exemplification. But the relationship between them is that of causal analysis. The first paragraph presents the cause, the second the effect.

As most of us know, the farm population of the United States has declined steadily for two decades, reaching a low of 10 million in 1965, slightly less than five percent of the nation's total population. Using a different yardstick, the total number of farms in the U.S. declined to just over 3 million in 1968, and commercial farms, those grossing over $10,000 per year (which, incidentally, produce 85 percent of the product sold), declined to approximately 1 million. There is no indication that this dwindling process is coming to an end: The annual rate of outmigration from farms stood at 6.3 percent during the period 1965–68—the highest rate in U.S. history.

With these dwindling numbers, of course, has come diminished political power. The term "farm bloc" has been forgotten. Neither candidate in the 1968 presidential election gave a major farm speech. Farm people and rural areas still maintain a respectable degree of power in the Senate, but the political strength of farm people in the House of Representatives has declined to an extremely low level. In 1969, for example, Congressional districts claiming 25 percent or more farm population stood at only 31: the East contained none, the West 1, the South 14, and the Midwest 16. And this handful of farm Congressmen (by this generous definition) are concentrated in two tiny, embattled enclaves: the Committee on Agriculture and the Subcommittee on Agricultural Appropriations.[12]

Both paragraphs begin with a generalization as a central statement. In the first paragraph, the writer says, "As most of us know, the farm population of the United States has declined steadily for two decades. . . . "He then gives statistics illustrating the trend. The paragraph is unified because all the supporting data relate to the central statement. Unity is achieved in the second paragraph in the same way.

Many traps exist in causal analysis for the unwary writer. Avoid a rush to either conclusion or judgment. Take your time. Don't draw inferences from insufficient evidence. Don't assume that just because one event follows another, the first caused the second—a fallacy that logicians call *post hoc, ergo propter hoc* (that is, *after this, therefore, because of this*). You need other evidence in addition to the time factor in order to establish a causal relationship.

For example, in the sixteenth century tobacco smoking was introduced into Europe. Since that time, the average European's life span has increased severalfold. It would be a fine example of the *post hoc* fallacy to infer that smoking has increased the life span, which in fact probably stems from better housing, nutrition, and medical care.

Another common error is applying a syllogism backwards.

The following syllogism is valid:

1. All dogs are mammals.
2. Jock is a dog.
3. Therefore, Jock is a mammal.

But if you reverse statements (2) and (3), you have an invalid syllogism:

1. All dogs are mammals.
2. Jock is a mammal.
3. Therefore, Jock is a dog.

Jock, of course, could be a cat, a whale, a Scotsman, or any other member of the mammal family.

You can often find flaws in your own reasoning or that of others if you break the thought process down into the three parts of a syllogism.

EFFECTIVE EXPOSITION

Exposition is a common approach to a technical report. You will constantly be called upon to explain aspects of your work. Writing good expository prose is no easy chore. But you will be well on your way if you remember that persuasion, clarity, and completeness are your chief goals. You will be better equipped to achieve these goals if you master all the expository devices: topical arrangement, exemplification, definition, classification and division, comparison, and causal analysis.

Master also the ability to move up and down the abstraction ladder. The writer who stays high on the ladder produces little more than windy generalizations unrelated to reality. On the other hand, the writer who deals in facts and facts alone leaves the reader with too many unresolved questions. What is the point of all this information? What does it mean? What can I do with it? It is the writer's task to draw the needed inferences and generalizations for the reader. Just be sure that your inferences and generalizations are tied to fact, not floating like hot-air balloons that have slipped their cables.

Rich and subtle variation in use of the expository devices is one of the chief distinguishing marks of the good writer. Following is a portion of a report called "Satellites of the Outer Planets."[13] The entire report has a simple topical arrangement consisting of an introduction, called "Planets and Satellites," followed by four sections: "Satellites of Jupiter," "Satellites of Saturn," "Satellites of Uranus," and "Satellites of Neptune." To develop the exposition, the writer uses several rhetorical devices within the overall pattern. To illustrate the interlocking nature of rhetorical devices used for support and development, we have identified the devices used in the introduction.

A satellite is a world that revolves around a planet as planets revolve around the Sun. Earth has a single large satellite, the Moon; Mars has two tiny satellites, Deimos and Phobos. But most of the satellites of the Solar System orbit the outer planets beyond Mars: 30 known objects, compared with the three satellites of Earth and Mars.

Definition

Comparison

Learning about these many other satellites is vital to our understanding of the origin, development, phenomena, and processes of the Solar System and its planets, including the Earth. However, satellites of the outer Solar System are so far away (and so hard to study) that scientists neglected them until recent years when new ways to observe them and new means to analyze the observations became available. NASA's close views of the Martian satellites, information about the Moon, and some NASA spacecraft pictures of Jupiter's large satellites created a surge of interest in the many satellites of the giant outer planets.

Causal Analysis

Satellites can be grouped into three types: objects about the size of the Earth-like planets, those about the size of larger asteroids, and small, irregularly shaped

Classification

objects. Examples of each type are

(1) Saturn's largest satellite, Titan, a world larger than the planet Mercury.

(2) Uranus' satellite, Miranda, about the size of the asteroid Ceres.

(3) Leda, a recently discovered satellite of Jupiter that is about the size of the Martian satellite Deimos.

Exemplification

Satellites can also be grouped in terms of the shape and inclination of their orbits and their orbital motion. Regular satellites are those with orbits that are nearly circular, close to the equatorial plane of the planet, and *posigrade* in direction; i.e., the satellite moves around the planet in the same direction the planets move around the Sun. Irregular satellites have orbits that are highly elliptical or are inclined to the plane of the planet's equator, or are in a *retrograde* direction (a direction opposite to that of the planet's revolution).

Classification

Definition

Regular satellites probably originated from the same processes that formed the planets. Irregular satellites were probably captured later.

Causal Analysis

EXERCISES

1. Reproduced here is a portion of a report called *The Planet Venus.*[14] The portion reproduced explains some of the facts currently known about the planet and speculates about some of its mysteries. It's a good piece of expository writing. Write an analysis of it that shows which rhetorical devices the author has used and where and why he has used them.

VENUS AS A PLANET

The mass, diameter, and density of Venus are all only slightly less than those of the Earth. But its surface is hotter, its atmosphere much denser, and its rotations much slower than those of the Earth. Venus is one of the three planets of the solar system that do not have satellites; the others are Mercury and Pluto. Also Venus does not have a significant magnetic field.

The diameter of Venus is 12,100 km (7519 mi) and its average density is 5.25 times that of water. This high density may imply that Venus has a core of nickel and iron like the Earth.

The surface of Venus cannot be observed visually from Earth because Venus is shrouded in a thick blanket of clouds. However, radar waves from Earth stations penetrate to the surface and astronomers have used them to map the surface of Venus. In addition, the surface emits radio waves—microwaves—which penetrate the clouds and can be received on Earth. Astronomers use these radio waves to estimate the temperature of the planet's surface.

In the early 1960's when these measurements were first made, astronomers were surprised to discover that the surface of Venus is extremely hot. The microwaves carry the message that it is about 430°C which is hot enough to melt zinc. They also told us that the temperature is the same both day and night on the surface of Venus.

The surface temperature of Venus is much higher than that of the Earth. The difference is attributed to the dense atmosphere of Venus which traps incoming solar radiation and prevents heat from being re-radiated into space. This is termed the "greenhouse effect."

A telescope reveals no details in the yellowish clouds of Venus. However, if these clouds are photographed in ultraviolet light, dark shadings can be seen that rotate in a period of about 4 days. These shadings have also been seen more clearly in ultraviolet pictures of Venus taken from a passing spacecraft.

But the planet itself rotates much more slowly than its cloud tops. Again the discovery was made by radio waves. Transmitted from the Earth by a big antenna, these waves are reflected by Venus. Radar astronomers can tell from the form of the radio echo that Venus rotates once in 243.1 days, in a direction opposite to the rotation of the Earth. This is called a retrograde direction. A day (the time from one sunrise to the next) on Venus lasts about 117 Earth days (see figure).

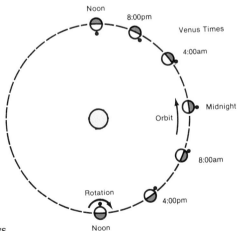

Axial Rotation: 243.1 Earth days

Orbital period: 225 Earth days (= Venus year)

Noon to Noon: 116.8 Earth days (= Venus day)

Venus year: 1.93 Venus days

A combination of retrograde rotation about its axis is 243.1 days and orbital revolution about the Sun in 225 days make a day on Venus 116.8 Earth days long, as shown in this diagram.

Another peculiar fact about the rotation of Venus is that it seems to be linked to the Earth. Each time Venus passes closest to the Earth at inferior conjunction,* the same face of Venus is turned towards the Earth. Most astronomers believe this is just a coincidence, but it may not be.

Astronomers do not know why Venus rotates so slowly. The other planets (except Mercury) rotate quickly on their axes. Mercury has been slowed by tidal friction because of its closeness to the Sun; but Venus is not only rotating slowly, it is also rotating in the "wrong" direction. The Sun could not have caused this. One possibility, proposed some years ago, is that Venus once had a big satellite that revolved around it in the retrograde direction. This satellite crashed into Venus and stopped the planet's rotation, or perhaps even pushed it a little into rotating the opposite way.

Astronomers looked at the spectrum of light reflected from the clouds of Venus and were able to find that its atmosphere contains much (97%) carbon dioxide. This again is a contrast to Earth which only has a small amount (0.03%) of carbon dioxide in its atmosphere. The reason there is so much carbon dioxide in the atmosphere of Venus is thought to be because Venus is a dry planet. The presence of oceans of water on Earth allowed Earth's early atmosphere of carbon dioxide to react with the rocks and form carbonates. Today, most of Earth's carbon dioxide is bound up as carbonates in rocks. Also living things on Earth used carbon dioxide from the atmosphere to provide carbon for their bodies. Oxygen was released by plants to form Earth's oxygen-rich atmosphere of today.

Another big difference between Earth and Venus, as mentioned earlier, is that Venus does not have a magnetic field of any significance. As a consequence its atmosphere has no protection from the blizzard of high-energy particles—protons and electrons—known as the solar wind. These particles are held away from the Earth by our magnetic field. But they speed directly into the upper atmosphere of Venus.

*Defined earlier in this way: "When Venus passes between Earth and Sun, and thus comes closer to the Earth, it passes through what astronomers call inferior conjunction. On the far side of the Sun it passes through superior conjunction."

2. Write a definition of some term in your academic discipline. Extend the definition with such devices as exemplification and comparison. Use a graphic if it will aid the reader.

3. Choose a topic and devise an organizational plan for it. Use one of these rhetorical devices for your organizing principle: topical arrangement, classification or division, or comparison.

4. Write a paper in which you find, explore, and explain thoroughly some trend in the contemporary world. The possibilities are limitless, but you may be happier exploring some trend in your own field. For example, are you in physics or engineering? Then you might be interested in the latest trends in nuclear power plant construction. Are you in forestry? Are there trends in the use and kinds of wood products? Your best sources for this paper are books of facts such as almanacs and yearbooks. Examples of such books are *Information Please Alma-*

nac, The New York Times Encyclopedic Almanac, The World Almanac and Book of Facts, Statistical Abstract of the United States, Canada Yearbook, Agricultural Statistics, Americana Yearbook, The Yearbook of Labour Statistics, The Municipal Yearbook, The United Nations Demographic Yearbook, The United Nations Statistical Yearbook.

5. Write a causal analysis of some trend or trends, perhaps using the same sources and facts that you used in Exercise 4. Your causal analysis can take various forms:

Trend A causes event B.
Both trends A and B are caused by event X.
If trends A and B continue, surely Y will result.

Your exploration of causality may be a matter of showing the significance of some trend to a specific audience. For example, of what significance is the rising average wage of engineers in industry to engineering schools? Be very factual and don't push your conclusions beyond what is warranted by your evidence.
Either Exercise 4 or 5 could well be written as a memorandum to an executive. See pages 311–313.

Chapter 7

Technical Narration, Description, and Argumentation

IN CHAPTER 6, we discuss technical exposition. In this chapter, we turn to narration, description, and argumentation. You will have frequent use of all four modes of discourse, sometimes in pure form but more often in combination. You might, for example, use argumentation as the dominant organizational plan for a report supporting the feasibility of some proposed research. But within your argument, you would probably use such expository devices as definition, exemplification, and causal analysis. Further, you might describe equipment to be used and narrate the history of past research to support your argument. The modes are means to ends, not ends in themselves. The ends, always, are to support your purpose and to satisfy the needs of your material and audience.

NARRATION

Narration tells a story—in technical writing, a factual story. Narration relates a series of events in an ordered sequence, usually chronological. The second paragraph of the following passage, an introduction to an article that describes a nuclear reaction that took place over two billion years ago, is a typical example:

> In 1942, when Enrico Fermi and his associates started up their nuclear-fission reactor at Stagg Field in Chicago, there was every reason to believe it was the first such reactor on the earth. The record book must now be corrected. In an open-pit uranium mine in the southeastern part

of the Gabon Republic, near the Equator on the coast of West Africa, are the dormant remains of a natural fission reactor. Within a rich vein of uranium ore the natural reactor once "went critical," consumed a portion of its fuel and then shut down, all in Precambrian times. The experiment at Stagg Field had been anticipated by almost two billion years.

The history of the natural reactor is an extraordinary sequence of seemingly improbable events. First, uranium from an entire watershed accumulated in concentrated local deposits, including one at a place now called Oklo. Then the conditions necessary to sustain the fission chain reaction were established; these included constraints on the concentration of uranium in the ore, on the size and shape of the lode and on the amount of water and other minerals present. After the reactor had shut down, the evidence of its activity was preserved virtually undisturbed through the succeeding ages of geological activity. Finally, the discovery of the reactor involved an investigative tour de force worthy of the best sleuths in detective fiction.[1]

Narration is frequently used in technical writing to provide a historical overview that will aid the reader in understanding some exposition that is to come. The passage just quoted serves such a use. For the most part, such narration is related in a businesslike way. Sometimes, however, the writer may choose to use narration more dramatically—perhaps in brochures or advertisements for a lay audience—to simplify complex ideas and provide human interest. The following introduction to a National Aeronautics and Space Administration brochure reporting the flight of the spacecraft *Voyager 2* for the general public fulfills both these purposes:

The year is 1610. Shakespeare's plays are the glory of England. The first newspapers have just been published in Europe. The Spanish are exploring the New World and feverishly harvesting its riches. Scholars are compiling the King James version of the Bible that will be published next year.

It is 67 years since a rebel Polish priest named Copernicus published his heretical description of the universe—a description that is shunned by the church even in this enlightened year of 1610.

On cold, clear winter evenings this year, an Italian scientist named Galileo Galilei is peering at the heavens with a new instrument he has built. He learned about this device, which he calls a telescope, from a Dutch spectacle maker. Among the celestial objects Galileo studies is Saturn, the most distant planet he knows about. Galileo turns his tiny, crude telescope toward Saturn. He is ill-prepared for the surprise the instrument provides.

Galileo has found other surprises in his surveys of the heavens, but Saturn appears, unbelievably, to have knobs protruding from its sides!

Or are they knobs? Could they be—his telescope is primitive and small—"cup handles"? His next conclusion: Two or more satellites must lie so close to Saturn that his telescope can't separate them. Later he decides even that conclusion is wrong.

Time passes and more pressing things—including his own trial by the

Inquisition on charges of heresy—occupy Galileo's time and thoughts. Because he is the most respected scientist of his time, and by recanting the heresy that the Sun is the center of the universe, Galileo survives the trial.

Years later, old and isolated from the world, Galileo again turns his telescope to Saturn. As he looks, his hands tremble; he shakes his head in disbelief, blinks his eyes, and stares again. The "cup handles," or whatever they were, have disappeared!

"God has tricked me," the deeply religious scientist writes that night in his journal.

Today we know that Galileo wasn't tricked, but that every 15 years Saturn's rings—the "cup handles"—turn edge-on to Earth and disappear from our view.[2]

The writer of the above passage followed it with a narration of the completed portion of the *Voyager 2*'s flight:

Voyager 2's Saturn encounter began on June 5, 1981, at a distance of 77 million kilometers from Saturn. The instruments began a four-month-long scrutiny of the Saturnian system. . . . By July 31, when Voyager 2 was 24.7 million kilometers from Saturn, the narrow-angle camera could no longer capture all Saturn's disk in a single picture. Then the camera began photographing small segments of the disk; the pictures fit together in mosaics to reveal details in the banded clouds.

By August 10, even four-photograph mosaics of Saturn did not capture the entire disk. Both the wide-angle cameras were directed, instead, to concentrate on atmospheric phenomena that deserve special attention. Mosaics and single photographs covered progressively smaller sections of the planet. The infrared instrument began to map the temperatures of Saturn and the satellites. . . .[3]

Narratives don't have to be restricted to past events. They can be used to forecast future events, as in this segment where the writer projects the future flight plan of the *Voyager 2*:

As it speeds toward Saturn, Voyager 2 will make distant observations of Titan. Finally the spacecraft will fly past the planet at 8:25 P.M. (PDT) on August 25, only 101,000 kilometers above the cloud tops and, a few minutes later, will skim within 36,000 kilometers of the A-ring. . . . Peering backward as it streaks away, Voyager 2 will continue to monitor the Saturnian system until the evening of September 28. The spacecraft will be on its way to Uranus, arriving there in January 1986.[4]

Narration also plays an important role in relating the details of an experiment. Often the events of an experiment may best be described in chronological order: "Placing the bunsen burner under the test tube, we brought the solution to 98° C," and so forth. Notice the narrative movement in this excerpt from an account of an experiment with a new haying system.

The baling chamber of a standard 14- by 18-inch baler was reduced to 12- by 12-inch size. Other parts of the baler were modified to produce well-formed, cube-shaped bales. Few problems were encountered in baling hay with high moisture content. The bale thrower functioned well with the small cubic bales. Most experimental bales were produced at a rate of 4 to 6 tons of dry hay per hour. Higher rates caused greater variation in the length and density of bales.

Whenever possible, bales were handled by mechanical means throughout the haying operation: A bale thrower loaded bales into trailing wagons. Hay was then unloaded from the rear of the self-unloading forage wagon into the hopper of an ordinary chain-and-flight farm elevator. The elevator delivered bales to a distributing conveyor above the drying compartment of the bin. Here, a plow that moves the length of the distributing conveyor discharged bales, alternately to either side, uniformly filling the drying compartment. The handling rate of bales in and out of the drier was normally about 12 tons per hour.

Hand labor was required to rake bales from the trailing wagons into the elevator hopper. Some experimental work was done on a mechanical device to perform this function.

After the bales have been dried, the front wall of the bin can be swung open. The conveyor inside the bin then moves the bales to the front, where a man rakes them directly into an elevator hopper. At the University's Rosemount Station, dry bales were removed from the bin, conveyed to self-unloading wagons, and hauled to a storage area. In a farm setup, the drier should be located near storage areas so the bales can be conveyed directly from the drier to storage.[5]

In narrating the experiment, the writer is actually describing process, and process description is probably the chief use of narration in technical writing. By process we mean a sequence of events that in chronological order progresses from a beginning to an end and results in a change or a product. The process may be humanly controlled, such as the manufacture of an automobile. Or it may be natural—the metamorphosis of a caterpillar to a butterfly, for example. In either case the key phrase is *chronological order*. It is the chronological order of the events that provides the organizational structure for process description. Figure 7-1 illustrates the idea nicely.

Process descriptions are written in one of two ways:

- *For the doer*—to provide instructions for performing the process.
- *For the interested observer*—to provide an understanding of the process.

A cake recipe provides a good example of instructions for performing a process. You are told when to add the milk to the flour, when to reserve the whites of the eggs for later use. You are instructed to grease the pan *before* you pour in the batter, and so forth. Writing good performance instructions is such an important application of technical writing that we have devoted all of Chapter

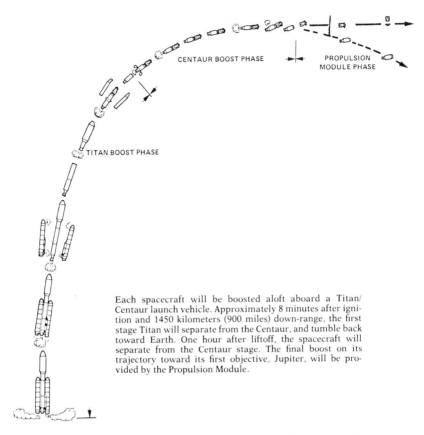

CENTAUR BOOST PHASE

PROPULSION
MODULE PHASE

TITAN BOOST PHASE

Each spacecraft will be boosted aloft aboard a Titan/
Centaur launch vehicle. Approximately 8 minutes after igni-
tion and 1450 kilometers (900 miles) down-range, the first
stage Titan will separate from the Centaur, and tumble back
toward Earth. One hour after liftoff, the spacecraft will
separate from the Centaur stage. The final boost on its
trajectory toward its first objective, Jupiter, will be provided by the Propulsion Module.

FIGURE 7-1 A Verbal and Pictorial Description of a Process [From
U.S., National Aeronautics and Space Administration, *Voyager—Journey to
the Other Planets*, prepared by the Jet Propulsion Laboratory (Pasadena,
Calif., 1977), 3.]

14 to it. In this chapter we'll explain only the second type—providing
an understanding of the process.

Insofar as you have grasped the proper order of events, the organization of a process description will not present any particular problem to you. You simply follow the order of events as they occur. However, the amount of detail you include is likely to be a problem.

In writing a set of instructions, your goal is to have the reader
perform the process. Therefore, you must be complete in every
detail. In writing process descriptions to provide understanding,
you'll find that extensive detail is not always necessary or even desirable. When *Time* magazine, for example, publishes an article about
open-heart surgery, its readers do not expect a complete set of
instructions on how to perform such an operation. Rather, they

expect their curiosity to be satisfied in a general way. In selecting detail, be guided by your purpose and audience.

Our next example comes from a book called *Wood Handbook.* Its subtitle is *Wood as an Engineering Material,* and its preface makes clear that its primary audience is composed of architects and engineers. One small section of it deals with shipworms, marine organisms that attack wood immersed in salt water:

> Shipworms are the most destructive of the marine borers. They are mollusks of various species that superficially are wormlike in form. The group includes several species of *Teredo* and several species of *Bankia,* which are especially damaging. These are readily distinguishable on close observation but are all very similar in several respects. In the early stages of their life they are minute, free-swimming organisms. Upon finding suitable lodgement on wood they quickly develop into a new form and bury themselves in the wood. A pair of boring shells on the head grows rapidly in size as the boring progresses, while the tail part or siphon remains at the original entrance. Thus, the animal grows in length and diameter within the wood but remains a prisoner in its burrow, which it lines with a shell-like deposit. It lives on the wood borings and the organic matter extracted from the sea water that is continuously being pumped through its system. The entrance holes never grow large, and the interior of a pile may be completely honeycombed and ruined while the surface shows only slight perforations. When present in great numbers, the borers grow only a few inches before the wood is so completely occupied that growth is stopped, but when not crowded they can grow to lengths of 1 to 4 feet according to species.[6]

Notice that no attempt is made to give the full information about the shipworm that an entomologist might desire. We don't learn, for example, how the shipworm reproduces, nor do we even learn very clearly what it looks like. For the intended readers of the *Wood Handbook,* such information is not needed. They do need to know what is presented—the process by which the shipworm lodges on the wood and how it bores into it. Knowing this, they can be alert to the potential for damage represented by the shipworm. Also they will understand the need for the kinds of preventive measures, such as chemical treatment and sheathing, explained later in the text.

In another section of the book, the process of heat transfer through a wall is described:

> Heat seeks to attain a balance with surrounding conditions, just as water will flow from a higher to a lower level. When occupied, buildings are heated to maintain inside temperature between inside and outside. Heat will therefore be transferred through walls, floors, ceilings, windows, and doors at a rate that bears some relation to the temperature difference and to the resistance to heat flow of intervening materials. The transfer of heat takes place by one or more of three methods—conduction, convection, and radiation (see figure).
> Conduction is defined as the transmission of heat through solid mate-

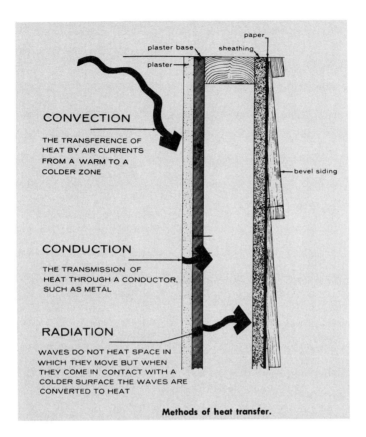

CONVECTION

THE TRANSFERENCE OF
HEAT BY AIR CURRENTS
FROM A WARM TO A
COLDER ZONE

CONDUCTION

THE TRANSMISSION OF
HEAT THROUGH A CONDUCTOR,
SUCH AS METAL

RADIATION

WAVES DO NOT HEAT SPACE IN
WHICH THEY MOVE BUT WHEN
THEY COME IN CONTACT WITH A
COLDER SURFACE THE WAVES ARE
CONVERTED TO HEAT

Methods of heat transfer.

rials; for example, the conduction of heat along a metal rod when one end is heated in a fire. Convection involves transfer of heat by air currents; for example, air moving across a hot radiator carries heat to other parts of the room or space. Heat also may be transmitted from a warm body to a cold body by wave motion through space, and this process is called radiation because it represents radiant energy. Heat obtained from the sun is radiant heat.

Heat transfer through a structural unit composed of a variety of materials may include one or more of the three methods described. Consider a frame house with an exterior wall composed of gypsum lath and plaster, 2- by 4-inch studs, sheathing, sheathing paper, and bevel siding. In such a house, heat is transferred from the room atmosphere to the plaster by radiation, conduction, and convection, and through the lath and plaster by conduction. Heat transfer across the stud space is by radiation and convection. By radiation, it moves from the back of the gypsum lath to the colder sheathing; by convection, the air warmed by the lath moves upward on the warm side of the stud space, and that cooled by the sheathing moves downward on the cold side. Heat transfer through sheathing, sheathing paper, and siding is by conduction. Some small air spaces will be found back of the siding, and the heat transfer across these spaces is principally by radiation. Through the studs from gypsum lath to sheathing, heat is transferred by conduction and from the outer surface of the wall to the atmosphere, it is transferred by convection and radiation.[7]

The readers are furnished far more detail in this description than they were in the description of the shipworm. Even an illustration is provided to make the process more understandable. The process of heat transfer, while still background information, is more useful to the architects and engineers than is the life cycle of a shipworm. Many factors of design depend upon a complete understanding of the process. Again, audience and purpose have determined the amount of detail presented.

Nothing we have said makes the selection of the content for any process description easy. You must exercise a good deal of judgment in the matter. But we have given you some guidelines to aid your judgment. As in all writing, you must decide what your audience needs to know to satisfy its purpose and yours.

The matter of sentence structure is also of some importance in process description. In writing performance instructions, you will commonly use the active voice and imperative mood—sentences like "Clean the threads on the new section of pipe" and "Add pipe thread compound to the outside threads." In following your instructions, it's the reader, after all, who is the doer. With its implied *you*, the imperative voice directly addresses the reader. But in the generalized process description for understanding, where the reader is not the doer, the use of the imperative mood would be inappropriate and even misleading.

In writing a process for understanding, therefore, you'll ordinarily use the indicative mood in both active and passive voice:

- *Active voice*—The size of the cover opening controls the rate of evaporation.
- *Passive voice*—The rate of evaporation is controlled by the size of the cover opening.

In the active voice the subject does the action. In the passive voice the subject receives the action. Use of the passive emphasizes the receiver of the action while deemphasizing or removing completely the doer of the action. (Incidentally, as the foregoing examples illustrate, neither doer nor receiver has to be a human being or even an animate object.) In situations where the doer is unimportant or not known, you should choose the passive voice. Conversely, in situations where the doer is known and important, you should choose the active voice. Because the active is usually the simpler, more direct statement of an idea, choose the passive voice only when it is clearly indicated. We have more to say on this subject on pages 180–181.

DESCRIPTION

You may have had instruction in writing description before. But probably it was geared to imaginative writing in which you were asked to evoke a feeling for some place or object. Technical description may be imaginative, but it rarely has anything to do with evoking feeling. Rather, technical description is used to provide a rational understanding of the thing described, whether that thing is a mechanism, a place, or even, for example, a book.

The physical description of some mechanism is perhaps the most common kind of technical description. It is a commonplace procedure with little mystery attached to it. For example, we see a friend smoking a pipe and endangering his thumb by using it to tap down hot tobacco coals. When he finishes his smoke, he risks cracking his pipe by banging it on an ashtray to shake out the ashes.

"Look," we say to him, "you ought to buy a pipe tamper and cleaner for yourself."

"What's that?"

"Well, it's a tool about the size of a penknife. In fact, it's made like a penknife. It's a narrow steel case with a dull knife blade and a thin pick folded into it that can be pulled out like penknife blades. One end of the case is round and flattened, about the size of a dime. That end of the case is the tamper. You use that to pack down the tobacco so that it burns more evenly.

"The pick is a thin piece of steel about the diameter of a toothpick and about half-an-inch longer. You use the pick to loosen the ashes in your pipe and turn them out into an ashtray. That way you don't have to bang your pipe and get it all dented."

"That so? What's the dull knife blade for?"

"You use the blade to ream out the cake that builds up in your pipe after you've smoked it awhile. The blade has a round tip, rather than a sharp one like most knife blades, so that you don't dig into the bottom of your pipe. The blade is sharp enough to cut through the cake but not sharp enough to cut into your pipe."

"Sounds like a handy gadget. I'll have to get one."

In this brief passage, we have used most of the techniques of good technical description. Our purpose has been to make you *see* the object and understand its function. We have described the overall appearance of the pipe cleaner and tamper and named the material with which it is made, steel. We have broken it down into its component parts. We have described the appearance of the parts and given their functions. We have used comparisons to familiar objects such as penknives, dimes, and toothpicks to clarify and shorten the description.

The use of comparison to familiar objects and the use of familiar terminology in general will greatly aid your audience in visualizing the thing described. We visualize things in essentially five ways—by shape, size, color, texture, and position—and have access to a wide range of comparisons and terms to describe all five.

For *shape*, you can describe things with terms such as *cubical, cylindrical, circular, convex, concave, square, trapezoidal,* or *rectangular*. You can use simple analogies such as *C-shaped, L-shaped, Y-shaped, cigar-shaped,* or *spar-shaped*. You can describe things as *threadlike* or *pencillike* or as *sawtoothed* or *football-shaped*.

For *size*, you can give physical dimensions, but you can also compare objects to coins, paper clips, cigarette packages, bread boxes, and football fields.

For *color*, you can obviously use familiar colors such as red and yellow and also, with some care, such descriptive terms as *pastel, luminous, dark, drab,* and *brilliant*.

For *texture*, you have many words and comparisons at your disposal, such as *pebbly, embossed, pitted, coarse, fleshy, honeycombed, glazed, sandpaperlike, mirrorlike,* and *waxen*.

For *position*, you have *opposite, parallel, corresponding, identical, front, behind, above, below,* and so forth.

As we have pointed out, technical description is rational, but nothing bars it from being imaginative so long as you apply your imagination accurately.

The following brief passage shows you how a combination of dimensions and a few words indicating shape and position can help you accurately visualize a canal in cross section:

> What was a barge canal like? Engineering experience had settled the overall geometry of canals well before the end of the 18th century. They were trapezoidal in cross section; at the bottom their width was 20 to 25 feet and at water level it was 30 to 40 feet. The water in them was only three to four feet deep, so that the angle of the sides was quite flat. This gradual slope helped to minimize erosion due to wave action and slumpage due to seepage.[8]

A description of a mechanism may be quite short, as in this description of a camera:

> *Camera*, lightproof box or container, usually fitted with a lens, through which an image of the scene being viewed is focused and recorded on film or some other light-sensitive material contained within.[9]

Mechanism descriptions may also be lengthy pieces of writing. Although the degree of complexity of the thing described may influence length, purpose is perhaps an even more significant factor. If all that is wanted is an overview of the mechanism and some notion of its function, the description can be short. On the other hand, the description must grow in length if the purpose is to describe fully the mechanism and all its parts. In their longer form, mechanism descriptions often follow a fairly standard pattern consisting of an introduction and a body.

In the *introduction* you give the function of the mechanism. You either describe its overall appearance or provide a graphic to fulfill that function. You divide the mechanism into its component parts and give an overview of how they work together. The following introduction to an encyclopedia description of a basic camera is typical:

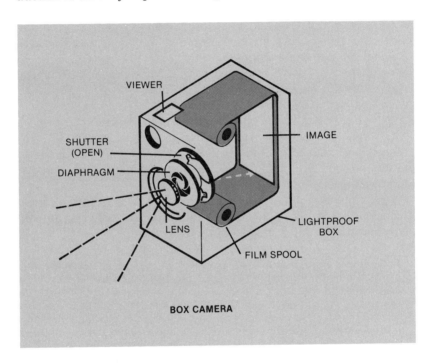

BOX CAMERA

Camera, an instrument for taking photographs. Every camera has four basic parts: a light-tight box which is called the camera body, a lens, a shutter, and a film holder. When an object is photographed, the shutter opens to admit rays of light from the object, the lens gathers the rays and focuses them to produce a latent image on a light-sensitive film, and the shutter then closes. For an explanation of the process by which a photograph is produced from the latent image on the exposed film, see PHOTOGRAPHY: Photographic Process.[10]

In the *body* of the description, you describe the parts and give their function. (If the parts are complex, they, in turn, may be divided into *their* component parts.) You again use graphics when appropriate. Here is how the same encyclopedia article describes the body and lens of the basic camera:

Body. The camera body serves as a rigid framework on which all other parts of the camera are mounted, and it also serves to protect the film from exposure to light except at the instant a photograph is taken. The bodies of most modern cameras are made of plastic or metal.

Lens. The lens is mounted over an opening at the front of the camera body. Its function is to produce an image on the film at the back of the camera by gathering and focusing the rays of light from the object. In simple box cameras the lens is usually a single piece of plastic or glass. In more elaborate cameras, the lens usually consists of several pieces of glass mounted in a metal cylinder called the lens barrel.

The light-gathering power of a lens is indicated by its *f-number,* or *relative aperture,* which is usually marked on the lens barrel. Lenses with low f-numbers have high light-gathering power and are called fast lenses, while lenses with high f-numbers have low light-gathering power and are called slow lenses. The light-gathering power of most camera lenses can be controlled by means of a *diaphragm.* The diaphragm resembles the iris of the eye. (See illustration.) When the diaphragm is open wide it allows a maximum amount of light to pass, and when it is partially closed it reduces the amount of light that is admitted. Since closing the diaphragm reduces the light gathering power of the lens, it has the effect of "slowing" the lens and thus of increasing its f-number.

Diaphragm. In many cameras the amount of light that strikes the film can be controlled by opening or closing the diaphragm. The diaphragm illustrated here is designed for an f/4 lens. It is shown at its smallest aperture, f/22, opened three stops, f/8, and opened two more stops to f/4, the diaphragm's largest aperture.

Usually, no ending is necessary in a mechanism description. You stop when you have described the last part. The camera description we've been working with, for example, continues on to describe the shutter and the film holder and then simply stops. Sometimes, however, you may want to round off your description with a summary or a complimentary close (see pages 222–223 and 225-226.)

The camera description we have shown you appears in a general encyclopedia and is intended for a lay audience. Therefore, it has the attributes of being simply written and fairly complete in its description of parts and functions. Its purpose is primarily instructional. In writing for an audience more knowledgeable about a mechanism, you might allow your description to have more professional terminology and to be more abbreviated. If your purpose is to point out to a potential customer the good features of your mechanism, your description of the functions might be sales oriented. All these differences are evident in the catalog description of Trane Process Air Handlers shown in Figure 7-2.

PROCESS AIR HANDLERS

Trane Process Air Handlers are designed to meet the most stringent indoor and outdoor applications in clean or contaminated environments. Units are also designed to operate under high and low temperature applications. Air handlers provide cooling and heating of air and air mixtures, air filtering, vapor condensation, heat recovery and humidification. Spark proof and/or noncorrosive construction is available when required.

Some of the common features available on Trane Process Air Handlers include:

• Dampers — face and bypass, leak-tight, high pressure, noncorrosive materials, operators as required and special bearings.
• Filters — inertia, roll, bag, cartridge and charcoal.
• Heat Exchangers — plate fin, fin and tube and electric coils.
• Fans — centrifugal, with FC, BI, AF wheels, and vane axial.
• Access doors and panels.
• Bird screens.
• Motors and electrical controls as required.
• Humidifiers.

FIGURE 7-2 Trane Catalog Description [From *Trane Process Machinery, Heat Transfer and Combustion Equipment* (La Crosse, Wisc.: The Trane Corporation, 1976). Copyright © The Trane Corporation, 1976. Used by permission.]

Up to this point, we have dealt with static mechanisms, that is, mechanisms that have limited or no movement. Obviously, many mechanisms—think of an internal combustion engine—are dynamic. When you describe the functions of a dynamic mechanism and its components, you're describing motion. The following example concerning a four-stroke-cycle engine illustrates this point:

> *The Four-Stroke Cycle.* In the four-stroke cycle an intake valve is opened as the piston is at the top of the cylinder, and the charge of fuel and air is drawn into the cylinder by the partial vacuum created by the piston's traveling down the cylinder. When the piston reaches the bottom of its stroke, the intake valve is closed and the piston, traveling upward, compresses the charge. When the piston reaches the top again, the mixture is ignited, and the resulting expansion of the hot gases forces the piston down. When the piston is at the bottom of its stroke again, the exhaust valve opens and the rising piston forces out the exhaust gases, leaving the cylinder clear for the introduction of the next charge of fuel and air. The entire process requires two complete down-and-up motions of the piston or two revolutions of the crankshaft.
>
> A flywheel stores up enough energy from the power stroke to carry the piston through the other three trips before the next power stroke.[11]

When you describe a dynamic mechanism, you move into the process description and the use of narration. This is a good example of what we have several times emphasized about the manner in which the modes of discourse are mutually supportive rather than mutually exclusive. For a more elaborate example of a description of a dynamic mechanism, complete with illustrations, see pages 327–329.

In mechanism description, as in all writing, you must consider your purpose and the needs of your audience. Therefore, the organization and content of mechanism descriptions are variable. But they are all likely to include statements about function, some sort of division of the mechanism into its component parts, and physical descriptions in words or graphics or some combination of the two.

Many of the principles found in mechanism description have nonmechanical applications. For example, we don't normally think of wood as a mechanism, but the same rational principles we have been discussing are found in the following passage about wood. The subject is divided, objective physical details are described, and function is discussed:

> Dry wood is made up chiefly of the following substances, listed in decreasing order of amounts present: Cellulose, lignin, hemicelluloses, extractives, and ash-forming minerals.
>
> Cellulose, the major constituent, comprises approximately 50 percent of wood substance by weight. It is a high-molecular weight linear polymer that, on chemical degradation by mineral acids, yields the simple

sugar glucose as the sole product. During growth of the tree, the linear cellulose molecules are arranged into highly ordered strands called fibrils, which in turn are organized into the large structural elements comprising the cell wall of wood fibers. The intimate physical, and perhaps partially chemical, association of celluose with lignin and the hemicelluloses imparts to wood its useful physical properties. Delignified wood fibers have great commercial value when reconstituted into paper. Moreover, they may be chemically altered to form synthetic textiles, films, lacquers, and explosives.

Lignin comprises . . .[12]

The author completes the description by describing in order lignin, hemicelluloses, extractives, and ash-forming minerals.

The same techniques can even be applied to something as non-mechanical as a book. In the following passage, a public-relations device known as a capabilities booklet is described. The booklet is broken down into its component parts, and each part is described in terms of content and function:

No single list of contents can possibly apply to all capabilities books. Here, however, are some of the major subjects you may find helpful in building your own booklet. This list is not necessarily in sequential order.

Brief corporate profile: 200 to 600 words, usually up front, defining the company's new situation.
Company creed: Short philosophic credo—if meaningful.
Short history: Growth, expansion, scope, etc. with much more emphasis on the past five years than on the previous 20.
Photos of facilities: Brief captions with outstanding photographs explaining advantages of using this equipment or this procedure or this service.
Photos or drawings of new products: (Where applicable.) Range, versatility, or unique results of the facilities illustrated.
Proofs of performance: Photos of current installations or applications with brief comment of where, what, and the *results* if available.
Differences or advantages: Diagrams or copy underscoring major differences between your current capabilities compared with former methods or with those of competitors (without mentioning names.)[13]

Writers of technical material must often describe places as well as mechanisms and other objects. Place descriptions appear in research reports when the location of the experiment has some bearing on the research. Highway engineers must describe the locations of projected highways. Surveyors have to describe the boundaries of parcels of property. Police officers have to describe the scene of an accident. In the following passage, an archaeologist reconstructs through imagination and scientific knowledge a Roman outpost in second-century Britain:

> A new fort, built of stone, rose on the eastern part of a level plateau. The enclosure was oblong; at the north and south ends the fort's massive walls, broken by central gates, ran 85 meters; on the east and west sides the length of the walls is just under 150 meters.
>
> The overgrown ruins of the last wood fort at Vindolanda, abandoned 40 years earlier, stood just beyond the west wall of the new fort. Here the garrison engineers must have dumped many carloads of clay in the process of covering up the earlier structures and preparing a level site for the stone foundations of what we call Vicus I. On both sides of the main road leading from the west gate were erected a series of long barrackslike buildings; the masonry was dressed stone bound with lime mortar. The largest barracks was nearly 40 meters long, and all of them seem to have been about five meters wide. We deduce that the structures were one story high and probably served as quarters for married soldiers. They were divided into single rooms, one to a family, by partitions spaced about seven meters part. There was probably storage space under the roof, and a veranda outside each room would have provided cooking space. The four barracks we have located so far could have housed 64 families. The British army in India offered almost identical housing to the families of its Indian soldiers. Smaller buildings, possibly housing for the families of noncommissioned officers, stood nearby.[14]

When describing a place, you must be aware of your point of view—that is, where you are positioned to view the place being described. In the passage just cited, the author begins with an aerial point of view. He looks down upon the fort and describes it in general outline—its walls, roads, and buildings. That job done, he returns to ground level, and his point of view is now that of a well-informed guide walking you through the barracks. Notice the masonry, he says. Here are the rooms. One family lived in each. See how the rooms were partioned off and where the storage space and cooking areas were. The writer uses the descriptive language of size, shape, and position. Physical description and function are gracefully combined.

When describing a place, if you, in your mind's eye, position yourself physically before you begin to write, you'll avoid confusing and annoying your readers with an inconsistent jumping about in your description. You can shift your point of view as our archaeologist has done, but be aware when you are making the shift. Have a good reason for doing it, and don't do it too often. Most often the reason will be to allow yourself to give an overview and then proceed to a close-up of selected details. That is evident in the foregoing passage.

The following passage concerns the details photographed by two spacecraft on Venus. The scene is visualized from the point of view of the spacecrafts' vantage points on the surface:

> Details on the surface of Venus were virtually unknown until two Soviet Venera spacecraft landed on it in 1975 and returned pictures. Earlier speculations had ranged from steaming swamps to dusty deserts, from carbonated seas to oceans of bubbling petroleum. All were wrong, though the desert concept seems closest to what is now known. Photographs from the Venera spacecraft revealed a dry rocky surface fractured and changed by unknown processes. There are rocks of many different kinds and a dark soil. The two spacecraft landed about 2000 km (1200 mi) from each other. One landed on an ancient plateau or plains area. At this site there are rocky elevations interspersed with a relatively dark, fine-grained soil. This soil seems to have resulted from a weathering of the rocks, possibly by a chemical action. The rocky outcrops are generally smooth on a large scale, with their edges blunted and rounded. The dark soil fills some of the cavities in the rocks. By contrast the other Venera landed at a site where there are rocks which look younger and less weathered, with not much evidence of soil between them.[15]

In this passage, the use of textured language is striking: *dry rocky surface fractured and changed by unknown processes; fine-grained soil; weathered; smooth; blunted and rounded.* Color is not mentioned, but the shade *dark* is a dominant note.

Remember that in any description you are selecting details. You would find it quite impossible to describe everything. Very often, the details chosen support a thesis, perhaps an inference or chain of inferences. Notice how the selected details in the following passage about Jupiter's ring support the inferences:

> Voyager 2 took pictures of Jupiter's ring on the inbound leg, but more interesting were the pictures it took while behind Jupiter, looking back at the ring. Where Voyager 1's pictures were faint, the ring now stood out sharp and bright in the newest photos, telling scientists instantly that the ring's particles scattered sunlight forward more efficiently than they scattered it backward, and therefore were tiny, dust-like motes. (Large particles backscatter more efficiently.)
>
> While the dust particles of the ring appeared to extend inward toward Jupiter, probably all the way to the cloud tops, the ring had a hard outer edge, as if cut from cardboard.
>
> Close examination of Voyager photos after the encounter revealed two tiny satellites, orbiting near the outer edge of the ring and herding the particles in a tight boundary. The source of the ring's dust probably lies within the bright portion of the ring itself. The dust may be due to micrometeorites striking large bodies in the ring.[16]

To summarize place description: choose one or several points of view. Your choices should help to support whatever you are trying to do. To help your reader visualize the place, use the descriptive language of size, shape, color, texture, and position. Select details that support your purpose.

ARGUMENTATION

In argumentation, you deal with propositions that lie somewhere between the bounds of verifiable fact on one hand and pure subjectivity on the other. Your purpose is to convince your audience of the probability that the propositions you are putting forth are correct. Let us illustrate.

Imagine for the moment that you are the planning engineer for a new housing subdivision called Hawk Estates. Hawk Estates, like many such new subdivisions, is being built close to a city but not in a city. The problem at issue is whether Hawk Estates should build its own sewage disposal plant or tap into the sewage system of the nearby city. (You have already ruled out individual septic tanks because Hawk Estates is built on nonabsorbent clay soil.) The city will allow the tap-in. You have investigated the situation and have decided that the tap-in is the most desirable alternative. The heads of the company planning to build Hawk Estates are not convinced. It is their money, so you must write a report to convince them.

First, you must state your major proposition: Hawk Estates should tap into the sewage system of the city of Colorful Springs.

Now you must support your major proposition. Your support normally will consist of a series of minor propositions, which will in turn be supported by verifiable facts and statements from recognized authorities. In this case your support may consist of the following minor propositions and facts.

- *Minor proposition:* The city sewage service can handle Hawk Estates' waste. *Support:* To support this proposition you give the estimated amount of waste that will be produced by Hawk Estates, followed by a statement from the Colorful Springs city engineer that the city system can handle this amount of waste. In an appendix put the actual figures that support the city engineer's statement. This first minor proposition is your most important one. Questions of cost, convenience, and so forth would be irrelevant if Colorful Springs could not furnish adequate service.
- *Minor proposition:* The overall cost to Hawk Estates taxpayers will be only slightly more if they are tapped into the city rather than having their own plant. *Support:* For facts you would list the initial cost of the plant versus the cost of the tap-in. You would further state the yearly fee charged by the city versus the yearly cost of running the plant. You might anticipate the opposing argument that the plant will save the taxpayers money. You could break down the cost per individual taxpayer, perhaps

showing that the tap-in would cost an average taxpayer only an additional $10 a year, a fairly nominal amount.

- *Minor proposition:* The proposed plant, a sewage lagoon, will represent a nuisance to the homeowners of Hawk Estates. *Support:* Because the cost for the tap-in is admittedly higher, your argument will probably swing on this minor proposition. State freely that well-maintained sewage lagoons do not smell particularly. But then point out that authorities state that sewage lagoons are difficult to maintain. Furthermore, if not maintained to the highest standards, they emit a distinct odor. To clinch your argument, you show that the only piece of land in Hawk Estates large enough to handle a sewage lagoon is upwind of the majority of houses, during prevailing winds. With the tap-in, of course, all wastes are carried away from Hawk Estates and represent no problem of odor or unsightliness whatsoever.

In your conclusions you point out that in cost and the ability to handle the produced wastes, the proposed plant and the tap-in are essentially equal. But, you point out, the plant will probably become an undesirable nuisance to Hawk Estates. Therefore, you recommend that the builder choose the tap-in over the plant.

Throughout any argument you appeal to reason. In most technical writing situations an appeal to emotion will make your case immediately suspect. Never use sarcasm in an argument. You never know whose toes you are stepping on or how you will be understood. Support your case with simply stated verifiable facts and statements from recognized authorities. In the preceding example, a statement from potential buyers that "sewage lagoons smell" would not be adequate. But a statement to the same effect from a recognized engineering authority from a nearby university would be acceptable and valid.

The intention of argument is to convince your audience that your opinions are correct. The major technique of argument is the use of a major proposition followed by a series of supported minor propositions. Within your propositions, anticipate and answer opposing arguments when you can, as in this example where the writer casts doubt upon certain projections:

> In view of these changes, most nuclear power projections, official and unofficial, are outdated. If we count plants to which there is now a firm and reasonable commitment, we arrive at a nuclear capacity of about 115,000 to 120,000 megawatts sometime in the 1990's. And no new construction permits are on the horizon.
>
> Currently, 72 plants are licensed for commercial operation, representing electric generating capacity of about 55,000 megawatts. There are also 77 large plants (with a capacity of roughly 1,000 megawatts each) in various states of completion, some almost finished, others hardly begun. I expect that at least 20 of the plants under construction will be cancelled, and only 1 or 2 of the 10 plants under construction-permit

review have much chance to see the light of day (for economic reasons). What this adds up to is no more than 50-odd plants (about 60,000 megawatts) headed for completion.[17]

Because their conclusions present probabilities rather than verifiable facts, the discussion sections of many scientific and technical reports are often rather subtle and understated arguments. Such is the case in the following discussion section from a report on an experiment with a tree-harvesting process known as strip thinning

About 200,000 acres of red pine 30–100 years old in northern Minnesota are expected to be available for thinning during the next 10 years. Strip thinning is a method of making mechanized harvesting adaptable to the management of these stands. The first cut, using the tree length or full tree harvesting, can be made with minimum disturbance to the forest. These systems provide for efficient operation and removal of the harvest from the forest as tree length logs for processing at the mill. This strip thinning will leave the forest in a condition adaptable to future thinnings by individual tree selection or additional strips.

The layout of strips and landings can be used for the life of this stand, thus reducing future operation costs. If carefully planned, strips and landings can make the forest more adaptable to multiple use. Landings have potential use for wildlife openings, overflow campgrounds, and hunting sites. Strips can be oriented to increase snow accumulation and delay snowmelt for watershed purposes.

Although a systematic layout of straight strips was used, the system has enough flexibility to meet the varied conditions in the forest. For example, the 16-foot width is not mandatory for maneuverability of skidders on the strip. A 12-foot width would have been acceptable over a large part of the area. Strips need not be perfectly straight: slight curving can be used to take advantage of openings. Some selective cuttings may be possible in the area adjacent to strips.

Mechanized harvesting that permits removal of full trees, coupled with strip cutting that limits disturbance to 25 percent or less of the forest area, can be used for harvesting timber in sensitive areas, such as scenic zones and recreational areas. Careful harvesting under winter conditions would cause less disturbance than summer logging.

Further trials are planned to determine the full potential of these harvesting systems. Different age classes and species mixtures as well as additional harvesting patterns will be included.

These systems offer the forest manager a powerful tool in controlling forest conditions. Besides prescribing the trees to be cut, he also prescribes the removal method, slash placement, and clearing of landings. He can control forest conditions for tree growth, esthetics, wildlife, water, or any special use for which required forest conditions can be prescribed. Of course, skill in selecting landings and laying out strips that can be effectively used in all future thinnings is required.[18]

Notice how the authors move easily from their major proposition—"Strip thinning is a method of making mechanized harvesting adaptable to the management of these stands"—through a series of minor propositions.

Within the framework of argument, you can use all the other modes: exposition, narration, and description.

PERSUASIVE STRATEGY

At the beginning of Chapter 6, "Technical Exposition," we speak of the modes of discourse as strategies. We'll close this chapter by exploring briefly the proper choice of a rhetorical strategy. The strategy or combination of strategies you choose depends in large part upon the objectivity or subjectivity of the information you are working with. All information ranges along a continuum from complete objectivity to complete subjectivity.

	Area of	
Objectivity	Reasonable Argument	Subjectivity

To illustrate this continuum in the classroom, we often conduct an experiment. We choose a dozen students and tell them that we are going to give them three situations and three corresponding questions to which they will respond with one-word answers. In the first situation, a piece of chalk is held up without identification, and each student is asked to write one word that will identify the object observed.

In the second situation, we ask each student to write down one word that describes his or her reaction to modern art, the work of Picasso or Jackson Pollock, for example.

In the third situation, each student is asked to write a one-word reaction to the way the president of the United States is handling whatever happens to be a controversial issue at the moment—the domestic economy, for example.

Then the answers for each situation are tallied. In the first situation, all the students inevitably—well, almost inevitably—write "chalk." In the second situation, we'll get a variety of one-word answers, such as "dislike," "junk," "interesting," "indifferent," "innovative," and "exciting." The reactions to the president's handling of the nation's economic problems may include "satisfactory," "unimpressive," "antiquated," "poor," and so forth.

The first situation, with the chalk, illustrates a response to a totally objective situation. Language is being used to record something that is completely verifiable: close examination and analysis of the object shown would clearly demonstrate whether or not it was a piece of chalk. Most of us understand how verifiable facts are applied, both in science and technology and in everyday life. If someone tells you that a room measures 25 feet by 60 feet, and you disagree, you don't argue philosophically to prove your point. Rather, you get a tape measure and verify which of you is closer to the truth. In this objective zone, then, you may describe, narrate, and explain, but you rarely need to argue in order to be persuasive.

The second situation, the reaction to modern art, takes us to the subjective end of the continuum—from the single answer "chalk" to many answers. The answers, in turn, range from favorable words like "interesting" and "innovative" to unfavorable words like "dislike" and "junk." In the objective zone, there is no reason to argue because we can verify the correct response. But subjective reactions to modern art are considerably less verifiable. Can we *prove* that the person who says "junk" is more correct than the person who says "interesting," or vice versa? Obviously, some works of art (regardless of their period) are superior to others, for fairly objective reasons. But one's *reaction* to any artistic trend—primitive, realistic, abstract, and so on—is purely a matter of individual taste: "I don't care if it's the greatest piece of sculpture in the world; to me, it looks like a hard-boiled egg."

In this subjective zone, when we do not share an opinion, its merits cannot be tested by objective scientific means. Although you can use description, narration, and exposition in an effort to challenge another's viewpoint, direct argument would probably not be persuasive because it is likely to butt up against deeply held personal values. You can, however, argue indirectly: "You should eat your spinach. I know you don't like its taste, but it's good for you."

In the last situation, the president's handling of the economic situation, the responses again range from favorable to unfavorable. But the different opinions here are not purely subjective, as in matters of taste. They are based on the students' best analysis of whatever hard evidence they have concerning the president's actions. Many of our communications fall in this zone of the continuum, between pure objectivity and pure subjectivity. In this zone, argument is permissible—in fact, inevitable. And when we argue, we have at our command all the other modes: exposition, description, and narration.

Although scientific evidence is available on issues that arise in this middle zone, it does not provide indisputable answers. Therefore, we have to evaluate whatever evidence we find and use it as a basis for the conclusions on which we act. If you were an agronomist who had planted corn with 30 inches of space between rows in one field and 40 inches in another, and you found that the first method produced the best yield, what would you conclude? Assuming that other variables in the experiment (such as moisture, fertilizer, and drainage) were carefully controlled, you would probably infer that the 30-inch spacing *caused* the greater yield. But your scientific caution would not allow you to state your findings with any great certainty until you had repeated the experiment several times with the same results. Even then, you would present the causal connection of spacing to yield as a conclusion: that is, as an opinion, not as a fact.

Henry David Thoreau once observed that if you find a trout in the milk can you can infer that someone has been watering the milk. He was right. But don't rush into conclusions. Gather your evidence. Weigh all the evidence and decide how much of it *really* supports your conclusions. Do not twist evidence to suit a preconceived result.

The whole process of gathering information, weighing evidence, drawing conclusions, and testing conclusions is, or course, the scientific method at work. Science depends upon the collection of facts and the projection of accurate conclusions based upon those facts. You, as a writer reporting on science and technology, must accurately record both facts and conclusions, using all four modes of discourse to do the job. Locating a particular topic on the objective-subjective continuum helps you to select the most appropriate strategy.

EXERCISES

1. Write a narration of two or three paragraphs that is intended to serve as a historical overview for some larger exposition. Write the overview for a lay audience. Choose as a subject some significant event in your own professional field such as Mendel's work on heredity, Salk's work on polio vaccine, the building of the Panama Canal, the establishment of the structure and function of DNA, the laying down of the first successful transatlantic cable, or Nightingale's development of modern nursing.

2. Write two versions of a narration intended to provide an understanding of a process. The first version is for a lay audience whose interest will be chiefly curiosity. The second version is for an expert or technical audience that has a professional need for the knowledge. The process might be humanly controlled such as buying and selling stocks, writing computer programs, fighting forest fires, giving cobalt treatments, or creating legislation. It could be the manufacture of some product—paint, plywood, aspirin, digital watches, maple syrup, fertilizer, extruded plastic. Or you might choose to write about a natural process—thunderstorm development, capillary action, digestion, tree growth, electron flow, hiccuping, the rising of yeast dough.

3. Choose three common household tools such as a can opener, vegetable scraper, pressure cooker, screwdriver, carpenter's level, or saw. For each write a one-paragraph description such as you might find in an encyclopedia. Then write a second version of each paragraph that could serve as a catalog description for a particular brand of the product, such as a General Electric can opener or Stanley saw.

4. Write a description of a dynamic mechanism. You may choose either a manufactured mechanism—such as a farm implement, electric

motor, or seismograph—or a natural mechanism, such as a human organ, an insect, or a geyser. Consider your readers to have a professional interest in the mechanism. They use it or work with it in some way. Divide the mechanism into its component parts, and using both words and graphics, describe its functions and physical appearance.

5. Describe some place that could be of importance to you in your professional field—a river valley, a power plant, a nurse's station, a business office, a laboratory. Have a definite purpose and audience for your description, such as supporting some inference about the place for an audience of your fellow professionals. Or you might emphasize some particular aspect of the place for a lay audience. Use at least two points of view, one that allows you to give an overview of the place and one that allows you to examine it in some detail.

6. Argue pro or con for a proposition such as the following:

- Camera X is superior to camera Y.
- Organic foods are worth the additional cost.
- Nurses should be allowed to prescribe therapy and medication.
- Health care should be nationalized.
- Agricultural chemicals are necessary and safe.
- Nuclear power plants are the answer to power needs in the United States for the next 30 years.

State your major proposition and support it with three to five minor propositions that are, in turn, supported by factual evidence. Anticipate opposing arguments and answer them. Your audience is composed of college-educated lay people.

Chapter **8**

Organizing Your Report

IN A SURVEY, a thousand engineers were asked what bugs them the most about writing. Leading the list of answers at 28 percent was "organizing and outlining."[1] Most writers would agree that organizing is a critical and difficult step. Poor organization leads to many other writing problems such as a lack of clarity and conciseness. Good organization smoothes the entire process. In this chapter we discuss the three steps of successful report organization. The first is the process of organizing the body of your report, that is, the main discussion of your report exclusive of introductions, summaries, and so forth. Second, we discuss the construction of a formal outline for your report for those occasions when a formal outline is required. Third, we cover planning your report format, that is, deciding what form your complete report will finally have. Will it be a memo or a more formal report? If a more formal report, what elements of reports such as tables of contents and abstracts should it contain?

THE ORGANIZATIONAL PROCESS

By stressing organization in this chapter we don't mean to imply that it is a static process performed in separation from all the other steps in writing such as gathering information and analyzing your audience. Rather, organization, as we indicate frequently, is a dynamic process that begins the moment you start thinking about a writing project. In Chapter 2, "Getting Started," we discuss your decision making about topic, purpose, audience, and your role as a writer. All influence organization. From the very beginning you have been thinking about what you want your readers to know and to be able to do as a result of reading your report. In Chapter 3, "Analyzing Your Audience," we advise you to be always thinking about your audience's purposes as well as your own. Audience factors—what readers already know, what they don't know, what they want to know, what you want them to know—govern organization in many ways. In Chapter 5, "Gathering and Checking Information," we give you ten steps to follow (page 74). The first one, "Calling upon Your Memory," suggests that you begin by taking inventory of the information you already possess on a subject. This inventory is not only a start on gathering subject matter; it is also a start on the organizational process. The jotted-down thoughts, random though they may be, begin to give form to the paper you will ultimately write. As you gather information from whatever source, you will take notes under headings that will suggest organizational units to you.

At the step labeled "Generalizing from Particulars and Particularizing from Generalizations" (page 80), we urge you to "lean back in your chair and take a long hard look at what you have learned so far." In your long hard look you apply the reasoning processes of induction and deduction. Both are organizational processes that may develop an entire report for you, or at least major sections of it. The last step of the ten is "Reviewing the Information Already Gathered." In this step we urge you to take stock frequently of your material as you gather it. We suggest that this periodic checking will keep you from going off on a tangent. Gradually, as through these reviews your purpose, audience, and information take shape, the overall organizational plan may come clear. And, of course, the four modes of discourse explained in Chapters 6 and 7—exposition, narration, description, and argumentation—are all methods of organization.

Because organization is this continuing, dynamic process involving all the prewriting steps, by the time you reach the moment of getting a plan on paper, you will often find that the basic problem has already been solved. For example, your task may be to give an account of some happening—perhaps an accident. You see at once that a narrative approach is the best. In a rough outline you line up your facts in chronological order, and you're ready to write.

Our restaurant example from Chapter 5, "Organizing and Checking Information," provides another instance in which the final organization may grow out of the prewriting steps. As our young restaurateur, you have reached the point where you are persuaded by your research data that remodeling a restaurant for teenagers into one for young adults would be a profitable venture. It now remains to make this recommendation to your boss and to argue for it. To argue seems to call for the argumentation mode. Looking at your information, you can pick out several premises that will support your recommendation:

- Demographic data support the idea of a restaurant for young adults.
- Well-run restaurants catering to young adults are quite successful.
- Your research has given you a restaurant design that you feel will appeal to young adults.
- The project seems to be financially feasible.

With these premises as umbrella headings, you can now sort out your data in a rational, organized way. As you do so, you remember you are operating more as a scientist than you are as a salesperson. Though you are convinced that remodeling is a good idea, you want to be sure to include information on the risks involved. You want your boss to have full information when she makes her decision. Working in this way, you reach a final organizational plan rather quickly and simply.

Unfortunately, the right organization is not always so immediately obvious. On such occasions, most practicing writers would agree that a period of meditation about the material on hand must precede not only the writing of a rough draft but even the making of an outline. An outline constructed before you have thought about a wide range of possibilities may restrict you so tightly that you will either have to ignore it later or mutilate your material to fit it. This period of meditation can take many forms. For some writers it involves a lot of pacing around; for others, long periods of starting at pads of paper, typewriters, or walls. Others shuffle and reshuffle their stacks of note cards, trying first one arrangement, then another. Some writers doodle little notes or write out networks of related ideas. However it is done, what is involved is taking stock of the information you possess. We suggest for such moments a technique we have personally used and have also taught our students for many years—a modified form of brainstorming. It seems to work particularly well at those moments when you are blocked and need to get something—anything—down on paper.

Brainstorming

What exactly is brainstorming? True brainstorming is a process in which you uncritically, without thought of organization, jot down every idea about a subject that pops into your head. The key to successful brainstorming is that you do not attempt to evaluate or organize your material at the first stage. These processes come later. Evaluation or organization at the first critical stage may cause you to discard an idea that may prove valuable in the context of all the ideas that the brainstorming session produces. Also, avoiding evaluation at this point prevents the self-censorship that often blocks a writer.

You can apply a modified form of brainstorming to organizing your report. We use the term *modified* because at the organizational stage you are not starting from scratch. Rather, you are working with the large body of information you have already gathered.

Take all your research materials—your interviews, your letters, your note cards, your questionnaires, your experimental data—and sit down at a table, preferably a large table. Place your materials in a stack before you. Now take a large pad of paper, the bigger the better. Thumb through your materials and write on the pad highly condensed versions of all the facts and ideas in your material. Also write down any ideas that your research materials suggest to you. These ideas that you gain through association may be facts you knew before researching; they may be organizational ideas; they may be analogies to explain difficult concepts; they may be additional thoughts about your audience; they may be almost anything. Do not evaluate them. Write them down. As you proceed, draw lines between ideas to separate them.

Obviously, this is no five-minute exercise. Brainstorming a report that will run 1,500 words when finished may take you several hours. If you are dealing with a complex subject, you may spend all day or several days simply revolving ideas around in your head and on paper.

We illustrate this process with excerpts from a brainstorming session on an essay you can see in finished form—Chapter 3, "Analyzing Your Audience." We have put the ideas down here exactly as they appear on the brainstorming pad—abbreviations, fragmented ideas, and all:

Purpose vs. audience analysis

----

purpose or writing—purpose of reading

----

on the job vs. in school

----

need long known
    Aristotle defined generation gap
    Roman water commissioner
    typographical experiments

Swain quote—use!

Ping-Pong balls and mousetraps

----

technicians, laymen (executives), operators, engineers, scientists, jr. engineers, 1st echelon technicians, 2nd ech.—how to break down?

----

why do I read scientific articles/what's in it for me? What does development mean for me? practical interest

----

instructions can be written for laymen or technician—or operator

----

Westinghouse stuff—Jim Souther

The brainstorming session for Chapter 3 filled four sheets of 22-by-17-inch newsprint with closely packed writing. We paid no attention to organization at this point. Neither did we evaluate. Some of the ideas made it through to the finished product; some did not. (Some show up in other chapters.) All we tried to do was to put our ideas and facts out in the open where we could use them and think about them.

What do you do after the brainstorming session? First read over everything you have written and then, if possible, do something else for a while. Read a book, watch TV, work on another project, have a chat with your friend down the hall, sleep on your ideas if you can. Let your subconscious mind work on your material. Many scientists and writers report that solutions to problems have come to them while they were asleep or doing something totally unrelated to solving the problem. Ernest Hemingway, telling about his early days in Paris, said that he had to learn not to think about his writing when he was not at work on it. In that way he allowed his subconscious mind to work on it.[2]

Probably you have had experiences yourself in which a problem has solved itself overnight. Of course, the subconscious mind must have material to work on if it is to solve the problem. Your research and your brainstorming session supply that material.

Organizing Your Material

When you return to the material that your brainstorming session produced, you must begin to organize and evaluate. For your first step, reexamine your purpose. You will have had your purpose in mind while you were researching, but perhaps your research has modified it to some degree. At this point write it out in precise terms. Nothing clarifies fuzzy thought as quickly as having to express it on paper. State your purpose with your audience in mind. You are never merely writing, say, an essay on "how to build an FM tuner" but rather "how to build an FM tuner if you are a high school student with a limited knowledge of electronics."

With your purpose and audience clearly stated, examine the results of your brainstorming session. You are now about to bring order out of the chaos before you.

Your first step should be to look for key ideas and organizational patterns that directly support your purpose. Find major ideas under which you can group less important ones. Let us use our audience analysis chapter again to illustrate. Looking at our four pages of information, we saw that types of audiences seemed to be the dominant idea. This suggested classification by audience type as the organizational scheme. First ideas may not always work, but in this case ours did. Once we had hit upon the proper rhetorical mode, our organization went well but not without a good deal of trial-and-error adjustment.

First, we had to decide how to classify our audience by type. We took a large sheet of paper and wrote the various audiences on it:

technicians
lay people (executives)
engineers
scientists
jr. engineers
operators

As an afterthought we added "history of audience analysis" to the list.

Next we took a large sheet of paper and wrote "Scientists" across the top of it. Under this heading we listed all the things from our four pages of ideas that seemed to apply to the scientist. Part of the list read like this:

definition by education
definition by experience
limited background information needed
OK for theoretical calculations
show all facts
OK to use math, formulas, scientific terms
tables OK
be honest

We did the same for "engineers" and for "lay people (executives)." We looked at our lists for a while. They didn't work as they should. Some of the things listed under "Scientists" fit just as well under "Engineers." We didn't want a lot of needless repetition. Some of the things we had to say about lay people and executives did not seem to belong in the same section. Some of the things we had to say did not fit under any of the headings.

We had to look for new headings. After several more trial-and-error runs, we ended up with these:

History of Audience Analysis
A Lay Audience
Executives
Experts
Technicians
The Combined Audience

Under these headings we made a more complete outline, filling in facts and ideas under the major headings. No one but ourselves was to see the outline, so we skipped the formality of roman numerals and other outline apparatus.

When we looked our outline over, we were pretty satisfied, but we had a nagging doubt about beginning with the history of audience analysis. For one thing, we were violating our basic classification scheme. We could probably justify that, but were we also violating our own rules of audience analysis? Would you be as fascinated by Aristotle's ideas on audience analysis as we were? Or by the fact that Sextus Julius Frontinius, water commissioner of Rome in A.D. 97, invented the information report? Reluctantly, we decided that you wouldn't be. We scrapped that section. We kept just a few examples to illustrate the introduction to the chapter.[3]

When our rough but thorough outline was finished, we organized our research notes into piles corresponding to the headings and sub-headings of the outline and were ready to write our first draft of the chapter. The planning of the chapter from brainstorming through organizing the research notes took 15 hours scattered over 4 days. In the hours not devoted to planning, we walked, read novels, did our other work, watched TV, slept, and tried not to think about the chapter.

The process we have just described is our way of organizing an extensive piece of writing. It may not work for everyone, although we know it does for many of our students. Other methods work for other people. Some people work directly from their research notes, laying them out on a large table and reading them over carefully. They then organize their material into piles, trying first one arrangement and then another. Some people begin by talking their papers into a tape recorder.

Some of these approaches and others we have not discussed are described by Dr. Blaine McKee in an article entitled "Do Professional Writers Use an Outline When They Write?"[4] McKee's article, based on research conducted with members of the Society for Technical

Communication, contains many interesting insights into the organizational practices of professional technical writers. Figure 8-1 is a table from his article showing the types of outlines used by the professionals.

Only 5 percent used no outline at all, and only 5 percent reported using the elaborate sentence outline. Most reported using some form of the topic outline or a mixture of words, phrases, and sentences. Most kept their outlines flexible and informal, warning against getting tied to a rigid outline too early. But all but 5 percent did feel the need to think through their material and to get some sort of organizational pattern down on paper before beginning the first draft of a report.

However you approach the organizational process, keep these points in mind:

- You can save yourself a lot of initial chaos and hard work if you remember that certain kinds of reports (or sections of reports) have fairly standard organizations. For example, a description of a mechanism almost always involves focusing on each of its component parts. Comparison papers and physical research reports also follow a fairly definite pattern. We describe these organizational patterns in Chapters 6 and 7 and in Part 2, "Applications."
- Writing things down is a wonderful discipline. It clarifies thought, keeps your mind from wandering, and records key points.
- Stay relaxed. Don't rush the organizational process. As many times as need be, go over your information, objectives, and audience analysis. Be willing to try a trial-and-error approach until you find the right plan. It's much easier to rearrange and discard information at this stage than it is later.

Type		Percent
Topic outline		60
Word	8.75 percent	
Word and phrase	51.25 percent	
Combination of word, phrase, and sentence		30
Sentence outline		5
No outline		5
		Total 100

FIGURE 8-1 Types of Outlines Used

THE FORMAL OUTLINE

An experienced writer seldom goes beyond the thorough but informal outlining procedure described in the previous section. Your instructors, however, will probably require you to submit formal outlines from time to time. They have several good reasons for this.

Reasons for Outlining

A formal outline is a good exercise in and of itself. You may be assigned to read an essay and outline it. By reading your outline, the instructor can quickly see if you have the ability to abstract the major ideas from a piece of writing—a necessary skill in research. Or the instructor may ask you to go through a body of research material and construct the outline for a paper that you will never actually write. This exercise allows both you and the instructor to concentrate on organizational skills without getting into writing skills.

In industry you would need an outline in cases of multiple authorship, such as a report prepared by a team or on a production-line basis. With a formal outline the writers know where their parts fit into the whole. And, of course, formal reports often require a table of contents, which closely resembles a formal topic outline (see pages 207–209).

Most often, however, instructors will ask for a formal outline of a paper that you are to write. Usually they will ask for the outline a week or two before the paper is due. They have two reasons for this early turn-in date. First, they get you to work on the paper early enough for you to think the paper through before you have to write it. Second, they allow themselves enough time to help you with your organization.

Writers often digress in reports. They have not always learned how to stay close to the main subject. Many writers hate to throw away any research material. They have worked hard to get their facts, and whether the facts really fit or not, they will attempt to shoehorn them in. Because of this tendency, writers sometimes include too much background material. A formal outline with its neat divisions will show a student writer and teacher alike when the student plans to develop irrelevant material at the expense of needed material. Look at the faulty outline in Figure 8-2. A paper written from this outline would never fulfill the stated purpose. The student read a good deal of the history of distillation and desalination and no doubt found it fascinating. (And he by no means wasted his time in absorbing all the history. The best reports, like the tips of icebergs, often suggest but do not display their underpinnings.) But most of the history is irrelevant to the purpose of the report and should not be included. The discussion of modern methods belongs in a promi-nent place. It is from these methods that the student must choose. But in the faulty outline, this discussion is the fifth subdivision under one of the subdivisions of the history of desalination—three times removed from prominence.

The student was correct in making the choice of a method a major division, but the outline as a whole is much too one-sided. In this instance, the senior officers designated as the audience do not need the ancient history. They are concerned with the application of mod-ern desalination methods. They are concerned with theory only inso-far as they need it to understand the application. The student has mislaid the purpose of the report and misjudged the needs of the audience. The outline clearly showed the teacher and the student where the organization failed.

The student's revised outline appears in Figure 8-3. In the revised outline, the student has discarded the irrelevant material. The first division of the outline states the problem, the solution of which is the purpose of the paper. The remaining divisions concentrate on the cri-teria that govern the choice and on the alternatives available. By the time division VII is reached, both the student and the audience have sound reasons for the choice.

Outlining Conventions

As illustrated in Figures 8-3 and 8-4, the outline has four parts: (1) title, (2) purpose statement, (3) audience statement, and (4) body. Observe the following conventions in the body of your outline:

- Make all entries grammatically parallel. Do not mix complete sentences with incomplete sentences. Do not mix noun phrases with verb phrases, and so forth. A formal outline with a hodgepodge of different grammat-ical forms will seem to lack—and, in fact, may lack—logic and consistency.

Incorrect	*Correct*
I. The overall view	I. The overall view
II. To understand the terminal phrase	II. The terminal phrase
III. About the constant-bearing concept.	III. The constant-bearing concept

- Never have a single division. Things divide into two or more; so, obvi-ously, if you have only one division, you have done no dividing. If you have a "I" you must have a "II." If you have an "A" you must have a "B," and so forth.

DESALINATION METHODS FOR AIR FORCE USE

Purpose: To choose a desalination method for Air Force bases located near large bodies of salt water.

Audience: Senior officers.

I. History of desalination
    A. Distillation among ancient peoples
        1. Greeks
            a. Aristotle
            b. Pliny the Elder
        2. Arabians
    B. Distillation in the Middle Ages
    C. Distillation in modern times
        1. France
            a. Cellier–Blumenthal
            b. Derosne
        2. Gt. Britain
            a. Fitzgerald
            b. Hales
        3. Shipboard equipment
        4. The Middle East today
        5. Modern methods
            a. Electrodialysis
            b. Reverse osmosis
            c. Multistage flash
            d. Long tube vertical
II. Choice of a method
    A. Theory of desalination
    B. Long tube vertical
    C. Justification of choice

FIGURE 8-2 A Faulty Outline

DESALINATION METHODS FOR AIR FORCE USE

Purpose: To choose a desalination method for Air Force bases located near large bodies of salt water.

Audience: Senior officers.

I.   Statement of the problem

   A.   Need for a choice

   B.   Choices available

   C.   Sources of data

II.   Explanation of criteria

   A.   Cost

   B.   Purity

   C.   Quantity

   D.   Ease of maintenance

III.   Electrodialysis

   A.   Theory of method

   B.   Judgment of method

      1.   Cost

      2.   Purity

      3.   Quantity

      4.   Ease of maintenance

IV.   Reverse osmosis

   A.   Theory of method

   B.   Judgment of method

      1.   Cost

      2.   Purity

      3.   Quantity

      4.   Ease of maintenance

[V, VI.   Same as III and IV for remaining methods]

VII.   Choice of a method

FIGURE 8-3   A Revised Outline

TITLE

Purpose: ▨▨▨▨▨▨▨▨▨▨▨▨▨▨▨▨▨▨▨▨▨▨▨▨▨
Audience: ▨▨▨▨▨▨▨▨▨▨▨▨▨▨▨▨▨▨▨▨▨▨▨

  I. ▨▨▨▨▨▨▨▨▨▨▨▨▨▨▨▨▨▨▨▨▨▨▨▨
    A. ▨▨▨▨▨▨▨▨▨▨▨▨▨▨▨▨▨▨▨▨▨▨▨
      1. ▨▨▨▨▨▨▨▨▨▨▨▨▨▨▨▨▨▨▨▨▨
        a. ▨▨▨▨▨▨▨▨▨▨▨▨▨▨▨▨▨▨▨▨
          (1) ▨▨▨▨▨▨▨▨▨▨▨▨▨▨▨▨
            (a) ▨▨▨▨▨▨▨▨▨▨▨▨▨▨▨
            (b) ▨▨▨▨▨▨▨▨▨▨▨▨▨▨▨
          (2) ▨▨▨▨▨▨▨▨▨▨▨▨▨▨▨▨▨
        b. ▨▨▨▨▨▨▨▨▨▨▨▨▨▨▨▨▨▨
        c. ▨▨▨▨▨▨▨▨▨▨▨▨▨▨▨▨▨▨
      2. ▨▨▨▨▨▨▨▨▨▨▨▨▨▨▨▨▨▨▨▨▨
      3. ▨▨▨▨▨▨▨▨▨▨▨▨▨▨▨▨▨▨▨▨▨
    B. ▨▨▨▨▨▨▨▨▨▨▨▨▨▨▨▨▨▨▨▨▨▨▨
      1. ▨▨▨▨▨▨▨▨▨▨▨▨▨▨▨▨▨▨▨▨▨
      2. ▨▨▨▨▨▨▨▨▨▨▨▨▨▨▨▨▨▨▨▨▨
 II. ▨▨▨▨▨▨▨▨▨▨▨▨▨▨▨▨▨▨▨▨▨▨▨▨
    A. ▨▨▨▨▨▨▨▨▨▨▨▨▨▨▨▨▨▨▨▨▨▨▨
    B. ▨▨▨▨▨▨▨▨▨▨▨▨▨▨▨▨▨▨▨▨▨
    C. ▨▨▨▨▨▨▨▨▨▨▨▨▨▨▨▨▨▨▨▨

FIGURE 8-4 An Outline Format

Incorrect	*Correct*
I. Visual capabilities	I. Visual capabilities
A. Acquisition	A. Acquisition
II. Interception and closure rate	B. Interception and closure rate
III. Braking	II. Braking

- Capitalize only the first letter of the entries and proper nouns.
- Do not include entries for your report's introduction or conclusion. Outline only the body of the report. Of course, the information in your purpose statement belongs in your introduction, and perhaps the information about audience belongs there as well.
- Use substantive statements in your outline entries. That is, use entries such as "Reverse osmosis" or "Judgment of method" that suggest the true substance of your information. Do not use cryptic expressions such as "Example 1" or "Minor premise."
- Use either a topic or a sentence outline.

Topic Outlines. The entries of a topic outline are either single words or phrases, all grammatically parallel. Do not use periods or any other marks of punctuation after the entries in a topic outline. The two outlines on desalination in Figures 8-2 and 8-3 are both topic outlines. Students generally prefer topic outlines to sentence outlines, finding them easier to handle. McKee's research indicates that professional technical writers prefer them. They have two great advantages: (1) The table of contents that you must construct for more formal papers is essentially a modified topic outline. You are ahead of the game, then, when you have a good topic outline from the start. (2) Similarly, the headings in formal technical papers are lifted right from the outline. Like the entries in outlines, headings must be grammatically parallel.

Sentence Outlines.   In a sentence outline all entries are complete sentences with proper punctuation. Sentence outlines have one clear advantage over topic outlines. When a writer has constructed a sentence outline, most of the topic sentences for the paragraphs should be well in hand. Writing the paper then becomes a matter of filling in the sentences containing supporting details. Figure 8-5 is an example of a sentence outline.

REPORT FORMAT

Planning the format of your report is the final step before you actually write it. Formats can range from the exceedingly simple— a brief pencilled note on plain paper—to the elaborate complexity of a large report with many divisions, subdivisions, headings, graphics, and so forth, all bound in a fancy cover. These elements of reports are discussed in Chapters 11, "Formal Elements of Reports," and 12, "Graphical Elements of Reports." What we discuss here is the strategy of choosing a proper format. Apply the following general principles when planning your format.

Using a Functional Format

Provide no more complexity in your format than the situation calls for. Your goal is to make your report pleasing and accessible to your readers. If your report is short, say, under five pages, a simple memorandum format will usually do (see pages 311–313). If your report will be beyond memorandum length, you'll need to provide such formal elements of reports as a table of contents and a title page. In a report with many graphics, you may need a list of figures. Supplementary data may be provided in an appendix. Definitions may be supplied in a glossary.

Weigh the value of each of these additions carefully. Add something only if you can justify it on the basis of functionality. For instance, a report that contains only two or three words that need to be defined for the readers does not need a glossary. Define the words as you use them. If you use many terms unfamiliar to your readers,

SAFE MANNED INTERPLANETARY TRAVEL DURING SOLAR FLARES

Purpose: To demonstrate that manned interplanetary travel is safe during periods of solar flares providing NASA understands the problem and protects the astronauts with the latest prediction techniques and shielding methods.

Audience: Undergraduate college students with a scientific or engineering background are the intended audience.

I.   Solar flares are associated with sunspots and produce dangerous radiation.
  A. Sunspots are regions of lower temperature than the surrounding areas of the sun's surface and appear as dark spots that are readily visible to an observer on the earth.
  B. Solar flares appear as sudden, intense brightenings of the areas around sunspots.
  C. Solar flares produce both corpuscular radiation and electro–magnetic radiation.

II. Biological effects constitute the greatest hazard to astronauts from solar radiation.
  A. Even small amounts of radiation will damage body cells.
  B. Depending on the radiation dose, the results can range from no obvious injury to death within 30 days.

III. Manned interplanetary travel will be possible with protection from solar flares.
  A. Predicting solar flares provides one means of protection.
    1. Short–range prediction is currently possible.
    2. Long–range prediction provides some protection but is not accurate enough for total protection.
  B. Shielding combined with short–range prediction will adequately protect the astronaut.
    1. Active shielding causes charged particles to miss the space capsule entirely.
    2. Passive shielding absorbs charged particles before they reach the astronauts.

FIGURE 8-5   A Sentence Outline

a glossary may be a good idea. Before including an appendix, ask yourself if your readers really need or want the supplementary information provided. If you honestly think not, don't include it. Perhaps, instead, tell your readers what supplementary material is available in your files and that you'll be happy to send it upon request. In writing, as in many things, form follows function.

Revealing Your Organization to Your Reader

You see your report as you have organized it. That is, in your mind, you see clearly all the divisions and subdivisions of your organizational plan. Your readers see your report page by page; therefore, they may not grasp your organization or understand its significance. Give them some help. Even a two- or three-page memo will profit from a few well-placed headings that reveal your organization. Longer reports definitely need headings and a table of contents based upon the headings.

Using Standard Formats When Appropriate

Many types of reports have a standard format that their readers expect the writer to follow. For example, empirical research reports written for scientists almost always include an introduction, literature review, materials and methods section, results section, and a discussion section. You would be wise when reporting research for scientists to use this standard format. Most of the reports we discuss in Part 2, "Applications," have formats that, while not rigid, are fairly standard. Use these formats unless you have good reason to deviate from them.

Providing for Selective Reading

Technical and business reports differ from literature in a very significant way. The writer of a short story or a novel hopes and expects that the reader will read every word from start to finish. The writers of technical or business reports know that many of their readers will not want or need the entire report. Rather, the readers will read selectively, scanning a report looking for those segments pertinent to their needs and interests. For instance, scientists reading through a professional journal will want to read thoroughly only those articles that relate closely to their specialities. They will want only summaries of the others. Therefore, abstracts are provided with each article to allow the scientific readers to extract key points quickly without reading the entire article. Furthermore, headings within the article that label the key parts such as "Materials and Methods" and "Results and Discussion" allow the scientists to scan the article, looking for the sections of interest.

In like manner, reports for executives should be given formats that allow for selective reading. As we point out in Chapter 3, "Analyzing Your Audience," not all readers of a report will be interested in the same parts of a report (see pages 45–47). Through the use of headings and a table of contents, you allow your executive readers to find the sections they need. Executives also appreciate summaries and clearly labeled conclusions and recommendations being placed early in the report.

Integrating Graphics

While still in the planning stage, review your material looking for places where you can integrate graphics into your report. Plan to lift masses of statistics out of densely packed paragraphs and to present them in simple tables and bar graphs. Visual ideas should be presented visually. Don't describe, for example, the keyboard of a word processor. Show it in a photograph or drawing. Plan to integrate these graphics in your report through proper placement and by referring to them at key points in your report.

Example Formats

As examples only, we show you here a few of the possible patterns using the report elements explained in Chapter 11, "Formal Elements of Reports." See pages 200–201 for page references to the elements. For an example of a full-scale report, see Appendix A, "A Student Report."

- *An internal report of 9 pages—informational, for executives:*
  Memorandum of transmittal
  Title page
  Informative abstract
  Introduction
  Discussion
  Complimentary close
- *An external report of 35 pages—decision making, for executives and experts:*
  Letter of transmittal
  Title page
  Table of contents
  List of figures
  Glossary
  Introduction
  Summary
  Conclusions
  Recommendations
  Discussion
  List of references
  Appendixes
- *Internal report—research report for scientists:*
  Title page
  Informative abstract
  Table of contents
  List of figures
  Introduction
  Discussion

Summary
Conclusions
Bibliography

EXERCISES

1. When asked to brainstorm about why the United States birthrate has been decreasing,* a group of students produced the reasons listed below. Obviously, there are enough ideas on the list for several books. Looking at the list, see how many topics and subtopics you can break it down into. You may have to supply some "umbrella" topic headings. Select one of the topics (or subtopics). Decide what your purpose in handling the topic is and who your audience will be. Perhaps, for example, you might wish to show a lay audience how some development listed is *one* of the causes for the declining birthrate. Using the material on the list and ideas of your own, organize a possible essay. If your instructor wishes you to, turn your organizational plan into a formal outline. If you were to write the essay, what facts would you need to support the opinions stated? Where would you look for such facts?

first child born later
divorce rate up
husbands and wives both working
increased family planning
National Organization of Women
end of 1950s and 1960s boom
world food shortages
popular books on population problems
a mobile society
average age at marriage increasing

inflation
rising age of U.S. population
more women in colleges and universities
legalized abortion
higher costs for health care
better contraception
women in work force increasing
shift from rural to urban living
changing sexual attitudes
the youth movement of the 1960s

*According to the *Statistical Abstract of the United States, 1981,* the birthrate per thousand population for the given years was as follows: 1950 (24.1), 1955 (25), 1960 (23.7), 1965 (19.4), 1970 (18.4), 1975 (14.8), 1979 (15.9).

2. For about 15 minutes, jot down everything you can think of about some technical or semitechnical subject. Choose a subject that you know well, so you can put down a good deal of information without further research—perhaps a subject related to a hobby or some school subject that you enjoy.

Do not evaluate the material as you put it down. Brainstorm the subject. After you complete the 15-minute exercise, examine and evaluate what you have. Narrow down to the most specific subject and purpose that you can. Choose your audience and role as a writer. Keeping your specific subject, purpose, audience, and role in mind, organize your jottings into a rough outline. Do not worry overmuch about outline format, such as roman numerals, parallel headings, and so forth.

3. Turn the informal outline you constructed for Exercise 2 into a formal outline. Use either a topic or sentence outline format. Your instructor may want to specify the outline type.

4. If you have been planning a report as called for in the exercises following Chapters 2 and 5, now prepare an organizational plan for that report. Use whatever outline type your instructor specifies.

5. Prepare a format plan for the report you outlined in Exercise 4. Your plan should look like one of the example plans we provide on pages 164–165. Include with your plan a paragraph explaining what graphics you intend to integrate into the report.

Chapter **9**

Achieving a Clear Style

EXAMPLES OF UNCLEAR writing style are all too easy to find, even in places where we would hope to find clear, forceful prose. Read the following sentence:

> While determination of specific space needs and access cannot be accomplished until after a programmatic configuation is developed, it is apparent that physical space is excessive and that all appropriate means should be pursued to assure that the entire physical plant is utilized as fully as feasible.

The murky sentence quoted comes from a report issued by a state higher education coordinating board. Actually, it's better than many examples we could show you. While difficult, the sentence is probably readable. Others are simply indecipherable. When you have finished this chapter, you should be able to analyze a passage like the one just cited and show why it is so unclear. You should also know how to keep your own writing clear, concise, and vigorous. Here are the principles that underlie everything we will tell you in this chapter:

- Write as honestly and objectively as you can. Objectivity comes from facts correctly reported and honest inferences drawn from them.
- Do not be mysterious. Tell your readers as clearly as you can what you want them to understand, know, or do.
- Remember that you have invited your readers to come to you. The burden of being clear rests upon you.

- Do not strike a pose when you write. You are not trying to impress readers with your superiority. You are trying to get ideas from your brain to theirs by the shortest, clearest route.

We discuss paragraphs, lists, clear sentence structure, specific words, and avoiding pomposity. We have broken our subject into five parts for simplicity's sake. But all the parts are closely related. All have but one aim—clarity.

THE PARAGRAPH

In Chapters 6 and 7, we discuss the rhetorical modes of exposition, description, narration, and argumentation. These modes may be used not only to develop reports but also to develop paragraphs within reports. Thus, paragraphs will vary greatly in organization and length, depending upon their purpose. In technical writing, however, the central statement of a paragraph more often than not appears at the beginning of the paragraph. This placement provides the clarity of statement that good technical writing must have. However, in a paragraph aimed at persuasion the central statement may appear at the close, where it provides a suitable climax for the argument. No matter where a central statement is placed, unity is achieved by relating all the other details of the paragraph to the statement, as in this paragraph of inference and speculation:

Mariner 10 confirmed the earlier findings of Mariner 5 and Venera 4 on their flights to Venus in October 1967 and discovered a bow shock—a wave in front of the planet like the bow wave of a ship. Somehow the ionosphere of Venus forms this bow shock in the solar wind and stops the solar wind from plunging directly into the atmosphere of the planet. How and why this bow shock occurs is not fully understood. Certainly the effect is very different from that of the Earth, Moon, Mars, Mercury and Jupiter. The Venus bow shock might be a direct interaction of the solar wind with the atmosphere of Venus, or with just the ionosphere. It may alternatively arise because the solar wind induces magnetic fields and produces thereby a pseudo or false magnetopause as though Venus had

We have italicized the central statement concerning a "bow shock" in front of Venus. The rest of the paragraph presents facts and analysis in support of the central statement.

a magnetic field like the Earth. Or the solar wind might act on the atmosphere of Venus as it does on a comet to produce the tail. Any of these explanations might be confirmed or disproved by another mission to Venus.[1]

How Long Should a Paragraph Be? Examination of well-edited magazines such as *Scientific American* reveals that their paragraphs seldom average more than 100 words in length. The editors of magazines know that paragraphs are for the reader. Paragraphing breaks the materials into related subdivisions for the reader's better understanding. When paragraphs run on too long, the central statements that provide the generalizations needed for reader understanding are either missing or hidden in the mass of supporting details.

In addition to considering the need of the reader for clarifying generalizations, editors also consider the psychological effect of their pages on the reader. They know that large blocks of unbroken print have a forbidding appearance that intimidates the reader. If you follow the practices of experienced editors, you will break your paragraphs whenever your presentation definitely takes a new turn. As a general rule, paragraphs in reports and articles should average 100 words or fewer. In letters and memorandums, because of their page layout, you should probably hold average paragraph length to fewer than 60 words.

How Do You Provide Transition? Most often, a paragraph presents a further development in a continuing sequence of thought. When this is true, the paragraph's opening central statement will be so closely related to the preceding paragraph that it provides most of the transition you need. When a major transition between ideas is called for, consider using a short paragraph to guide the reader from one idea to the next.

The following five paragraphs provide an excellent example of paragraph development and transition:

Hyperactivity is essentially a symptom, which may be the result of a child's basic personality, a temporary state of anxiety, or subclinical seizure disorders; or it may reflect a true hyperkinetic state. It may also, according to the Council on Child Health of the American Academy of Pedi-

Hyperactivity is the key word. It or *hyperactive* appears six times. Its repetition provides a major transitional device. The second sentence leads into the central theme of the five

atrics, be strictly "in the eyes of the beholder" (AAP Council on Child Health, 1975).

Descriptions of hyperactivity are generally given in behavioral terms, such as motor activity, attention span, frustration tolerance, excitability, impulse control, irritability, restlessness, and aggressiveness. Although these behaviors are measurable, they may not adequately reflect the kind of problems that different children have. Objective measures of attention span have been developed and have been used in some studies to help make the diagnosis, usually during tasks requiring continuous performance by the child. Standard questionnaires have also been designed, to be used by parents and teachers as they make observations at different times; such questionnaires provide useful evaluation information.

Researchers, however, acknowledge the difficulties of measuring these attributes and of interpreting the measurements. For example, in a study at the University of North Carolina designed to measure the motor activity and interest span of children diagnosed as hyperkinetic, Routh (1975) reported that of 78 referrals from physicians, teachers, and parents, only 47% of the children were judged to be overactive—despite the fact that all of the children were considered to be "problem children" by those referring them to the testing service. In a study of teacher ratings of hyperactivity conducted at the University of Iowa, older teachers rated more children hyperactive than did young teachers (Johnson, 1974).

Clinical experience also indicates that many factors alter the activity patterns of such children or the perception of the patterns by parents. Such factors range from the presence or absence of breakfast,

paragraphs—that hyperactivity may sometimes be "in the eye of the beholder." The central statement in the paragraph shows its relationship to the first paragraph through the use of the words *hyperactivity* and *descriptions*. The second half of the statement leads into the subject of the paragraph—the behavioral activities that provide a measure of hyperactivity. Many central statements provide transition by looking both backward and forward.

Researchers provides the new element that will be considered in this paragraph. The words *measuring* and *interpreting* alert the reader that this paragraph continues the central theme of all five paragraphs.

Facts to support the difficulty of measurement are given.

Clinical experience shifts the reader from *researchers* as a source of information to a new source. The word *also*

weather conditions and resultant seasonal cycles that alter activities during cold seasons, to the interpersonal family relationships or existence of disruptive family problems.

Age of the child appears to be another determinant of detection of the syndrome of hyperactivity. The majority of cases are noted at school, and there is usually a gradual diminution in the hyperactivity at puberty or slightly thereafter. While some observers cite isolated cases of identifiable hyperkinesis in older individuals, it is unusual. Also, in some areas of the world, the syndrome is not recognized as a problem by school authorities—again, an example of differences in perception or of occurrence.[2]

shows that the central idea of the previous paragraph is still being pursued. The clause beginning *many factors* introduces the subject matter of the paragraph. *Age of the child* introduces the central subject of this paragraph. The words *another, determinant, detection,* and *hyperactivity* all announce the relationship of the age of the child to the central theme of the five paragraphs.

The final sentence rounds off the five paragraphs by returning, in more formal language, to the central theme that hyperactivity may sometimes be "in the eye of the beholder."

The five paragraphs illustrate that you will develop paragraphs coherently when you keep your mind on the central theme. If you do so, the words needed to provide proper transition will come rather naturally. More often than not, your transitions will be repetitions of key words and phrases supported by such simple expressions as *also, another, of these four, because of this development, so, but,* and *however.* When you wander away from your central theme, no amount of artificial transition will really wrench your writing back into coherence.

LISTING

One of the simplest things you can do to ease the reader's chore is to break down complex statements into lists. Visualize the printed page. When it appears as an unbroken mass of print, it intimidates readers and makes it harder for them to pick out key ideas. Get important ideas out into the open where they stand out. Lists help to clarify introductions and summaries. You may list by (1) starting each separate point on a new line, leaving plenty of white space around it, or

(2) using numbers within a line as we have done here. Examine the following summary from a student paper, first as it might have been written and then as it actually was:

> The exploding wire is a simple-to-perform yet very complex scientific phenomenon. The course of any explosion depends not only on the material and shape of the wire but also on the electrical parameters of the circuit. In an explosion the current builds up and the wire explodes, current flows during the dwell period, and "post-dwell conduction" begins with the reignition caused by impact ionization. These phases may be run together by varying the circuit parameters.

Now, the same summary as a list:

> The exploding wire is a simple-to-perform yet very complex scientific phenomenon. The course of any explosion depends not only on the materials and shape but also on the electrical parameters of the circuit.
> An explosion consists primarily of three phases:
>
> 1. The current builds up and the wire explodes.
> 2. Current flows during the dwell period.
> 3. "Post-dwell conduction" begins with the reignition caused by impact ionization.
>
> These phases may be run together by varying the circuit parameters.

The first version is clear, but the second version is clearer, and readers can now file the information in their minds as "three phases." They will remember it longer.

Some writers avoid using lists even when they should, so we hesitate to suggest any restrictions on the technique. Obviously, however, there are some subjective limits. Lists break up ideas into easy-to-read, easy-to-understand bits, but too many can give your page the appearance of a laundry list. Also, journal editors sometimes object to lists in which each item starts on a separate line. Such lists take space, and space costs money. So use lists when they clarify your presentation, but use them discreetly.

CLEAR SENTENCE STRUCTURE

The basic English sentence structure appears in two patterns, *subject-verb-object* (SVO) and *subject-verb-complement* (SVC):

    John (S) hit(V) the ball(O).
    The baby(S) cried(V) lustily(C).

Around such simple sentences as "John hit the ball," the writer can hang a complex structure of words, phrases, and clauses that serve to modify and extend the basic idea. In this section on clear sentence structure, we discuss how to go about that task. In order, we discuss sentence length, sentence order, sentence complexity and density, active verbs, active and passive voice, and first-person point of view.

Sentence Length

How how should a sentence be? Some research[3] indicates the following scale of length versus reading ease:

very easy	8 words or fewer
easy	11
fairly easy	14
standard	17
fairly difficult	21
difficult	25
very difficult	29 words or more

Another study notes that professional writers' sentences average in the low 20s.[4]

A great many formulas to measure the readability of writing are available, some exceedingly complicated. However, in an article reviewing the use of such formulas, Dr. George R. Klare of Ohio University concludes "that counts of the two simple variables of word length and sentence length are sufficient to make relatively good predictions of readability."[5] That is, the use of shorter words and shorter sentences usually correlates positively with ease in reading. But Klare cautions that shorter sentences and words simply *indicate* that a piece of writing will be more readable. Whether they *cause* the increased readability is another and more complicated question.

Klare's research suggests that while present readability studies do not give any one the right to be arbitrary about formulas for writing, they do furnish us with some useful guides. And probably the best ones are the simplest. Hold your sentence length down. Technical concepts are hard enough to grasp even when described in simple sentences. Bring your skills in audience analysis to bear on the problem. Suppose you have an audience that includes a good many people with only an eighth-grade reading ability. In this case, sentences that average 14 words, "fairly easy" on the preceding scale, would be suitable. For most business correspondence, the "standard" 17-word average would be a good choice. For people with high reading skill—many executives, for example—a 20-word average would be appropriate. Stay away from the "very difficult" average of 29 words or more no matter who is in your audience. Remember, too, that you're dealing with *average* sentence length. Don't write sentences all of one length. For an average of 14 words, you would probably range from 5- to 25-word sentences. Klare also mentions word length as a factor. We will say more about that in a moment.

Sentence Order

What is the best way to order a sentence? Is a great deal of variety in sentence structure the mark of a good writer? One writing teacher, Dr. Francis Christensen of the University of Southern California, looked for the answers to those two questions. He examined large samples from 20 top writers, among them John O'Hara, John Steinbeck, William Faulkner, Ernest Hemingway, Rachel Carson, and Gilbert Highet. In his samples, he included 10 fiction writers and 10 nonfiction writers.[6]

What Christensen discovered seems to disprove any theory that good writing requires extensive sentence variety. The professionals whose work was examined depended mostly on basic sentence patterns. They wrote 75.5 percent of their sentences in plain subject-verb-object (SVO) or subject-verb-complement (SVC) order, as in these two samples:

> Doppler radar increases capability greatly over conventional radar. (SVO)
>
> Doppler radar can be tuned more rapidly than conventional radar. (SVC)

Another 23 percent of the time, the professionals began a sentence with a short adverbial opener:

> As with any radar system, Doppler does have problems associated with it.

These adverbial openers are most often simple prepositional phrases or single words such as *however, therefore, nevertheless,* and the other conjunctive adverbs. Generally they provide the reader with a transition between thoughts. Following the opening, the writer usually continued with a basic SVO or SVC sentence.

These basic sentence types—SVO(C) or adverbial + SVO(C)—are used 98.5 percent of the time by the professional writers in Christensen's sample. What did the writers do with the remaining 1.5 percent of their sentences? For 1.2 percent, they opened the sentence with verbal clauses based upon participles and infinitives such as *"Breaking* ground for the new church," or *"To see* the new pattern more clearly." The verbal opener was again followed most often with an SVO or SVC sentence, as in this example:

> Looking at it this way, we see the radar set as basically a sophisticated stopwatch that sends out a high energy electromagnetic pulse and measures the time it takes for part of that energy to be reflected back to the antenna.

Like the adverbial opener, the verbal opener serves most of the time as a transition.

The remaining .3 percent of the sentences (1 sentence in 300) are inverted constructions in which the subject is delayed until after the verb, as in this sentence:

> No less important to the radar operator are the problems caused by certain inherent characteristics of radar sets.

What can we conclude from Christensen's valuable study? Simply this: the professionals were interested in getting their content across, not in tricky word order. They conveyed their thoughts in a clear container, not clouded by extra words. You should do the same.

Sentence Complexity and Density

The complexity and density of a sentence are closely related to sentence length and sentence order. Research indicates that sentences that are *too* complex in structure or *too* dense with content are difficult for many readers to understand.[7] Basing our observations on this research, we wish to discuss four particular problem areas: too many words in front of the subject, too many words between the subject and the verb, noun strings, and multiple negatives.

Words in Front of the Subject. As the Christensen research indicates, professional writers open with something before their subjects about 25 percent of the time. When these openers are held to a reasonable length, they create no problems for readers. The problems occur when the writer stretches such openers beyond a reasonable length. What is *reasonable* is somewhat open to question and depends to an extent on the reading ability of the reader. However, most would agree that the 27 words and 5 commas before the subject in the following sentence make the sentence difficult for many readers:

> *Because of their ready adaptability, ease of machining, and aesthetic qualities that make them suitable for use in landscape structures such as decks,*

fences, and retaining walls, preservative-treated timbers are becoming increasingly popular for use in landscape construction.

The ideas contained in this too dense sentence become more accessible when spread over two sentences:

Preservative-treated timbers are becoming increasingly popular for use in landscape construction. Their ready adaptability, ease of machining, and aesthetic qualities make them highly suited for use in landscape structures such as decks, fences, steps, and retaining walls.

The second version has the additional advantage of putting the central idea in the sequence before the supporting information.

The conditional sentence is a particularly difficult version of the sentence with the subject too long delayed. You can recognize the conditional by its *if* beginning:

If heat (20°–35° C or 68°–95° F optimum), moisture (20% + moisture content in wood), oxygen, and food (cellulose and wood sugars) are present, spores will germinate and grow.

To clarify such a sentence, move the subject to the front and the conditions to the rear. Consider the use of a list when you have more than two conditions:

Spores will germinate and grow when the following elements are present:

- Heat (20°–35° C or 68°–95° F optimum)
- Moisture content (20% + moisture content in wood)
- Oxygen
- Food (cellulose and wood sugars)

Words Between Subject and Verb. In the following sentence, too many words between the subject and the verb cause difficulty:

Creosote, *a brownish-black oil composed of hundreds of organic compounds, usually made by distilling coal tar, but sometimes made from wood or petroleum,* has been used extensively in treating poles, piles, crossties, and timbers.

The sentence becomes much easier to read when it is broken into three sentences and first things are put first:

Creosote has been used extensively in treating poles, piles, crossties, and timbers. It is a brownish-black oil composed of hundreds of organic compounds. Creosote is usually made by distilling coal oil, but it can also be made from wood or petroleum.

You might wish to break down the original sentence into only two sentences if you had an audience that you thought could handle denser sentences:

Creosote, a brownish-black oil composed of hundreds of organic compounds, has been used extensively in treating poles, piles, crossties, and timbers. It is usually made by distilling coal tar, but it can also be made from wood or petroleum.

Noun Strings. Noun strings are another way that writers sometimes complicate and compress their sentences beyond tolerable limits. A noun string is a sequence of nouns that serves to modify another noun: for example, *multichannel microwave radiometer*, where the nouns *multichannel* and *microwave* serve to modify *radiometer*. Sometimes the string may also include an adjective, as in *special multichannel microwave radiometer*.

Nothing is grammatically wrong with the use of nouns for modifiers. Such use is an old and perfectly respectable custom in English. Expressions such as *fire fighter* and *creamery butter* in which the modifiers are nouns go unremarked and virtually unnoticed. The problem occurs when writers either string many nouns together in one sequence or use many noun strings in a passage. Both tendencies are evident in this paragraph:

We must understand who the initiators of *water-oriented greenway efforts* are before we can understand the basis for *community environment decision making processes*. State government planning agencies and commissions and *designated water quality planning and management agencies* have initiated such efforts. They have implemented *water resource planning and management studies* and have aided *volunteer group greenway initiators* by providing *technical and coordinative assistance*.[8]

In many such strings, the reader has great difficulty in sorting out the relationships among the words. In *volunteer group greenway initiators* does *volunteer* modify *group* or *initiators?* The reader has no way of knowing.

The solution to untangling difficult noun strings is to include the relationship clues such as prepositions, relative pronouns, commas, apostrophes, and hyphens. For instance, placing a hyphen in *volunteer-group* would clarify that *volunteer* modified *group*. The strung-out passage above was much improved by the inclusion of such clues:

> We must understand who the initiators of efforts *to* promote water-oriented greenways are before we can understand the process *by which* a community makes decisions *about* environmental issues. Planning agencies and commissions *of the* state government and agencies *which* have been designated *to* plan and manage water quality have initiated such efforts. They have implemented studies *on* planning and managing water resources and have aided volunteer groups that initiate efforts *to* promote greenways by providing them with technical advice and assistance *in* coordinating their activities.[9]

The use of noun strings in technical English will no doubt continue. They do have their uses, and technical people are very fond of them. But perhaps it would not be too much to hope that writers would hold their strings to three words or fewer and not use more than one per paragraph.

Multiple Negatives. Writers can introduce excessive complexity into their sentences through the use of multiple negatives. By *multiple negative*, we do not mean the grammatical error of the *double negative:* for instance, "He does *not* have *none* of them." We are talking about perfectly correct constructions that include two or more negative expressions: for example, "We will *not* go *unless* the sun is shining." Other examples would be "We will *not* pay *except* when the damages exceed $50" and "The lever will *not* function *until* the power is turned on."

The positive statements of all of these are better and clearer than the negative versions:

- We will go only if the sun is shining.
- We will pay only when the damages exceed $50.
- The lever functions only when the power is turned on.

Most research shows that readers have difficulty sorting out passages that contain multiple negatives. If you doubt the research, try your hand at interpreting this government regulation (italics for negatives are ours):

§928.310 Papaya Regulation 10.

Order. (a) *No* handler shall ship any container of papayas (*except* immature papayas handled pursuant to §928.152 of this part):

(1) During the period January 1 through April 15, 1980, to any destination within the production area *unless* said papayas grade at least

Hawaii No. 1, *except* that allowable tolerances for defects may total 10 percent *Provided*, that *not* more than 5 percent shall be for serious damage, *not* more than 1 percent for immature fruit, *not* more than one percent for decay: *Provided further*, that such papayas shall individually weight *not* less than 11 ounces each.[10]

Active Verbs

The verb determines the structure of an English sentence. Many sentences in technical writing falter because the finite verb does not function properly: that is, (1) comment upon the subject, (2) state a relationship about the subject, or (3) relate an action that the subject performs. Look at the following sentence:

Sighting of the gound was accomplished by the pilot at 7 A.M.

English verbs can easily be changed into nouns, but sometimes—as we have just seen—the change can lead to a faulty sentence. The writer has put the true action into the subject and subordinated the pilot and the ground as objects of prepositions. The sentence should read:

At 7 A.M. the pilot *saw* the ground.

The poor writer can ingeniously bury the action of a sentence almost anywhere. With the common verbs *make, give, get, have,* and *use,* the writer can bury the action late in the sentence in an object:

The punch card operator *has* the job of translating language symbols into machine symbols.

Or,

The speaker did not *give* a satisfactory explanation of his technique.

Properly revised, the sentences read:

The punch card operator *translates* language symbols into machine symbols.
The speaker did not adequately *explain* his technique.

The poor writer can even bury the action in an adjective:

A new discovery produces an *excited* reaction in a scientist.

Revised:

A new discovery *excites* a scientist.

There is constructions may trap the writer into inefficient sentences. For example:

There are some technical editors who prefer the passive voice.

Revised:

Some technical editors prefer the passive voice.

When writing, and particularly when rewriting, you should always ask yourself: "Where's the action?" If the action does not lie in the verb, rewrite the sentence to put it there, as with this sample:

Music therapy is the scientific *application* of music to accomplish the *restoration, maintenance,* and *improvement* of mental health.

This sentence provides an excellent example of how verbs are frequently turned into nouns: by the use of the suffixes *ion, ance* (or *ence*), and *ment.* If you have sentences full of such suffixes, you may not be writing as actively as you could be. Rewritten to put active ideas into verb forms, the sentence reads this way:

Music therapy *applies* music scientifically *to restore, maintain,* and *improve* mental health.

The rewritten sentence defines "music therapy" in one-third less language than the first sentence, without any loss of meaning or content.

Active and Passive Voice

We discuss active and passive voice sentences on page 131, but let us quickly review the concept again here. In an active voice sentence the subject performs the action and the object receives the action, as in "Mary hit the ball." In a passive voice sentence, the subject *receives* the action, as in "The ball has been hit." If you want to include the doer of the action, you must add this information in a prepositional phrase as in "The ball has been hit *by Mary.*"

We urge you to use the active voice more than the passive, but we do not wish to imply that you should ignore the passive altogether. The passive voice is often useful. You can use the passive voice to emphasize the object receiving the action. The passive voice in "Influenza may be caused by any of several viruses" emphasizes *influenza.* The active voice in "Any of several viruses may cause influenza" emphasizes the *viruses.*

Often the agent of action is of no particular importance. When such is the case, the passive voice is appropriate because it allows you to drop the agent altogether:

Edward Jenner's work on vaccination was published in 1796.

Be aware, however, that inappropriate use of the passive voice can cause you to omit the agent when knowledge of the agent may be vital. Such is often the case in giving instructions: "All doors to this building will be locked by 6 P.M." may not produce locked doors until it is rewritten in the active voice: "The night manager will lock all doors to this building by 6 P.M."

While the passive voice has its uses, most editors feel that too much of it produces lifeless and wordy writing. They therefore advise against using it except when it is clearly appropriate. The *Council of Biology Editors Style Manual* succinctly expresses the reasons for this advice:

> Use the active voice except where you have good reason to use the passive. The active is the natural voice, the one in which people usually speak and write, and its use is less likely to lead to wordiness and ambiguity. Avoid the "passive of modesty," a device of writers who shun the first-person singular. "I discovered" is shorter and less likely to be ambiguous than "it was discovered." When you write "Experiments were conducted," the reader cannot tell whether you or some other scientist conducted them. If you write "I" or "we" ("we" for two or more authors, never as substitute for "I"), you avoid dangling participles, common in sentences written in the third person passive voice.[11]

As the biology editors point out, the passive voice does, indeed, cause many dangling participles, as in "While conducting these experiments, the chickens were seen to panic every time a hawk flew over." Chickens conducting experiments? Not really. The active voice straightens out the matter: "While conducting these experiments, we saw that the chickens panicked every time a hawk flew over." (See also "Dangling Modifier," page 459.)

Write in the active voice whenever appropriate. When rewriting, find and revise inappropriate passive voice sentences into the active.

First-Person Point of View

Once, many reports and scientific articles were written in the third person—"This investigator has discovered"—rather than first person—"I discovered." Because this practice introduces needless complexity, it is now much less common. Along with the *Council of Biology Editors Style Manual,* many other style manuals for scientific journals now advise against the use of the third person on the grounds that it is wordy and confusing. We agree with this advice.

The judicious use of *I* or *we* in a technical report is entirely appropriate. Incidentally, such usage will seldom lead to a report full of *I's* and *we's.* After all, there are many agents in a technical report other than the writer. In describing an agricultural experiment, for example, researchers will report how *the sun shone, photosynthesis occurred, rain fell, plants drew nutrients from the soil,* and *combines harvested.* Only occasionally will researchers need to report their own actions. But when they must, they should be able to avoid such roundabout expressions as "It was observed by this experimenter." (See also pages 411-412)

A Caution

We must caution you before we leave this section on clear sentence structure. We are not urging upon you an oversimplified primer style, one often characterized by such sentences as "Jane hit the ball" and "See Dick catch the ball." Mature styles have a degree of complexity to them. Good writers, as Christensen's research shows, do put information before the subject. Nothing is wrong with putting information between the subject and verb of a sentence. You will find many such sentences in this book. What we have done, however, is alert you to the fact that research does show that sentences that are too long, too complex, too dense, for whatever reason, cause many readers difficulty. Keep in mind that despite increasingly good research into its nature writing is a craft and not a science. Be guided by the research available, but do not be simplistic in applying it.

SPECIFIC WORDS

Remember the semanticists' abstraction ladder that we describe for you on page 108: a ladder composed of rungs that ascend from very specific words such as *table* to abstractions such as *furniture, wealth,* and *factor?* The human ability to move up and down this ladder enabled us to develop language, on which all human progress depends. Because we can think in abstract terms, we can call a moving company and tell it to move our furniture. Without abstraction we would have to bring the movers into our house and point to each object we wanted moved. But like many helpful writing techniques, abstraction is a device you should use carefully.

Stay at an appropriate level on the abstraction ladder. Do not say "inclement weather" when you mean "rain." Do not say "overwhelming support" when you mean "62 percent of the workers supported the plan." Do not settle for "suitable transportation" when you mean "a bus that seats 32 people."

Writing that uses too many abstractions is lazy writing. It relieves writers of the need to observe, to research, and to think. They can speak casually of "factors," and neither they nor their readers really know what they are talking about. Here is an example of such lazy writing. The writer was setting standards for choosing a desalination plant to be used at Air Force bases.

- The cost must not be prohibitive.
- The quantity of water must be sufficient to supply a military establishment.
- The quality of the water must be high.

The writer here thinks he has said something. He has said little. He has listed slovenly abstractions when with a little thought and research he could have listed specific details. He should have said:

- The cost should not exceed $3 per thousand gallons.
- To supply an average base with a population of 5,000, the plant should purify 750,000 gallons of water a day (AFM 88-10 sets the standard of 150 gallons a day per person.)
- The desalinated water produced should not exceed the national health standard for potable water of 500 parts per million of dissolved solids.

Abstractions are needed for generalizing, but they cannot replace specific words and necessary details. Words mean different things to different people. The higher you go on the abstraction ladder, the truer this is. The abstract words—*prohibitive, sufficient,* and *high*—could have been interpreted in as many different ways as the writer had readers. No one can misinterpret the specific details given in the rewritten sentences.

Abstractions can also burden sentences in another way. Some writers are so used to thinking abstractly that they begin a sentence with an abstraction and *then* follow it with the specific word, usually in a prepositional phrase. They write,

The problem of producing fresh water became troublesome at overseas bases.

Instead of,

Producing fresh water became a problem at overseas bases.

Or,

The circumstance of the manager's disapproval caused the project to be dropped.

Instead of,

The manager's disapproval caused the project to be dropped.

We do not mean to say you should never use high abstractions. A good writer moves freely up and down the abstraction ladder. But when you use words from high on the ladder, use them properly—for generalizing and as a shorthand way of referring to specific details you have already given.

POMPOSITY

When writing, state your meaning as simply and clearly as you can. Do not let the mistaken notion that writing should be more elegant than speech make you sound pompous. Writing *is* different from speech. Writing is more concise, more compressed, and often better organized than speech. But elegance is not a prerequisite for good writing.

A sign at a service station where one of us gets his gasoline reads, "No gas will be dispensed while smoking." Would anyone in that service station speak that way? Of course not. He would say, "Please put out that cigarette" or "No smoking, please." But the sign had to be elegant, and the writer sounds pompous, and illiterate as well.

If you apply what we have already told you about clear sentence structure, you will go a long way toward tearing down the fence of artificiality between you and the reader. We want to touch on just three more points: empty words, elegant variation, and pompous vocabulary.

Empty Words

The easiest way to turn simple, clear prose into elegant nonsense is to throw in empty words, like these phrases that begin with the impersonal *it:* "It is evident," "It is clear that," or most miserable of all, "It is interesting to note that." When something is evident, clear, or interesting, readers will discover this for themselves. If something is not evident, clear, or interesting, rewrite it to make it so. When you must use such qualifying phrases, at least shorten them to "evidently," "clearly," and "note that." Avoid constructions like "It was noted by Jones." Why not simply say, "Jones noted"?

Many empty words are simply jargon phrases writers throw in out of sheer habit. You see them often in business correspondence. A partial list follows:

to the extent that	is already stated
with reference to	in view of
in connection with	inasmuch as
relative to	with your permission
with regard to	hence
with respect to	as a matter of fact

We could go on, but so could you. When such weeds crop up in your writing, pull them out.

Another way to produce empty words is to run an abstract word in tandem with a specific word. This produces such combinations as

ten in number *for* ten
wires of thin size *for* thin wires
red in color *for* red

When you have expressed something specifically, do not throw in the abstract term for the same word.

Elegant Variation

Elegant variation will also make your writing sound pompous. (H. W. Fowler and F. G. Fowler invented the term *elegant variation* in their book, *The King's English*. H. W. Fowler used it again in *A Dictionary of Modern English Usage*. Writers should own both books and use them often.) Elegant variation occurs when a writer substitutes one word for another because of an imagined need to avoid repetition. This substitution can lead to two problems:(1) The substituted word may be a pompous one. (2) The variation may mislead the reader into thinking that some shift in meaning is intended. Both problems are evident in the following example:

Insect damage to evergreens varies with the condition of the plant, the pest species, and the hexapod population level.

Confusion reigns. The writer has avoided repetition, but the reader may think that the words *insect*, *pest*, and *hexapod* refer to three different things. Also, *hexapod*, though a perfectly good word, sounds a bit pompous in this context. The writer would better have written,

> Insect damage to evergreens varies with the condition of the plant, the insect species, and the insect population level.

Remember also that intelligent repetition provides good transition. Repeating key words reminds the reader that you are still dealing with your central theme (see pages 169-171).

Pompous Vocabulary

Generally speaking, the vocabulary you think in will serve in your writing. Jaw-breaking thesaurus words and words high on the abstraction ladder will not convince readers that you are intellectually superior. Such words will merely convince readers that your writing is hard to read. We are not really telling you here that you must forgo your hard-won educated vocabulary. If you are writing for readers who would understand words like *extant* or *prototype*, then use them. But use them only if they are appropriate to your discussion. Don't use them to impress people. Remember, too, the conclusion reached by Klare (see pages 173-174) that excessive word length is a significant factor in reducing reader understanding.

Nor are we talking about the specialized words of your professional field. At times these are quite necessary. Just remember to define them if you feel your reader will not know them. What we are talking about is the desire some writers seem to have to use pompous vocabulary to impress their readers.

The following list is a sampling of heavy words and phrases along with their simpler substitutes.

accordingly: so	*in connection with:* about
acquire: get	*initiate:* begin
activate: begin	*in order to:* to
along the lines of: like	*in the event that:* if
assist: help	*in the interests of:* for
compensation: pay	*in this case:* here
consequently: so	*make application to:* apply
due to the fact that: because	*nevertheless:* but, however
facilitate: ease, simplify	*prior to:* before
for the purpose of: for	*subsequent to:* later, after
in accordance with: by, under	*utilize:* use

You would be wise to avoid the word-wasting phrases on this list and other phrases like them. You really don't need to avoid the words

shown. All are perfectly good words. But to avoid sounding pompous, don't string large clumps of such words together. Be generous
in your writing with simpler substitutes we have shown you. If you
don't, you are more likely to depress your readers than to impress
them. Don't be like the pompous writers who seek to bury you under
the many-syllabled words they use to express one-syllable ideas, as
in this example from the U.S. Department of Transportation:

> The purpose of this PPM [Policy and Procedure Memorandum] is to
> ensure, to the maximum extent practicable, that highway locations and
> designs reflect and are consistent with Federal, State and local goals and
> objectives. The rules, policies, and procedures established by this PPM
> are intended to afford full opportunity for effective public participation
> in the consideration of highway location and design proposals by high-
> way departments before submission to the Federal Highway Adminis-
> tration for approval. They provide a medium for free and open discus-
> sion and are designed to encourage early and amicable resolution of
> controversial issues that may arise.

We urge you to read as much good writing—both fiction and nonfiction—as time permits. Stop occasionally as you do and study the
authors' choice of words. You will find most authors to be lovers of
the short word. Numerous passages in Shakespeare are composed
almost entirely of one-syllable words. The same holds true for the
King James Bible. Good writers do not want to impress you with
their vocabularies. They want to get their ideas from their heads to
yours by the shortest, simplest route.

SUMMARY

A final example will summarize much that we have said. Insurance
policies have for so long been verbal bogs that most buyers of insurance have long since given up on finding one clearly written. However, the St. Paul Fire and Marine Insurance Company decided that
perhaps it was both possible and desirable to simplify the wording
of its policies. For a trial run, the company turned Dr. Rudolf Flesch
loose on one of its policies. He rewrote it, eliminating empty words
and using only words familiar to the average reader. He avoided
excessive sentence complexity and used predominantly the active
voice and active verbs. He broke long paragraphs into shorter ones.
The insurance company becomes *we* and the insured *you*. Definitions
are included where needed rather than segregated in a glossary.

The resulting policy is wonderfully clear. Compare a paragraph
of the old with the new.[12]

Old

Cancellation

This policy may be cancelled by the Named Insured by surrender thereof to the Company or any of its authorized agents, or by mailing to the Company written notice stating when thereafter such cancellation shall be effective. This policy may be cancelled by the Company by mailing to the Named Insured at the address shown in this Policy written notice stating when, not less than thirty (30) days thereafter, such cancellation shall be effective. The mailing of notice as aforesaid shall be sufficient notice and the effective date of cancellation stated in the notice shall become the end of the policy period. Delivery of such written notice either by the Named Insured or by the Company shall be equivalent to mailing. If the Named Insured cancels, earned premium shall be computed in accordance with the customary short rate table and procedure. If the Company cancels, earned premium shall be computed pro rata. Premium adjustment may be made at the time cancellation is effected or as soon as practicable thereafter. The check of the Company or its representative, mailed or delivered, shall be sufficient tender of any refund due the Named Insured. If this contract insures more than one Named Insured, cancellation may be effected by the first of such Named Insureds for the account of all the Named Insureds; notice of cancellation by the Company to such first Named Insured shall be deemed notice to all Insureds and payment of any unearned premium to such first Named Insured shall be for the account of all interests therein.

New

Can This Policy Be Canceled?   Yes it can. Both by you and by us.

If you want to cancel the policy, hand or send your cancellation notice to us or our authorized agent. Or mail us a written notice with the date when you want the policy cancelled. We'll send you a check for the unearned premium, figured by the short rate table—that is, pro rata minus a service charge.

If we decide to cancel the policy, we'll mail or deliver to you a cancellation notice effective after at least 30 days. As soon as we can, we'll send you a check for the unearned premium, figured pro rata.

Examples that substitute specific, familiar words for the high abstractions of the original policy are used freely. For instance:

You miss a stopsign and crash into a motorcycle. Its 28-year-old married driver is paralyzed from the waist down and will spend the rest of his life in a wheelchair. A jury says you have to pay him $1,300,000. Your standard insurance liability limit is $300,000 for each person. We'll pay the balance of $1 million.

Or:

> We'll defend any suit for damages against you or anyone else insured even if it's groundless or fraudulent. And we'll investigate, negotiate and settle on your behalf any claim or suit if that seems to us proper and wise.
>
> You own a two-family house and rent the second floor apartment to the Miller family. The Millers don't pay the rent and you finally have to evict them. Out of sheer spite, they sue you for wrongful eviction. You're clearly in the right, but the defense of the suit costs $750. Under this policy we defend you and win the case in court. The whole business doesn't cost you a penny.

Incidentally, there is no fine print in the policy. It is set entirely in 10-point type, a type larger than that used in most newspapers and magazines. Headings and even different colored print are used freely to draw attention to transitions and important information. You can clean up your own writing by following the principles discussed in this chapter and demonstrated in this new insurance policy.

If you exercise care, your own manner of speaking can be a good guide in writing. You should not necessarily write as you talk. In speech, you may be too casual, even slangy. But the sound of your own voice can still be a good guide—in this way. When you write something, read it over; even read it aloud. If you have written something you know you would not speak because of its artificiality, rewrite it in a comfortable style. Rewrite so that you can hear the sound of your own voice in it.

EXERCISES

1. You should now be able to rewrite the example sentence on page 167 into clear, forceful prose. Here it is again; try it:

> While determination of specific space needs and access cannot be accomplished until after a programmatic configuation is developed, it is apparent that physical space is excessive and that all appropriate means should be pursued to assure that the entire physical plant is utilized as fully as feasible.

2. Following are some expressions that the Council of Biology Editors believes should be rewritten.[13] Using the principles you have learned in this chapter, rewrite them.

> an innumerable number of tiny veins
> at this point in time

bright green in color
we conducted inoculation experiments on
due to the fact that
during the time that
fewer in number
for the reason that
goes under the name of
if conditions are such that
in the event that
in view of the fact that
it is often the case that
it is possible that the cause is
it would appear that
lenticular in character
oval in shape
plants exhibited good growth
prior to
serves the function of being
subsequent to
the fish in question
the treatment having been performed
throughout the whole of this experiment
the tube which has a length of 3 m
a process for the avoidance of waste
judging by present standards, these trees are

- If we interpret the deposition of chemical signals as initiation of courtship, then initiation of courship by females is probably the usual case in mammals.
- A direct correlation between serum vitamin B_{12} concentration and mean nerve conduction velocity was seen.
- It is possible that the pattern of herb distribution now found in the Chilean site is a reflection of past disturbances.
- Following termination of exposure to pigeons and resolution of the pulmonary infiltrates, there was a substantial increase in lung volume, some improvement in diffusing capacity, and partial resolution of the hypoxemia.
- Many biological journals, especially those that regularly publish new scientific names, now state in each issue the exact date of publication of the preceding issue. In dealing with journals that do not follow this practice, or with volumes that are issued individually, the biologist often needs to resort to indexes . . . in order to determine the actual date of publication of a particular name.
- Some in the population suffered mortal consequences from the lead compound in the flour.

3. Turn the following sentence into a paragraph of several sentences. See if listing might be a help. Make the central idea of the passage its first sentence.

If, on the date of opening of bid or evaluation of proposals, the average market price of domestic wool of usable grades is not more than 10 percent above the average of the prices of representative types and grades of domestic wools in the wool category which includes the wool required by the specifications (see (f) below), which prices reflect the current incentive price as established by the Secretary of Agriculture, and if reasonable bids or proposals have been received for the advertised quantity offering 100 percent domestic wools, the contract will be awarded for domestically produced articles using 100 percent domestic wools and the procedure set forth in (e) and (f) below will be disregarded.

4. Lest you think all bad writing is American, here are two British samples quoted in a British magazine devoted to ridding Great Britain of gobbledygook.[14] Try your hand with them.
   - The garden should be rendered commensurate with the visual amenities of the neighborhood.
   - Should there be any intensifiication of the activities executed to accomplish your present hobby the matter would have to be reappraised.

5. Write a short report, or rewrite one you did earlier, using the stylistic principles of this chapter.

Chapter **10**

Writing and Revising Your Report

IN THIS CHAPTER we come to the actual writing of your report. We discuss two subjects: (1) writing the rough draft and (2) revising the paper. We do not try to convince you that writing is an easy mechanical job. It is not. But we do give you suggestions that should make a tough job easier.

WRITING THE ROUGH DRAFT

Writing a rough draft is a very personal thing. Few writers do it exactly alike. Most write from an organized plan of some sort; a few do not. (See pages 154–155.) Some write at a fever pitch; others write slowly. Some writers leave revision entirely for a separate step. Some revise for style and even mechanics as they go along, working slowly, trying, in effect, to get it right the first time. All we can do is to describe in general the practices of most professional writers. Take our suggestions and apply them to your own practices. Use the ones that make the job easier for you and revise or discard the rest.

Probably our most important suggestion is to begin writing as soon after the planning stage as possible. Writing is hard work. Most people, even professionals, procrastinate. Almost anything can serve as an excuse to put the job off: one more book to read, a movie that has to be seen, anything. The following column by Art Buchwald describes the problem of getting started in a manner that most writers would agree is only slightly exaggerated.

MARTHA'S VINEYARD—There are many great places where you can't write a book, but as far as I'm concerned none compares to Martha's Vineyard.

This is how I managed not to write a book and I pass it on to fledging authors as well as old-timers who have vowed to produce a great work of art this summer.

The first thing you need is lots of paper, a solid typewriter, preferably electric, and a quiet spot in the house overlooking the water.

You get up at 6 in the morning and go for a dip in the sea. Then you come back and make yourself a hearty breakfast.

By 7 A.M. you are ready to begin Page 1, Chapter 1. You insert a piece of paper in the typewriter and start to type "It was the best of times . . ."
" Then you look out the window and you see a sea gull diving for a fish. This is not an ordinary sea gull. It seems to have a broken wing and you get up from the desk to observe it on the off chance that somewhere in the book you may want to insert a scene of a sea gull with a broken wing trying to dive for a fish. (It would make a great shot when the book is sold to the movies and the lovers are in bed.)

It is now 8 A.M. and the sounds of people getting up distract you. There is no sense trying to work with everyone crashing around the house. So you write a letter to your editor telling him how well the book is going and that you're even more optimistic about this one than the last one which the publisher never advertised.

It is now 9 A.M. and you go into the kitchen and scream at your wife, "How am I going to get any work done around here if the kids are making all that racket? It doesn't mean anything in this family that I have to make a living."

Your wife kicks all the kids out of the house and you go back to your desk. It occurs to you that your agent may also want to see a copy of the book, so you tear out the paper and start over with an original and two carbons: "It was the best of times. . ."

You look out the window again and you see a sailboat in trouble. You take your binoculars and study the situation carefully. If it gets worse you may have to call the Coast Guard. But after a half-hour of struggling they seem to have things under control.

Then you remember you were supposed to receive a check from the *Saturday Review* so you walk to the post office, pause at the drugstore for newspapers, and stop at the hardware store for rubber cement to repair your daughter's raft.

You're back at your desk at 1 P.M. when you remember you haven't had lunch. So you fix yourself a tuna fish sandwich and read the newspapers.

It is now 2:30 P.M. and you are about to hit the keys when Bill Styron calls. He announces they have just received a load of lobsters at Menemsha and he's driving over to get some before they're all gone. Well, you say to yourself, you can always write a book on the Vineyard, but how often can you get fresh lobster?

So you agree to go with Styron for just an hour.

Two hours later with the thought of fresh lobster as inspiration, you sit down at the typewriter. The doorbell rings and Norma Brustein is standing there in her tennis togs looking for a fourth for doubles.

You don't want to hurt Norma's feelings so you get your racket and

for the next hour play a fierce game of tennis, which is the only opportunity you have had all day of taking your mind off your book.

It is now 6 P.M. and the kids are back in the house, so there is no sense
trying to get work done any more for that day.

So you put the cover on the typewriter with a secure feeling that no
matter how ambitious you are about working there will always be somebody on the Vineyard ready and eager to save you.

Reprinted by permission of Art Buchwald

But you must begin, and the sooner the better. Find a quiet place
to work, one with few distractions. Choose a time of day when you
feel like working, and go to work. If you have followed our instructions on research and organization, much of your work is already
done. You should have before you a complete plan of your discussion. Whether your plan is a formal outline or an informal one, you
have your discussion divided into sections. You have a plan for your
format and know what elements you'll need, such as abstracts, conclusions, and recommendations. You know who your readers are and
why you are writing for them. You have thought about the vocabulary level suitable for them and how much detail they are likely to
need or want. Close at hand, you have your research material divided
into sections that match your organizational plan. Your cards are dis-
tributed into well-organized subpacks.

Where should you begin? Usually, it's a good strategy to begin not
with the beginning but with the section that you think will be the
easiest to write. If you do so, the whole task will seem less overwhelming. As you write one section, ideas for handling others will
pop into your mind. When you finish an easy section, go on to a
tougher one. In effect, you are writing a series of short, easily handled discussions rather than one long one. Think of a 1,500-word discussion as three short, connected, 500-word discussions. You will be
amazed at how much easier this attitude makes the job. We should
point out that some writers do prefer to begin with their introductions and even to write their summaries, conclusions, and recommendations, if any, first. They feel this sets their purpose, plan, and
final goals firmly in their minds. If you like to work that way, fine.
Do remember, though, to check such elements after you have written
the discussion to see if they still fit.

When we advise you to write your discussion in sections, we
assume that you plan to write a rough draft first, followed by a careful revision. Most writers find this, by far, the most efficient procedure. While you are writing your rough draft leave plenty of space
between the lines and at the margins for later revisions. Knowing
that you can smooth out your draft later, you will be more relaxed.
You will write faster and better.

How fast should you write? Again, this is a personal thing, but most professional writers write very rapidly, close to 1,000 words an hour. Probably you should plan on at least 500 words an hour. We advise you not to worry overmuch about phraseology or spelling in a rough draft. Proceed as swiftly as you can to get your ideas on paper. Later you can smooth out your phrasing and look up the words with doubtful spelling. However, if you do get stalled, reading over what you have written and tinkering with it a bit is a good way to get the flow going again.

Do not write for more than two hours at a stretch. This time span is one reason why you want to begin writing a long, important discussion at least a week before it is due. A discussion written in one long five- or six-hour stretch reflects the writer's exhaustion. Break at a point where you are sure of the next paragraph or two. When you come back to the writing, read over the previous few paragraphs to help you collect your thoughts and then begin at once.

Make your rough draft very full. You will find it easier to delete material later than to add it. Nonprofessional writers often write thin discussions because they think in terms of the writing time-span rather than the reading time-span. They have been writing on a subject for perhaps an hour and have grown a little bored with it. They feel that if they add details for another half-hour they will bore the reader. Remember this: at 250 words a minute, average readers can read an hour's writing output in about two minutes. Spending less time with the material than the writer must, they will not get bored. Rather than wanting less detail, they may want more. Don't infer from this advice that you should pad your discussion. Brevity is a virtue in professional reports. But the discussion should include enough detail to demonstrate to the reader that you know what you're talking about. The path between conciseness on one hand and completeness on the other is often something of a tightrope.

As you write your rough draft, indicate where your references and visual aids will go. And don't be a slave to your organizational plan. Often the writing process clarifies things in your mind—gives you insights into your work—in a way that organizing does not. Such clarification may call for a revised organization.

REVISING THE PAPER

After you have finished your rough draft, let it cool for a while. If you go on to your revision immediately after writing your draft, you will miss many errors. Try to wait at least a day. Several days are better. When you are ready, sit down to it with a pencil, a pair of scissors, additional paper, a glue stick, and an open mind. A glue stick, which comes under many different trade names, is like a lip-stick in which the stick of glue replaces the lip rouge. The solid glue is easy to spread and apply. It does not wrinkle the paper as mucilage does, and, unlike rubber cement, it does not give off toxic fumes.

If you are fortunate enough to be working on a word processor, you can skip the advice about scissors and glue. But the rest of the process may remain similar. We have done this fifth edition of the text you are reading on a word processor. We have found it an immensely powerful tool that speeded revision greatly. We did find that, working with the word processor, we revised more while we were composing than in the past. Otherwise, our practices and thought processes were similar to those in past revisions. Research is only now beginning to measure the effect of word processors on writing and revising, but it seems likely they will have a positive effect.

Read the draft first for organization and content. Try to put your-self in your reader's place. Does your discussion take too much for granted? Are questions left unanswered that the reader will want answered? Are links of thought missing? Have you provided smooth transitions from section to section, paragraph to paragraph? Do some paragraphs need to be split, others combined? Is some vital thought buried deep in the discussion when it should be put into an emphatic position at the beginning or end? Have you avoided irrele-vant material or unwanted repetitions?

In checking content, be sure that you have been specific enough. Have you quantified when necessary? Have you stated that "In 1984, 52 percent of the workers took at least 12 days of sick leave" rather than "In a previous year, a majority of the workers took a large amount of sick leave"? Have you given enough examples, facts, and numbers to support your generalizations? Conversely, have you gen-eralized enough to unify your ideas and to put them into the sharpest possible focus?

Is your information accurate? Don't rely on even a good memory for facts and figures that you are not totally sure of. Follow up any gut feeling you have that anything you have written seems inaccu-rate, even if it means a trip back to the library or laboratory. And check and double-check your math and equations. You can destroy an argument (or a piece of machinery) with a misplaced decimal point.

Be rigorous in your logic. Can you really claim that A caused B? Have you sufficiently taken into account other contributing factors? Examine your discussion for every conceivable weakness of organi-zation and content and be ready to pull it apart. All writers find it difficult to be harshly critical of their own work, but you will find it a necessary attitude.

If you have done your job thoroughly in the research and planning stage, you will find few weaknesses in content and organization. But it is an unusual and lucky writer who finds none at all. Now is when the generous spacing and margins and the scissors and glue stick come into play.

Write small additions and corrections between the lines and in the margins. Use the proofreader's marks on the inside of the back cover of this book as needed. When you have major additions and reorganizations to do, use your scissors. Suppose you decide that you need a paragraph of transition between two major sections. Write the needed paragraph. Cut your paper apart at the point where the paragraph is needed and glue the new paragraph into place between the sections. A professional writer always has scissors handy during revision. Is a heading botched with no room to write in a new one? Cut the botched heading out and insert a new one. Try to be neat. You or a typist will have to make sense of the revised draft.

After you have revised your draft for organization and content, read it over for style. (We treat this as a separate step, which it is. Of course, if you find a clumsy sentence while revising for organization and content, rewrite it immediately.) Use Chapter 9, "Achieving a Clear Style," to help you. Rewrite unneeded passives. Cut out words that add nothing to your thought. Cross out the pretentious words and substitute simpler ones. If you find a cliché, try to express the same idea in different words. Simplify; cut out the artificiality and the jargon. Be sure the diction is suitable to the occasion and the audience. Count the words in a few sentences here and there to see if you are close to the average you want. Break long involved sentences into two or even three shorter ones. Remember that you are trying to write effectively, not impressively. The final product should carry your ideas to the reader's brain by the shortest, simplest path.

Check your mechanics. Are you a poor speller? Check every word that looks the least bit doubtful. Develop a healthy sense of doubt and use a good dictionary. Some particularly poor spellers read their draft backwards to be sure that they catch all misspelled words. (Once again, if you are lucky enough to be working on a word processor, most have spelling programs to check the spelling.) Do you have trouble with subject-verb agreement? Be particularly alert for such errors. We have provided you with a handbook, pages 451–485, that covers some of the more common mechanical problems.

When you are through with checking mechanics, complete your documentation and finish your graphics.

Once you have revised the draft of your discussion, write the formal elements you need, such as the introduction, abstracts, and conclusions. You can now do the title page. You can also start such things as your table of contents and list of illustrations, though you will have to wait for the final draft in order to insert the proper page numbers.

When all is ready, you type your report or send it to the typist. No matter who types your report, proofread the final draft. The author of a report is responsible for any errors, even if someone else has done the typing.

We'll close here with a word of caution. We have been talking about your revising your own work. Therefore, we could cheerfully urge you to be ruthless with yourself. If someday you find yourself in the position of editing other people's work, you will find a gentler attitude more helpful. When editing, you are dealing with more than the paper lying in front of you. You are dealing with the human being who wrote the paper. It's your obligation to both author and reader to help make the work before you as clear and correct as possible, but you must have reasonable regard for the author's rights. So avoid arbitrary attitudes. Don't just gleefully slap labels like "confusing" on the page. Why is the offending passage confusing? Perhaps the comment might better be phrased as a specific question, such as, "Isn't A really the same as B?" Be particularly careful about changing a writer's sentences. Don't change them just because they are not written the way you would have written them. Be sure that you can back up any change you do recommend in sentence structure—or anthing else, for that matter—by demonstrating that the change really does clarify the thought or increase the efficiency or impact of the writing. If you can't justify a change, perhaps you had better allow the original to stand.

CHECKLIST

Some writers feel that, like airplane pilots, they need a checklist, that is, a brief list of all the vital steps they must perform. We provide such a checklist for you here. (We also reproduce it on the inside front cover.) Used conscientiously in revision, this list will help you avoid many embarrassing oversights in your work.

1. Do you have a good title? Is it short but informative?
2. Have you stated clearly and specifically the purpose of the investigation and the report?
3. Have you stated clearly and fully the outcome of the investigation— what was actually accomplished?
4. Have you clearly and fully described the methods, materials, and equipment you used in conducting the investigation?
5. Have you put into the report everything required?
6. Does the report contain anything that you would do well to cut out?
7. Does the report present the substance of your investigation in a way that will be clear and readily understandable to your intended readers?

8. Are your paragraphs clear, well organized, and of reasonable length?
9. Is your prose style (diction) clear and readable?
10. Have you included all the formal elements that your report needs?
11. Are your headings and titles clear, properly worded, and parallel?
12. Have you inserted the documentary reference numbers in your text?
13. Have you keyed the tables and figures into your text and have you sufficiently discussed them?
14. Are all parts and pages of the manuscript in the correct order?
15. Will the format of the typed or printed report be functional, clear, and attractive?
16. Does your manuscript satisfy stylebook specifications governing it?
17. Have you included required notices, distribution lists, and identifying code numbers?
18. Do you have written permission to reproduce extended quotations or other matter under copyright?
19. Have you proofread your manuscript for matters both large and small?
20. While you were composing the manuscript did you have any doubts or misgivings that you should now check out?
21. What remains to be done, such as proofreading your typist's final copy?

EXERCISES

1. Revise an earlier report for organization, content, style, and mechanics. Use scissors and glue stick where needed. Provide all needed formal elements. Add graphics or improve the ones you already have. Type the final draft or have it typed. Turn in to your instructor the revised draft and the final typed version.

2. This exercise is both a class and individual effort. Your instructor will provide you with a draft of a discussion and a working bibliography for it. The class will decide the following:

   - Audience and purpose for the report and the role of the writer.
   - The formal elements the report needs to make it a functional report.
   - Where some of the prose should be replaced or supplemented by graphics.

   After your class discussion, take your scissors and glue stick to the draft. Revise it for organization, style, and mechanics. Prepare the needed graphics and indicate where in the draft they should go. Write the needed formal elements. The exercise is complete when the report is ready for the typist.

3. If you have been planning and organizing a report as called for in the exercises following Chapters 2, 5, and 8, now write that report. Include all necessary graphics and formal elements.

Chapter 11

Formal Elements of Reports

IN CHAPTER 8, "Organizing Your Report," on pages 161–164, we discuss the strategies to follow in planning the format of your report. In this chapter, we discuss formal elements of reports—the tools you can use to carry out those strategies. We divide the elements into three groups: prefatory elements, main elements, and supplementary elements. In addition, we provide a section on documentation. To help you find needed elements, here is a list of them with reference to the page on which the explanation of each begins:

THE FUNCTIONS OF FORMAT

The formal elements of reports answer functional needs. For example, many professional reports are long, and the table of contents serves the functional purpose of sorting things out for the reader. Professional reports are likely to pass between people who do not know each other very well. The letter of transmittal serves as an introduction and breaks the ice, so to speak. Busy readers, who sometimes lack the time to read an entire report, must read professional reports selectively. They need an informative abstract to give them the major points in a hurry and headings to help them locate the parts they want. To a large extent, then, the degree of formality needed in a report's format depends on situation and audience.

Our approach to format is descriptive, not prescriptive; that is, we describe some of the more conventional practices found in modern technical reporting. We realize fully, and you should too, that many colleges, companies, and journals call for practices different from the ones we describe. Therefore, we do not recommend that you must follow at all times the practices in this chapter. If you are a student, however, your instructor may, in the interests of class uniformity, insist that you follow this chapter fairly closely.

Whether you follow the conventions described in this chapter or others, you should realize the importance of following some consistent, conventional plan when you put your reports together. Inconsistency in format confuses your readers. It is just one more thing for them to worry about. As in the rest of the reporting process, in planning format put the readers' need to understand above all else.

PREFATORY ELEMENTS

The prefatory elements help your readers to get into your report. The letter of transmittal or preface may be the readers' first introduction to the report. The table of contents reveals the structure of your organization. In the glossary readers will find the definitions of terms that may be strange to them. All the prefatory elements discussed in this section contribute to the success of your report. Take care with them and do them well.

Letter of Transmittal and Preface

We have placed the letter of transmittal and preface together because in content they are often quite similar. They usually differ in format and intended audience only. You will use the letter of transmittal when the audience is a single person or a single group. Many of your major reports in college will include a letter of transmittal to your professor, usually placed just before or after your title page. When on the job, you may handle the letter differently. Often it is mailed before the report, as a notice that the report is forthcoming. Or it may be mailed at the same time as the report but under separate cover.

Generally, you will use the preface for a more general audience where you may not know specifically who will be reading your report. The preface or letter of transmittal introduces the reader to the report. It should be fairly brief. Always include the following basic elements:

- Statement of transmittal or submittal (included in the letter of transmittal only)
- Statement of authorization or occasion for report
- Statement of subject and purpose

Additionally, you may include some of the following elements:

- Acknowledgments
- Distribution list (list of those receiving the report—used in the letter of transmittal but not in the preface)
- Features of the report that may be of special interest or significance
- List of existing or future reports on the same subject
- Background material
- Summary of the report
- Special problems (including reasons for objectives not met)
- Financial implications
- Conclusions and recommendations

How many of the secondary elements you include depends upon the structure of your report. If, for example, your report's introduction or discussion includes background information, there may be no point in including such material in the preface or letter of transmittal. See Figures 11-1 and 11-2 for a sample letter of transmittal and a sample preface.

If the report is to remain within an organization, the letter of transmittal will become a memorandum of transmittal. This changes nothing but the format. (See pages 311–313.)

```
Gatlin Hall
Weaver University
Briand, MA 02139

July 27, 1983


Dr. Ross Alm
Associate Professor of English
Weaver University
Briand, MA 02139

Dear Dr. Alm:

I submit the accompanying report entitled "Characteristics
of Venus and Mercury" as the final project for English 430,
Technical Writing.

The report discusses the characteristics of both Venus and
Mercury, covering size, mass, density, physical appearance,
and atmosphere.  I have made an effort to provide a base for
understanding the significance of the space probes to Venus
and Mercury carried on by NASA's Jet Propulsion Laboratory
(JPL).  Recent information about Mercury obtained by the
most recent probe, Mariner 10, is incorporated into the
report.

I am indebted to Ms. Mary Fran Buehler of JPL who has
allowed me to quote extensively from her unpublished work
on Mariner 10.

Sincerely,

Anne K. Chimato

Anne K. Chimato
English 430
```

FIGURE 11-1   Letter of Transmittal

Cover

A report's cover serves three purposes. The first two are functional and the third esthetic and psychological.

Common typewriter paper requires protection during handling and storage. Pages ruck up, become soiled and damaged, and may eventually be lost if they are not protected by covers. Because they are what readers first see as they pick up a report, covers are the appropriate place for prominent display of identifying information such as the report title, the company or agency by or for which the report was prepared, and security notices if the report contains proprietary or classified information. Incidentally, students should not

PREFACE

In recent years the National Aeronautics and Space Administration (NASA) has explored the inner planets of our solar system, Venus and Mercury, with unmanned space probes. This report, part of NASA's educational series for high school students, reports the information from the latest probe, Mariner 10. Characteristics of both Venus and Mercury—including size, density, physical appearance, and atmosphere—are discussed.

Of particular interest in this report is the surprising finding that Mercury, contrary to scientific expectation, has a magnetic field. This finding may cause present theories about the generation of magnetic fields within planets to be revised.

For lists of other reports on NASA's unmanned space probes write to NASA, Jet Propulsion Laboratory, California Institute of Technology, Pasadena, California 91125.

FIGURE 11-2 Preface

print this sort of information directly on the cover. Rather, they should type the information onto gummed labels readily obtainable at the college bookstore, and then fasten the labels to the cover. A student label might look like the one in Figure 11-3.

Esthetically and psychologically, covers bestow dignity, authority, and attractiveness. They help to convert a bundle of manuscript into a finished work that looks and feels like a report, and has some of the characteristics of a printed and bound book.

Suitable covers need not be expensive and sometimes should not be. Students, particularly, should avoid being pretentious. All three purposes are frequently well served by covers of clear plastic or light cardboard, perhaps of 30- or 40-pound substance. Students can buy such covers in a variety of sizes, colors, and finishes.

CHARACTERISTICS OF VENUS AND MERCURY

by

Anne K. Chimato

English 430   27 July 1983

FIGURE 11-3 Student Label

While you are typing your report, remember that when you fasten it into its cover about an inch of left margin will be lost. If you wish an inch and a half margin, you must leave two and a half inches on your paper. Readers grow irritated when they must exert brute force to bend open the covers in order to see the full page of text.

Title Page

Like report covers, title pages perform several functions. They dignify the reports they preface, of course, but far more important, they provide identifying matter and help to orient the report users to their reading tasks.

To give dignity, a title page must be attractive and well designed. Symmetry and balance are important, as are neatness and freedom from clutter. The most important items should be boldly printed; items of lesser importance should be subordinated. These objectives are sometimes at war with the objective of giving the report users all the data they may want to see at once. Here we have listed in fairly random order the items that sometimes appear on title pages. A student paper, of course, would not require all or even most of these items. The first four listed are usually sufficient for simple title pages.

- Name of the company (or student) preparing the report
- Name of the company (or instructor and course) for which the report was prepared
- Title and sometimes subtitle of the report
- Date of submission or publication of the report
- Code number of the report
- Contract numbers under which the work was done
- List of contributors to the report (minor authors)
- Name and signature of the authorizing officer
- Company or agency emblem and other decorative matter
- Proprietary and security notices
- Abstract
- Library identification number
- Reproduction restrictions
- Distribution list (A list of those receiving the report. If the letter of transmittal does not contain this information, the title page should.)

Understandably, placing all of these items on an 8½-by-11-inch page would guarantee a cluttered appearance. The moral is plain: if you cannot put down all that you would like, put down—clearly and boldly—all that you must. See Figures 11-4 and 11-5 for typical title pages.

Pay particular attention to the wording of your title. Titles should be brief but descriptive and specific. The reader should know from the title what the report is about. A title such as "Effects of Incuba-

CHARACTERISTICS OF VENUS AND MERCURY

Prepared for

Professor Ross Alm

English 430

Technical Writing

by

Anne K. Chimato

Abstract

This report describes the characteristics of both Venus
and Mercury, covering size, rotation, revolution about the
sun, density, physical appearance, and atmosphere. New
findings about Mercury's magnetic field are presented.
Conclusions about Earth's early history are drawn from
Mercury's cratered surface and Earth's ancient rocks.

27 July 1983

FIGURE 11-4 Title Page

tion Temperatures on Sexual Differentiation in the Turtle, *Chelydra sepentina*" is illustrative. From it you know specifically the research being reported. To see how effectively this title works, leave portions of it out and see how quickly your understanding of what the article contains changes. For example, "Sexual Differentiation in the Turtle" would suggest a much more comprehensive report than does the actual title. A title such as "Effects of Incubation Temperatures" could as well be about chickens as turtles. On the other hand, adding the words "An Investigation into" to the beginning of the title would add nothing really useful. The test of whether a title is too long or

```
        FINAL REPORT OF THE COMMITTEE ON

WORD PROCESSING ALTERNATIVES FOR THE OXFORD INSURANCE COMPANY


                   Prepared for

                 Dennis Colcombet

         Vice President for Information Services




                 Committee Members

              Betty Robinett, Chair

              Robin Brown

              Terry Collins

              Donald Ross




                  5 March 1983



              Distribution List

           Ann Bailly, Office Supervisor
           Donald Browne, Computer Technician
           Arthur Walzer, Purchasing Officer
```

FIGURE 11-5   Title Page

too short isn't in the number of words it contains but what happens
if words are deleted or added. Keep your titles as brief as possible,
but make sure they do the job.

Table of Contents

A table of contents (TOC) performs at least three major functions. Its
most obvious function is to indicate by number the page on which
discussion of each major topic begins; that is, it serves the reader as
a locating device. Less obviously, a TOC displays the extent and
nature of the topical coverage and suggests the logic of the organi-

zation and the relationship of the parts. Still earlier, in the prewriting stage, provisional drafts of the TOC enable the author to "think on paper"; that is, they act as outlines to guide the composition.

A system of numbers, letters, type styles, indentations, and other mechanical aids has to be selected so that the TOC will perform its intended functions. Figure 11-6 shows a TOC suitable for student reports.

As you can see in Figure 11-6, the heading CONTENTS identifies the page. The titles of prefatory parts, main sections, and main terminal parts are given in capitals. Divisions of main sections are given less prominence by means of indentation and a combination of capital and small letters. Beginning page numbers of parts are given at the right, lined up vertically on their right-hand digits. Subdivisions may be listed without page numbers. Spaced leader dots are sometimes used to carry the reader's eye from the end of each title to the beginning page number and, at the same time, to pull the page together visually. This practice is by no means universal. Some people feel the leader dots clutter the page and therefore do not use them. (For example, see Figure 17-1, page 386.) If you use a numbering system in your report (see pages 229–231), the TOC should reflect that system.

When you design your own TOC, as with the title page, avoid overcrowding. Seldom is there justification for listing parts subordinate to the subdivisions of sections; very shortly a point is reached where users have almost as much trouble locating items in the TOC as they have in locating them by flipping through the pages of the report. Be sure to prepare the final draft of your TOC by referring to the typed pages of the otherwise finished report. It is simply amazing to discover the differences between an early version of a TOC and the updated version. Whole sections may have been dropped or have been placed in a new order. Wordings of headings may have been drastically changed. Old page numbers may no longer apply. Remember that the TOC entries and the headings on the text pages must be worded exactly the same. Every entry in the TOC must also be in the report. However, we have already pointed out that every heading in the report need not be in the TOC.

List of Illustrations

If a report contains more than a few illustrations, say more than three or four, it is customary to list the illustrations either on a separate page or on the TOC page. Before we go into detail, however, it seems essential to explain what we mean by illustrations.

Illustrations are of two major types: tables and figures. A table is any array of data, often numerical, arranged vertically in columns

CONTENTS

ii

FIGURE 11-6 Table of Contents

and horizontally in rows, together with the necessary headings and notes. Any illustration that does not satisfy this definition of table is automatically a figure. Figures, therefore, include photographs, maps, graphs, organization charts, and flow diagrams—literally anything that does not qualify as a table by the preceding definition. (For further details, see Chapter 12, "Graphical Elements of Reports.")

If the report contains both tables and figures, it is customary to use the page heading ILLUSTRATIONS, a combining term, and to list all the figures first and then all the tables.

The titles of all illustrations should be as brief and yet as self-explanatory as can be. Avoid the cumbersome expression "A Figure Showing Characteristic Thunderstorm Recording." Say, simply, "Characteristic Thunderstorm Recording." On the other hand, do not be overly economical and write just "Characteristic" or "A Comparison." At best, such generic titles are only vaguely suggestive.

Figure 11-7 shows a simple version that should satisfy most ordinary needs. Notice in the figure that arabic numbers are used for figures and roman numerals for tables. This practice is common but by no means standard.

Glossary and List of Symbols

Reports dealing with technical and specialized subject matter almost invariably include abbreviations, symbols, and terms not known to the nonspecialist. Thus a communication problem arises. Technically trained persons have an unfortunate habit of assuming that what is well known to them is well enough known to others. This assumption is seldom justified. Terms, symbols, and abbreviations undergo changes in meaning with time and context. In one context, ASA may stand for American Standards Association; in another context, for Army Security Agency. The letter K may stand for Kelvin or some mathematical constant. The meaning given to Greek letters may change from one report to the next even though both were done by the same person. Furthermore, writers seldom have complete control over who will read their reports. A report intended by the author for an engineering audience may have to be read by members of management, the legal department, or sales. In doubtful instances, it is wise to play it safe by including a list of symbols or a glossary or both. Readers who do not need these aids can easily ignore them; those who do need them will be immeasurably grateful.

Figure 11-8 illustrates a list of symbols. Figure 11-9 illustrates a glossary. In Figure 11-9 notice that the terms to be defined are set up to stand out plainly from the definitions. Notice also that the definitions here are fragmentary sentences. Whatever kind of definitions

ILLUSTRATIONS

FIGURE 11-7   List of Illustrations

SYMBOLS

A	Mass number
A.W.	Atomic weight
c	Velocity of light (2.998×10^{10} cm/sec)
D	H^2 atom (deuterium)
E	Energy
e−	Electron
e	Electronic charge (1.602×10^{-10} abs. coulomb)
ev	Electron volt
F	Free energy
(g)	Gas phase
H	Heat content
h	Planck's constant (6.624×10^{-27} erg sec)
I_{sp}	Specific impulse
k	Boltzmann's constant (1.3805×10^{-16} erg/deg)
ln	Natural logarithm

*   *   *   *   *   *   *   *

α	Alpha particle
γ	Gamma ray
ζ	Bond energy
μ	Micro
ρ	Density

v

FIGURE 11-8 List of Symbols

212

GLOSSARY

Btu ------------------the amount of heat required to raise the temperature of one pound of water one degree Fahrenheit

degree day -----------a temperature standard around which temperature variations are measured

design temperature ----the maximum reasonable temperature expected during the heating or cooling season upon which the design calculations are based

heat transmission
    coefficient -------the quantity of heat in Btu transmitted per hour through one square foot of a building surface

infiltration ---------the air leaking into a building from cracks around doors and windows

sensible heat --------heat that the human body can sense

thermal conductivity - the quantity of heat in Btu transmitted by conduction per hour through one square foot of a homogeneous material for each degree Fahrenheit difference between the surfaces of the material

thermal resistance ---the reciprocal of thermal conductivity

vii

FIGURE 11-9 Glossary

you decide to write—full or fragmentary sentences—you should be consistent throughout your glossary.

Abstracts

Discussed here are two kinds of abstracts: informative and descriptive. Each kind is often set off by itself on a separate page or pages of a report. Sometimes the descriptive abstract is placed on the title page. (See Figure 11-4, page 206.) Because your full report contains complete documentation, you need not footnote or otherwise document the information in abstracts.

Never use "I" statements in either kind of abstract. Report your information impersonally as though it were written by someone else. The informative abstract in Figure 11-10 shows the style. This is not an arbitrary principle. If you were to publish an article, your abstract would likely be reprinted in an abstracting journal where the use of "I" would be inappropriate.

Frequently a report will contain both a descriptive and an informative abstract. When this is so, it is common practice to label the descriptive abstract as *Abstract* and the informative abstract as *Summary*.

Informative Abstract.   An informative abstract is a summary of the key information in your complete report. Good informative abstracts are difficult to write. At one extreme, they lack adequate information; at the other, they are too detailed. In the informative abstract you must pare down to material essential to your purpose. This can be a slippery business.

Suppose, for example, you are writing a report to explore the knowledge about the way the human digestive system absorbs iron from food. In the body of the report you discuss an experiment conducted with Venezuelan workers that followed isotopically labeled iron through their digestive systems. To enhance the credibility of the information presented, you include some details about the experiment. You report the conclusion that vegetarian diets decreased iron absorption. How much of all this should you put in your abstract? Given your purpose, the location and methodology of the experiment would not be suitable material for the abstract. You would simply report that in one experiment vegetarian diets have been shown to decrease iron absorption.

When you are selecting material for your informative abstract, remember that the abstract should be able to stand alone. In fact, for the busy reader it may serve as a substitute for the entire report.

ABSTRACT

Venus has a diameter of 7520 miles and a density slightly less than that of Earth. Because of Venus' persistent cloud cover, little is known about its surface features. Venus' day is the equivalent to 127 Earth days, and its year equals 225 Earth days. Its cloud layer is approximately 20-miles thick. Venus' surface temperature is approximately 475 C, and its surface pressure is equivalent to the pressure in Earth's ocean at a depth of 2400 feet. Mercury has a diameter of 3000 miles and a density greater than Earth's. Mercury has a period of revolution around the sun equal to 88 Earth days, and its day is equal to 176 Earth days. Therefore, Mercury's day is 2-years long by its time. Mercury's surface is cratered like the surface of Earth's Moon. Mercury has virtually no atmosphere but does show minute quantities of helium. Its surface temperature varies from 325 C to -125 C. An unexpected discovery of Mariner 10 was that Mercury has a magnetic field. Because Mercury rotates so slowly, scientists thought that it would have no magnetic field. This finding may cause present theories about the generation of magnetic fields within planets to be revised. The cratering of Mercury suggests that at some point in Earth's early history it may have been similarly pockmarked. Evidence for this conclusion can also be found in the most ancient of Earth's rocks.

iv

FIGURE 11-10 Informative Abstract

Therefore, you must include enough information to satisfy the purpose of your report. State major conclusions, recommendations, and decisions, if any, in the informative abstract.

A good rule of thumb for length, unless you have instructions otherwise, is that your abstract should be from 5 to 10 percent of the length of your body. For business reports, this will normally be an acceptable limit. However, most journals, because of rising publication costs, will set arbitrary limits of under 200 words for abstracts. The abstracts printed in Figure 4-11 on page 68 are good examples of such short abstracts.

Our advice for writing informative abstracts has necessarily been rather general. Depending upon where your abstract may appear, you may receive much more specific advice. For example, most professional journals or societies publish stylebooks that include specifications abut how to write an abstract. We have reprinted one such specification in Figure 11-11. Read it to gain more insight into abstract writing. Also, because abstract writing uses the same techniques as summary writing, you might want to read what we say about that on pages 222–226 and 395–398.

Descriptive Abstract. The main purpose of the descriptive abstract is to help busy readers decide if they need or want the information in the report enough to read it entirely. The descriptive abstract merely tells what the full report contains. It cannot serve as a substitute for the report itself. Many reports contain descriptive abstracts, and many abstracting journals print them. Here is a descriptive abstract based on the information in Figure 11-10:

> This report describes the characteristics of both Venus and Mercury, covering size, rotation, revolution about the sun, density, physical appearance, and atmosphere. New findings about Mercury's magnetic field are presented. Conclusions about Earth's early history are drawn from Mercury's cratered surface and Earth's ancient rocks.

Note that the descriptive abstract discusses the *report*, not the subject. After reading this abstract, you know that "New findings about Mercury's magnetic field are presented," but you do not know what those findings are. You must read the report for that information. Whether the report is ten or a thousand pages long, a descriptive abstract can cover the material in less than ten lines. In contrast, the informative abstract must grow longer as the report's discussion grows longer.

1.4
Abstract

An abstract is a brief summary of the content and purpose of the article. In APA journals the abstract is used in place of a concluding summary and appears directly under the by-line. All APA journals except *Contemporary Psychology* require an abstract.

The abstract allows readers to survey the contents of an article quickly. Because, like the title, it is used by *Psychological Abstracts* for indexing and information retrieval, the abstract should be self-contained and fully intelligible without reference to the body of the paper and suitable for publication by abstracting services without rewriting. Information or conclusions that do not appear in the main body of the paper should not appear in the abstract. Because so much information must be compressed into a small space, authors sometimes find the abstract difficult to write. Leaving it until the article is finished enables you to abstract or paraphrase your own words.

An abstract of a *research* paper should contain statements of the problem, method, results, and conclusions. Specify the subject population (number, type, age, sex, etc.) and describe the research design, test instruments, research apparatus, or data-gathering procedures as specifically as necessary to reflect their importance in the experiment. Include full test names and generic names of drugs used. Summarize the data or findings, including statistical significance levels, if any, as appropriate. Report inferences made or comparisons drawn from the results.

An abstract of a *review* or *theoretical* article should state the topics covered, the central thesis, the sources used (e.g., personal observation, published literature, or previous research bearing on the topic), and the conclusions drawn. It should be short but informative. For example, "The problem was further discussed in terms of Skinner's theory" is not an informative statement. The abstract should tell the reader the *nature* or *content* of the theoretical discussion: "The discussion of the problem centered on Skinner's theory and the apparent fallacy of determinism."

An abstract for a research paper should be 100–175 words; one for a review or theoretical article, 75–100 words. General style should be the same as that of the article.

Remember, to the degree that an abstract is succinct, accurate, quickly comprehended, and informative, it increases your audience.

FIGURE 11-11 Instructions for Abstracting [From *Publication Manual of the American Psychological Association*, 2nd ed. (Washington, D.C.: American Psychological Association, 1974), 15. Copyright © 1974 by the American Psychological Association. Reprinted by permission.]

MAIN ELEMENTS

The body of a report contains its detailed information and interpretation. The body needs to be introduced and, normally, finished off with an ending of some sort that may include a summary, conclusions, recommendations, or simply a graceful exit from the report. We discuss all these elements in this section on main elements.

Introduction

In your introduction you should announce four things immediately: (1) subject, (2) purpose, (3) scope, and (4) plan of development. Early in your paper, you also give any needed theoretical or historical background, and this is sometimes included as part of the introduction.

Subject.   Never begin an introduction with a superfluous statement. The writer who is doing a paper on core memory in computers and begins with the statement "The study of computers is a vital and interesting one" has wasted the readers' time and probably annoyed them as well. Announce your specific subject loud and clear as early as possible in the introduction, preferably in the very first sentence. The sentence "This paper will discuss several of the more significant applications of the exploding wire phenomenon to modern science" may not be very subtle, but it gets the job done. The reader knows what the subject is. Often, in conjunction with the statement of your subject, you will also need to define some important terms that may be unfamiliar to your readers. For example, the student who wrote the foregoing sentence followed it with these two:

> A study of the exploding wire phenomenon is a study of the body of knowledge and inquiry around the explosion of fine metal wires by a sudden and large pulse of current. The explosion is accompanied by physical manifestations in the form of a loud noise, shock waves, intense light for a short period, and high temperatures.

In three sentences the writer announced the subject and defined it. The paper is well under way.

Sometimes, particularly if you are writing for nonspecialists, you may introduce your subject with an interest-catching step. This step may be rather extended, as in this example:

> Traveling in orbit around the earth at an altitude of some 270 miles, the Skylab astronauts rotated their spacecraft to establish a bearing on one of their principal check points: a cluster of several hundred round green spots, each half a mile in diameter, arrayed in an orderly pattern on the earth's surface below them. What they were viewing was a dense

concentration of circular irrigated fields in north-central Nebraska, a pattern easily identified from space. Passengers on commercial jet airliners increasingly notice the same sight over many other areas of the continential U.S., including eastern Colorado, central Minnesota, the Texas Panhandle, the Pacific Northwest and northern Florida. Now the proliferating green circles can be seen even in the middle of the Sahara. What is being observed is perhaps the most significant mechanical innovation in agriculture since the replacement of draft animals by the tractor.

These circular green fields, most often found in arid or semiarid country, are being watered by the world's first successful irrigation machine.[1]

Or you may simply extract a particularly interesting fact from the main body of your theme. For example,

Last year, the Federal Aviation Administration attributed 16 aircraft accidents to clear air turbulence. What is known about this unseen menace that can cripple an aircraft, perhaps fatally, almost without warning?

In this example, the writer catches your interest by citing the accident rate caused by clear air turbulence and nails the subject down with a rhetorical question in the second sentence. Interest-catching introductions are used in brochures, advertisements, and magazine and newspaper articles. You will rarely see an interest-catching introduction in business reports or professional journals. If you do, it will usually be a short one.

Purpose. Your statement of purpose tells the reader *why* you are writing about the subject you have announced. By so doing you also, in effect, answer the reader's question "Should I read this paper or not?" For example, a writer who had word processors as her subject announced her purpose this way: "In this article I use illustrative narratives to argue that direct use of word processors can benefit any communicators in business, industry, or academe."[2] Readers who have no reason to be interested in such a discussion will know there is no purpose in their reading the report.

Another way to understand the purpose statement is to realize that it often deals with the *significance* of the subject. A writer who had the application of human engineering to technical writing as his subject announced his purpose this way:

Regardless of your writing specialty, however, you will be more effective as a technical writer if you become familiar with human engineering and learn to apply human-engineering principles.[3]

Scope. The statement of scope further qualifies the subject. It announces how broad and, conversely, how limited the treatment of the subject will be. Often it indicates the level of competence expected in the reader for whom the paper is designed. For example, a student who wrote, "In this report I explain the operation of the Cockcroft-Walton accelerator in a manner suitable for college undergraduates" declared her scope as well as her purpose. She is explaining the *operation* of the Cockcroft-Walton accelerator, and the audience she is designing her report for is not composed of high school students or graduate physicists but college undergraduates. She further qualified her scope and audience by stating, "This explanation will be comprehensive enough to allow sophomore students of physics to use and operate the accelerator without further research."

Plan of Development. Part of the introduction is telling the reader how you plan to develop your paper. The principle of psychological reinforcement is at work here. If you tell your readers what you are going to cover, they will be more ready to comprehend as they read along. The following, taken from the introduction to a paper on iron enrichment of flour, is a good example of a plan of development:

> This study presents a basic introduction to three major areas of concern in regard to iron enrichment: (1) questions on which form of iron is best suited for enrichment use; (2) potential health risks from super enrichment—cardiovascular disease, hemochromatosis, and the masking of certain disorders; and (3) the inconsistencies in basic knowledge as they relate to the definitions, extent, and causes of iron deficiency.

You need not necessarily think of the announcement of subject, purpose, scope, and plan of development as four separate steps. Often, subject and purpose or scope and plan of development can be combined. In a short paper, perhaps two or three sentences might cover all four points, as in this example:

> Concern has been expressed recently over the possible presence in our food supply of a class of chemicals known as "phthalates" (or phthalate esters). To help assess the true significance of this concern as to the safety and wholesomeness of food, this article discusses the reasons phthalates are used, summarizes their toxicological properties, and evaluates their use in food packaging materials in the context of their total use in the environment.[4]

Also, introductions to specialized reports may have peculiarities of their own. These will be discussed in Part 2, "Applications."

Theoretical or Historical Background. When necessary theoretical or historical background is not too lengthy, you can incorporate it into your introduction. In fact, when handled properly such information may well catch the interest of the reader, as in the following example:

> Climatologists attribute the warming trend to the furnaces of civilization which have been spewing forth increasing loads of carbon dioxide to the atmosphere. This colorless and odorless gas, exhaled by man and used by plants to make themselves green, restricts the escape into space of infrared radiation from the sun-warmed earth. Since increased CO_2 absorbs more of the infrared radiation than formerly, a larger amount of heat accumulates, causing a slight but significant increase in average global temperature. This impact of atmospheric CO_2 on climate—dubbed the "greenhouse effect"—has become more apparent in recent years because of the escalating rate at which power plants and industry throughout the world have burned coal, oil from shale, and synthetic oil and gas.[5]

If necessary background material is extensive, however, it more properly becomes part of the body of your paper.

Body

The body will be the longest section of your report. Your purpose and your content will largely determine the form of this section. Therefore, we can prescribe no set form for it. In presenting and discussing your information you will use one or more of the rhetorical forms described in Chapters 6 and 7 or the special techniques described in Part 2, "Applications." In addition to your text you will probably also use headings (see pages 228–229) and the various visual aids such as graphs, tables, and illustrations described in Chapter 12, "Graphical Elements of Reports."

When thinking about your discussion, remember that almost every technical report answers a question or questions: What is the best method of desalination to create an emergency water supply at an overseas military base? How are substances created in a cell's cytoplasm carried through a cell's membranes? What is the nature of life on the ocean floor? How does a hydraulic pump work? To answer such specific questions, general questions must be asked and answered. Often the reporter's old standbys—Who? What? When? Where? Why? How?—are good starting points. Add to these the always important "so-what?" What are the significant implications of your information. However you approach your discussion, project yourself into the minds of your readers. What questions do they need answered to understand your discussion? What details do they need to follow your argument? You will find that you must walk a narrow line between too little detail and too much.

Too little detail is really not measured in bulk but in missing links in your chain of discussion. You must supply enough detail to lead the reader up to your level of competence. You are most likely to leave out crucial details at some basic point that, because of your familiarity with the subject, you assume to be common knowledge. If in doubt about the reader's competence at any point, take the time to define and explain.

Many reasons exist for too much detail, and almost all stem from writers' inability to edit their own work. When you realize that something is irrelevant to your discussion, discard it. It hurts, but the best writers will often throw away thousands of words, representing hours or even days of work. You must always ask yourself questions like these: Does this information have significance, directly or indirectly, for the subject I am explaining or for the question I am answering? Does the information move the discussion forward? Does it enhance the credibility of the report? Does it support my conclusions? If you don't have a yes answer to one or more of these questions, the information has no place in the report, no matter how many hours of research it cost you.

Ending

Depending upon what sort of paper you have written, you ending can be (1) a summary, (2) a set of conclusions, (3) a set of recommendations, or (4) a graceful exit from the paper. Frequently, you'll need some combination of these. We'll look at the four endings and at some of the possible combinations. It's also possible in executive reports that the "ending" may actually be placed at the front of the report. See pages 36, 386–388, and 494–496.

Summary.   Many technical papers are not argumentative. They simply present a body of information that the reader needs or will find of interest. Frequently, such papers end with summaries. In a summary, you condense for your readers what you have just told them in the discussion. Each major section of your discussion should be restated in the summary. Sometimes you may wish to number the points for clarity. The following, from a paper of about 2,500 words, is an excellent summary:

> The exploding wire is a simple-to-perform yet very complex scientific phenomenon. The course of any explosion depends not only on the material and shape of the wire but also on the electrical parameters of the circuit.

An explosion consists primarily of three phases:

1. The current builds up and the wire explodes.
2. Current flows during the dwell period.
3. "Post-dwell conduction" begins with the reignition caused by impact ionization.

These phases may be run together by varying circuit parameters.

The exploding wire has found many uses: it is a tool in performing other research, a source of light and heat for practical scientific application, and a source of shock waves for industrial use.

Summaries should be concise and they should introduce no material that has not been covered in the report. You construct a summary as you construct an informative abstract. You read your discussion over, noting your main generalizations and your topic sentences. You smoothly blend these together into a paragraph or two. Sometimes you will represent a sentence from the discussion with a sentence in the summary. At other times you will shorten such sentences to phrases or clauses. The last sentence in the foregoing example actually represents a summary of four sentences from the writer's discussion. The four sentences, themselves, were the topic sentences from four separate paragraphs.

Conclusions. Some technical papers work toward a conclusion. They ask a question, such as "Are nuclear power plants safe?" present a set of facts relevant to answering the question, and end by stating a conclusion: "Yes," "No," or sometimes, "Maybe." The entire paper aims squarely at the final conclusion. In such a paper, you argue inductively and deductively. You bring up opposing arguments and show their weak points. At the end of the paper, you must present your conclusions. Conclusions are the inferences drawn from the factual evidence of the report. They are the final link in your chain of reasoning. In simplest terms, the relationship of fact to conclusion goes something like this:

Facts	*Conclusion*
Car A averages 25 miles per gallon.	On the basis of miles per gallon,
Car B averages 40 miles per gallon.	Car B is preferable.

Because we presented a simple case, our conclusion was not difficult to arrive at. But even more complicated problems present the same relationship of fact to inference.

In working your way toward a major conclusion, you ordinarily have to work your way through a series of conclusions. In answering the question of nuclear power plant safety, you would have to answer a good many subquestions concerning such things as security of the radioactive materials used, adequate control of the nuclear reaction, and safe disposal of nuclear wastes. The answer to each subquestion is a conclusion. You may present these conclusions in the body of the report, but it's usually a good idea to also draw them all together at the end of the report to prepare the way for the major conclusion.

Earlier (on page 220) we showed you the introduction to a report that questioned whether the class of chemicals known as phthalates endangered public health. Here are the conclusions to that report:

> Based on the observations made thus far, there is no evidence of toxicity in man due to phthalates, either from foods, beverages, or household products as ordinarily consumed or used.
>
> These observations, coupled with the limited use of phthalate-containing food packaging materials and the low rate of migration of the plasticizers from packaging material to food, support the belief that the present use of phthalates in food packaging represents no hazard to human health.[6]

All the conclusions presented are supported by evidence in the report.

In larger papers or when dealing with a controversial or complex subject, you would be wise to precede your conclusions with a summary of your facts. By doing so, you will reinforce in your reader's mind the strength and organization of your argument. For an example of such a combination, see Appendix A, pages 494–496. In any event, make sure your conclusions are based firmly upon evidence that has been presented in your report. Few readers of professional reports will take seriously conclusions based upon empty, airy arguments. Conclusions are frequently followed by recommendations.

Recommendations. A conclusion is an inference. A recommendation is the statement that some action be taken or not taken. The recommendation is, of course, based upon the conclusions and is the last step in the process. You conclude that Brand X bread, for example, is cheaper per pound then Brand Y and just as nutritious and tasty. Your final conclusion, therefore, is that Brand X is a better buy. Your recommendation is "Buy Brand X."

Many reports such as feasibility reports, environmental impact statements, and research reports concerning the safety of certain foods or chemicals are decision reports that end with a recommendation. For example, we are all familiar with the government recommendations that have removed certain artificial sweeteners from the market and that have placed warnings on cigarette packages. These recommendations were all originally stated at the end of reports looking into these matters.

Recommendations are simply stated. They follow the conclusions, often in a separate section, and look something like this:

Based upon the conclusions reached, we recommend that our company

- Not increase the present level of iron enrichment in our flour.
- Support research into methods of curtailing rancidity in flour containing wheat germ.

Frequently, you may have a major recommendation followed by additional implementing recommendations, as in the following:

We recommend that the Department of Transportation build a new bridge across the St. Croix River at a point approximately three miles north of the present bridge at Hastings.

- The department's location engineers should begin an immediate investigation to decide the exact bridge location.
- Once the location is pinpointed, the department's right-of-way section should purchase the necessary land for the approaches to the bridge.

You need not support your recommendations when you state them. You should have already done that thoroughly in the report and in the conclusions leading up to the recommendations. It's likely, of course, that a full-scale report will contain, in sequence, a summary, conclusions, and recommendations. For more information about this subject and for examples of summaries, conclusions, and recommendations, see pages 394–402 and 494–496.

Complimentary Close. A short, simple, nonargumentative paper often requires nothing more than a graceful exit, a complimentary close. As you would not end a conversation by turning on your heel and stalking off without a "good-bye" or a "see you later" to cover your exit, you do not end a paper without some sort of close. In a short informational paper that has not reached a decision, the facts should be still clear in the readers' minds at the end, and they will not need a summary. One sentence, such as the following that might end a short speculative paper on lasers, will probably suffice: "Because lasers seem to have almost unlimited uses, they cannot fail to receive increasing scientific attention in the years ahead."

Sometimes an appropriate quotation can be used to get you out of a report gracefully, as in the following:

> As Roger Revelle and H. E. Seuss stated in 1957: "Human beings are now carrying out a large-scale geophysical experiment of a kind that could not have happened in the past nor be repeated in the future. Within a few centuries we are returning to the atmosphere and oceans the concentrated organic carbon stored in the sedimentary rocks over hundreds of millions of years. This experiment, if adequately documented, may yield a far-reaching insight into the processes determining weather and climate."[7]

Combination Endings. We have treated summaries, conclusions, and recommendations separately. And, indeed, it's likely that a full-scale report leading to a recommendation will contain in sequence separate sections labeled *Summary, Conclusions, and Recommendations.* When such is the case, the summary will often be restricted to a condensation of the factual data offered in the body. The implications of the data will be presented in the conclusions, and the action to be taken in the recommendations. (See, for example, pages 394–402 and 494–496.)

However, in shorter reports, and in reports that do not have recommendations as their *primary* goal, all three—factual summary, conclusions, and recommendations—may be combined into one section. If this section is then presented at the end of the report it will probably be labeled *Summary.* If it is presented at the front of the report—in a position before even the introduction—it may be labeled *Abstract.* It will be, in fact, what you know as an informative abstract. (See pages 214–216.)

It's unfortunate, perhaps, that there is a slight confusion of terms when these elements are combined in different ways or placed in different locations. Don't let the confusion in terminology confuse the essence of what is involved here. In all but the simplest reports, you must draw things together for your readers. You must condense and highlight your significant data and present any conclusions and recommendations you may have. Notice how this summary of a scientific research report smoothly combines all these elements:

SUMMARY

In many turtles the hatchling's sex is determined by the incubation temperature of the egg, warm temperatures causing femaleness and cool temperatures maleness. Consequently, the population sex ratio depends upon the interaction of (i) environmental temperature, (ii) maternal choice of nest site, and (iii) embryonic control of sex determination. If environmental temperature differs between populations, then sex ratio selection is expected to adjust either maternal behavior or embryonic temperature-sensitivity to yield nearly the same sex ratio in the different populations.

To test this hypothesis in part, we have compared sex determining temperatures among embryos of emydid turtles in the northern and southern U.S. We predicted that embryos of southern populations should develop as male at higher temperatures than those of northern populations. The data offer no support for this prediction among the many possible comparisons between northern and southern species. The data actually refute the prediction in both of the North-South intraspecific comparisons. Further study is needed, in particular, of nest temperatures in the different populations.[8]

SUPPLEMENTARY ELEMENTS

Reports may relegate subordinate material to appendixes and annexes. Reports need a system of headings and numbers to show their order and sequence. Finally, information used in the report has to be documented. All these functions are fulfilled by the supplementary elements of reports.

Appendixes and Annexes

Appendixes and annexes have become increasingly important in modern reporting practices. Many report users want to extract the heart of a report with a minimum expenditure of time and energy. This development has led to several changes in reporting practices within recent years, including greater economy of language and the adoption of appendixes and annexes to hold subordinate materials. The "main report" may now consist of 12 to 20 pages whereas the same content in earlier days might have required readers to plow through a hundred or more pages.

A knowingly designed TOC, brisk language in the prefatory parts, meaty summaries of two to five pages bringing well forward the conclusions and recommendations, the insertion of colored separator pages to "seal off" detail in appendixes and annexes—all have combined to make the report user's job simple and the report maker's job more difficult and hazardous.

Here, however, we are concerned with a limited aspect of the whole effort leading to "reader's ease"—the use of appended materials to clean up the oversized body of the voluminous report of earlier days. Therefore, you should know that detailed data, analysis, and supporting evidence are now often removed from the body and relegated to annexes and appendixes. The student report reprinted as Appendix A, on pages 489–504, provides in its annexes and appendixes a good example of how they can be used to ease a busy reader's task.

During the final stages of major reorganization, you should determine whether the following materials should be shifted to appendixes or annexes:

- Case histories
- Supporting illustrations
- Detailed data
- Transcriptions of dialogue
- Intermediate steps in mathematical computation
- Copies of letters, announcements, and leaflets mentioned in the report
- Samples, exhibits, photographs, and supplementary tables and figures
- Extended analyses
- Lists of personnel
- Suggested collateral reading
- Anything else that is not essential to the sense of the main report

This tendency to reduce the body of the report to basic material has some disadvantages. In a drastic attempt to achieve conciseness and economy, some reports have been stripped of their circumstantial detail. Others have been made so brief and cryptic that they defy interpretation. Others have had their main parts excised and transferred beyond the colored separator page—the iron curtain of some modern reports. On the other hand, the motives behind these changes have a sound foundation. Reports have become too long and too detailed. Body discussions have become clogged with marginally relevant material. Reports have become tedious for the ordinarily busy and interested reader.

We therefore suggest that you be receptive to suggestions that certain front and body materials be shifted to the back. Consider each such suggestion on its merits, but consider also the effect on the whole. Do not be so eager to shift material to the appendixes and annexes that you prevent the reader from understanding the major points of the report.

Headings

Headings are used to display your report's coordination and subordination to the reader. The heading itself is a phrase that describes what is discussed in the paragraph or paragraphs that follow it. Coordination and subordination are shown through your headings by a consistent use of various type faces and positions for different level headings.

You may be surprised by the amount of typographical variation that you can obtain from a standard typewriter:

1. You can set up a heading entirely in capital letters.

`NEMATODES`

2. You can use a combination of capital and small letters.

`Nematodes and Fungi`

3. You can use "expanded capitals."

`N E M A T O D E S`

4. You can center headings as in 1, 2, and 3 above, or you can place them at the left margin.

`Pathogens`

`Fungi`

5. Of course, you can also indent them five or more spaces from the left margin.

    `Fungi`

6. Finally, you can choose between underlining and not underlining.

    <u>`Pathogens`</u>
    `Fungi`

As illustrated, you can use the possibilities in various combinations so that with ingenuity and foresight you can construct ten or more distinctive levels of headings using only a standard typewriter.

Because headings are used to show the organization of a paper, an obvious relationship exists between them and the TOC. All entries in the TOC must be reproduced verbatim as headings. In addition, you may also include headings to display lower levels of organization not included in the TOC. Rarely should a student paper include more than four levels of headings. Figure 11-12 illustrates a system of headings that might actually run over four or five pages.

The heading is not a part of the sentence following it. It stands outside the text. An example of misuse would be

Data systems. These are designed for a specific command function.

Note that "Data systems" is the subject of the sentence through the pronoun "These." While it seems natural, avoid such use of the heading. In the interests of clarity, repeat the key term somewhere in the first sentence or two that follow the caption.

Numbering System

Quite often, in order to facilitate reference back and forth in a report or book, a numbering system is combined with the headings. The

```
        C O N T R O L L I N G   S O I L - B O R N E

      P A T H O G E N S   I N   T R E E   N U R S E R I E S

    TYPES OF SOIL-BORNE PATHOGENS AND THEIR EFFECTS ON TREES

        Simply stated, the effects of soil-borne pathogens...

    ...........................................................

    The Soil-borne Fungi

        At one time it was thought that soil-borne fungi....

    ...........................................................

        Basidiomycetes.  The basidiomycetes are a class of

    fungi whose species....................................

        Phycomycetes.  The class phycomycetes is a very

    diversified type of fungi.  It is the ..................

    The Plant Parasitic Nematodes

        Nematodes are small, nonsegmented...................

    ...........................................................

        TREATMENTS AND CONTROLS FOR SOIL-BORNE PATHOGENS

    ...........................................................

    ...........................................................
```

FIGURE 11-12   A Four-Level System of Headings

three systems in most common use are the traditional outline system, the century-decade-unit system (often called the Navy System), and the multiple-decimal system. The three schematics that follow illustrate the three systems.

```
Traditional outline system:    T I T L E
                               I.  FIRST-LEVEL HEADING
                                   A.  Second-Level Heading
                                       1.  Third-level heading
                                       2.  Third-level heading
                                   B.  Second-Level Heading
```

```
                          II.  FIRST-LEVEL HEADING
                           A.  Second-Level Heading
                                1.  Third-level heading
                                2.  Third-level heading
                           B.  Second-Level Heading
Century-decade-unit system:  T I T L E
                          100  FIRST-LEVEL HEADING
                          110  Second-Level Heading
                                111  Third-level heading
                                112  Third-level heading
                          120  Second-Level Heading
                          200  FIRST-LEVEL HEADING
                          210  Second-Level Heading
                                211  Third-level heading
                                212  Third-level heading
                          220  Second-Level Heading
Multiple-decimal system:   T I T L E
                          1  FIRST-LEVEL HEADING
                          1.1  Second-Level Heading
                                1.1.1  Third-level heading
                                1.1.2  Third-level heading
                          1.2  Second-Level Heading
                          2  FIRST-LEVEL HEADING
                          2.1  Second-Level Heading
                                2.1.1  Third-level heading
                                2.1.2  Third-level heading
                          2.2  Second-Level Heading
```

Many companies or government agencies specify one of these systems; therefore, it pays to be familiar with them all. If you use a numbering system with your headings, you must also repeat the numbers before the entries in the TOC. Often, numbers will be used only with the discussion elements of your report, that is, those elements currently marked with roman numerals in Figure 11-6. However, it is also acceptable to use numbers with other major elements, such as introductions, conclusions, recommendations, and references.

Pagination

The numbering of pages, or pagination, is seldom a complicated problem. Brief manuscripts and reports having little prefatory matter are nearly always given arabic numbers, running in one continuous series from first page to last. It is the convention to place these numbers at the bottom of the page or at the upper right-hand corner.

When centered near the bottom of the page, each number may be given a hyphen on each side:

-17-

As reports grow longer and more complicated, the page-numbering system may also need to be more complex. In such reports, the prefatory pages are often given small roman numbers, as *i, ii,iii*. The pages of the body of the report are given arabic numbers. Sometimes the introduction is numbered with the prefatory material, sometimes with the body. In either case the arabic series begins at 1 and usually continues through to the last page of the report. A number is not placed on the title page. However, a number is allowed for it; that is, the following page is number *2* or *ii*.

The arabic series is not invariably retained throughout the report. Sometimes the appendixes or annexes, if any, are given capital letters, and then the pages of an appendix bear their distinctive letter plus their own series of arabic numbers: A1, A2; B1, B2, B3, and so on. Using the letter-plus-number system for appendixes and annexes possesses two advantages. For one thing, it is a distinctive system which clearly reveals that a page is part of an appendix or annex and not part of the body of the report. For another thing, use of a distinctive system makes it possible to print the appendixes and annexes before the pagination of the main report has been made firm. As a corollary, pages may be added to or removed from an appendix or annex without disturbing the pagination of the rest of the report.

What do you do if an appendix is a reprint of a document having its page number? The answer, quite simply, is to retain the original page numbers and then beneath them add the new page numbers in brackets:

128

[426]

Clarity for the reader is always a paramount objective of pagination. In any case, you should never be more complicated than you have to be.

DOCUMENTATION

There is a bewildering complexity of documenting systems from college to college, journal to journal, company to company. Therefore we cannot claim a universal application for the instructions that follow. Use them barring conflicting instructions from your instructor, college, employer, or the stylebook of the journal or magazine in which you hope to publish.

Before we go into the mechanics of documentation, it might be wise to discuss why and when you need to document.

First of all, documenting fulfills your moral obligation to give credit where credit is due. It lets your readers know who was the originator of an idea or expression and where his or her work is found. Second, systematic documentation makes it easy for your readers to research your subject further.

When do you document? Well, to begin with, give credit when you quote directly from the work of another. Beyond documenting direct quotation, it is not always easy to determine what should and what should not be documented. You do not document general information or what might be called common knowledge. For example, even if you referred to a technical dictionary to find that creatinine's more formal name is methyglycocyamidine, you would not be obligated to show the source of this information. It is general information, readily found in many sources. If on the other hand you should include in your paper an opinion that certain quantum effects can be explained without inferring that atomic particles behave like waves, you would need to document the source of this opinion.

Generally, you should document if the material involves speculation or if it is found in only a few sources. Do not clutter your pages with references to information readily found in many sources. If in doubt as to whether to document or not, play it safe and document. If you can pinpoint your obligation sentence by sentence or paragraph by paragraph then document closely and specifically. If you owe a more general obligation, then indicate that general obligation as in this example:

1. Much of the information in this section was provided by Dr. James Blake, Associate Professor of Chemistry, the Arden School of Technology.

We will explain two systems of documentation. The first involves using notes, the second a bibliography or reference list. In our explanation we follow the 13th edition of the widely used *The Chicago Manual of Style*, published by the University of Chicago Press.[9] Many of the major technical and scientific journals generally conform to the documentation style recommended by *The Chicago Manual*. If your work is in the humanities, you might prefer to use the style described in the *MLA Handbook*.[10]

Notes and Footnotes

Notes may be (1) displayed at the bottom of the page on which the material to be documented appears or (2) gathered together at the end of the report under the heading NOTES. In the first method, the notes are called footnotes and in the second, endnotes or simply notes. The first method is illustrated in Figure 11-13. The second is more common in student reports and journal articles. We use it in this book; see our "Chapter Notes" on pages 527–533. In both cases the note form is the same. In both cases the note number is indicated in the paper by a superscript number, that is, a number placed above the line of type, as you can see in the many note numbers we give in this book (for instance, on page 233) and in Figure 11-13.

In a list of notes, entries appear in the list in the order that the superscript note numbers occur in the paper. (See Figure 11-14 for a sample list.) The number corresponding to the note number is placed on-line, paragraph indented from the left margin, and followed by a period; the body of the note is then begun in the third space to the right of the number. (In the style of *The Chicago Manual*, a footnote is made in the same way. See Figure 11-13.)

In a student or company report, each line of an individual entry is single-spaced; double-spacing is used between separate entries. If you are preparing a report for publication, double space all lines.

Not everyone needs word-processing equipment. For some it may just be a frill that makes "a company more efficient at doing unimportant things."[1] Nonetheless, word processing is becoming increasingly important, and we need criteria to help us make appropriate choices. The criteria should cover both the equipment capabilities and the maintenance support the vendor provides.[2] Also, in making a choice, we must analyze carefully the tasks that we expect.

1. Jason A. Chamberlain, "Manager's Memo: Basic Steps in Evaluating New Technology," Word Processing & Information Systems, May 1981, 52.

2. Susan M. Briles, "What to Look for in Word-Processing Equipment, Technical Communication 29, no. 3 (1982): 20.

FIGURE 11-13 Footnotes on a Page

Though few authorities agree on every detail of an individual entry, all agree on the basic order of a note: (1) the author's or editor's name, (2) the title, and (3) the facts of publication. Following *The Chicago Manual,* we recommend that you place a comma between the three components of the note. Consult the samples that follow, the samples in Figure 11-14, and our notes on pages 527–533

NOTES

1. John Sterling Harris, "So You're Going to Teach Technical Writing: A Primer for Beginners," Technical Writing Teacher 2 (Fall 1974): 4.

2. Paul L. Briand, "The Nonsense about Technical Writing," Journal of Engineering Education 57 (1967): 507.

3. Donald H. Cunningham and Vivienne Hertz, "An Annotated Bibliography on the Teaching of Technical Writing," College Composition and Communication 21 (1970): 177-79.

4. Ibid.

5. United Nations, Economic and Social Council, Bibliography of Publications Designed to Raise the Standard of Scientific Literature (Paris, 1963), 204-206.

6. Briand, "Nonsense about Technical Writing," 508.

7. The Chicago Manual of Style, 13th ed. (Chicago: University of Chicago Press, 1982), 337.

8. Private communication with Thomas L. Warren, Department of English, Oklahoma State University, Stillwater, Oklahoma, 20 March 1983.

9. H.W. Fowler and F.G. Fowler, The King's English (New York: Oxford University Press, 1940), 46.

10. Juanita Williams Dudley, "Writing Skills of Engineering and Science Students," IEEE Transactions on Engineering Writing and Speech 14 (June 1971): 42.

11. Harris, "Technical Writing," 5.

12. Joseph M. Williams, Style (Glenview, Ill.: Scott, Foresman and Company, 1981), 62.

13. Ibid., 78.

14. Daniel B. Felker et al., Guidelines for Document Designers (Washington, D.C.: American Institutes for Research, 1981), 11.

FIGURE 11-14 List of Endnotes

to find the correct forms for most widely used types of entries. If you conform to the basic three-part pattern we have described, you should be able to construct most notes in a satisfactory way. We now describe the components of that pattern.

Author. In footnotes put the author's name in natural order, first name or initial, middle name or initial, and last name. Use the name as it appears on the title page. For multiple authors or editors, indicate first the name of the author or editor whose name appears first on the title page; if there are two or three authors or editors, list all names. If there are more than three authors or editors, list the first one and designate all authors or editors beyond the first one by "et al." (from the Latin *et alii*, "and others").
Examples:

Van Wyck Brooks, (an author)

P. L. Robinson and R. E. Dodd, (two authors)

W. B. Yeats, ed., (an editor)

Ludovico Ariosto and Paul L. Briand, eds., (two editors)

Hobard H. Willard, Lynne L. Merritt, Jr., and John A. Dean, (three authors)

Charles D. Hodgman et al., (more than three authors)

United Nations, Economic and Social Council, Commission on Human Rights, (author a named group)

Title. Underline book titles in typed or handwritten papers. This corresponds to the italics used in print to designate a title. Anything separately published is underlined. The title of an article from a book, magazine, encyclopedia, collection, anthology, or newspaper is placed within quotation marks. This title is then followed by the underlined title of the publication from which it has been taken. Give the title in full exactly as it is on the title page, but do not include subtitles and long explanatory phrases.
Examples:

Complete Field Guide to North American Wildlife, (book title)

"Moths and Ultrasound," *Scientific American,* (title of article in a periodical)

Review of *A Model of the Brain* by J. Z. Young, (book review)

"More Demand for Burley Predicted," Louisville *Courier-Journal,* (newspaper headline with the city not part of the paper's name)

"Magnetism Problems and the Analog Computer," (an unpublished the-sis or dissertation)

"Landscape Architecture," *Encyclopedia Britannica*, 15th ed., (an ency-clopedia article)

Interview Topic: The Air Force Role In Space, (an untitled interview)

"New Techniques in Making Nearly Perfect Mirrors," no. 2 in The Kitzh-aber Memorial Lectures in Geometrical Optics, (an unpublished lecture)

Facts of Publication.   Include those facts of publication neces-sary to identify the source clearly, so the reader may verify your cita-tion or pursue further the information taken from the source.

When citing a book, the title should be immediately followed by the number of the edition, if you are using an edition other than the first. Then, in parentheses, list the place of publication, followed by a colon; the publisher's name, exactly as it appears on the title page or copyright page, followed by a comma; and the date of publication. Unless the city's location is obvious (as with New York or San Fran-cisco) and it cannot be confused with another city of the same name, include the state (or country if applicable) when listing place of publication:

Middletown, Conn.: Wesleyan University Press
Washington, D.C.: American Chemical Society

When several places are listed on the title page, use only the first. For date of publication use the copyright date or the date of the latest edition. When no date is given, use the abbreviation "n.d." Place a comma after the end parenthesis, then cite the page or pages from which your information has come; for example, "24" or "312–16."

For periodical citations it usually is not necessary to indicate the publisher and place of publication. Place the volume number imme-diately after the name of the perodical (the periodical title is under-scored, but the volume number is not). Place the year of publication next in parentheses followed by a colon. After the colon, cite the page or pages from which your information has come.

If a journal renumbers its pages in each issue, the month or issue number must also be included. The month (or day, for weekly peri-odicals) is indicated before the year, and both are enclosed in paren-theses. Alternatively, the issue number may be given after the volume number, preceded by "no." If the periodical is published in one or more series, give the series designation just before the volume number.

Yearbooks and conference proceedings usually may be cited in the same fashion as periodicals. For bulletins, circulars, and other monographs that appear in numbered series, follow the citation form for books, inserting the title of the series and the number of the monograph, if any, after the title of the monograph.

Government publications that are books or monographs written by individuals are handled like other books or monographs. It is much more difficult to cite official reports, government bulletins, pamphlets, contract reports, and so forth, with or without known authors. Certainly the citation should include the name of the government (nation, state, city), the name of the agency, the author if known, the title of the report, and the date of publication. In the case of an anonymous document, the sponsoring body (such as a government agency) should be treated as the author. If the document has a number (for example, AFOSR TN-58-6, AERE-R-4705, AFM 10-4) this should be included. It is very helpful to indicate where a document can be located, for example, Office of Technical Services, U.S. Department of Commerce. You may have to consult a librarian for help in citing specific government publications.

Examples:

Album of Science (New York: Charles Scribner's Sons, 1982), 13–15. (a typical book citation)

A Textbook of Physical Chemistry, 2nd ed. (New York: Academic Press, 1979), 22. (second edition of a book)

Electrochemistry of Fused Salts, trans. Adam Peiperl (Washington, D.C.: The Sigma Press, Publishers, 1961), 21. (translated book)

J Anim Sci 54 (1982): 1132–37. (A typical periodical citation, to be preceded by the title of the article. It is common practice to abbreviate journal titles, as in this example. Most library reference desks will have a copy of Alkire's *Periodical Title Abbreviations*. With its yearly supplements it lists most journals and their abbreviations.)

Scientific American 247, no. 1 (1982): 40–51. (periodical that repaginates each issue)

Scientific American, July 1982, 40–51. (alternative way to cite a monthly periodical that repaginates each issue)

Can J Res, Sec F 26 (1948): 151–59. (periodical published in sections or series)

Florida Agricultural Experiment Station Bulletin 818 (1981): 36. (monograph published in numbered series, to be preceded by the title of the monograph)

Society for Technical Communication Proc 27 (1980): R-15. (annual proceedings of a society)

Describe any unpublished material as completely as you can:

Unpublished master's thesis, University of Denver, 1980. (unpublished thesis)

Interview with Professor Frederick H. MacIntosh, University of North Carolina, Chapel Hill, N.C., 20 April 1982. (interview)

Presented at the 36th annual meeting of the Colorado-Wyoming Academy of Science, Denver, Colorado, 30 April 1983. (lecture or paper presented at a meeting, to be preceded by title of lecture)

Private communication with John Galt, Colorado Springs, Colorado, 14 September 1980. (telephone conversation, brief meeting, or letter from an individual)

Subsequent References. If you refer to an entry more than once, you can choose among several ways of constructing your citation for the subsequent series. If you refer to the same entry for two or more notes in sequence with no intervening notes, the simplest citation is "Ibid." (from the Latin *ibidem*, "the same place"). If you are referring to the same entry, but to a different page of that entry, construct your citation as follows, "Ibid., 26."

When you are referring to the same entry more than once, but other notes intervene, your second and subsequent citations need repeat only the author's name and a shortened form of the title. When the page reference changes, follow the title with a comma and the proper page reference, for example, "Taylor, *Analysis*, 202."

Bibliographies and Reference Lists

Bibliographies can serve two purposes: they can be used in place of notes to document sources, or they can serve as general lists of references relating to the subject of a report. A bibliographical list can include works used for general background in preparing a report, as well as works specifically cited. Or the bibliography may be intended primarily as a guide to additional information for interested readers. (See our Appendix C, "A Selected Bibliography," pages 523–526, for an example of a bibliography intended as a guide.)

In reports where the subject matter is primarily scientific, the bibliography, rather than notes, is the most widely accepted way to cite specific sources. In this case, the title *References* or *Literature Cited* rather than *Bibliography* is used. In reports where the subject matter is drawn from the humanities, notes or footnotes are the accepted way to document sources, and a bibliography serves as a more general list of references.

The Chicago Manual recommends different formats for bibliographic entries in the sciences and humanities. We will discuss the scientific style here. For information on accepted style in the humanities, refer to *The Chicago Manual,* pages 439–483, or the *MLA Handbook.* (Also see our Appendix C, which follows *The Chicago Manual*'s humanities style.)

The format you follow for the entries in a scientific bibliography or reference list is similar to the format for notes, but with some significant differences (see Figure 11-15 for a model reference list):

1. Use periods instead of commas between the basic components of author, title, and facts of publication. Omit the parentheses around the facts of publication.
2. Transpose the author's name: last, first, middle name or initial. If you must list several names, transpose only the first author's name and punctuate between names in the style of Figure 11-15.
3. The year of publication follows the author's name and precedes the title as in Figure 11-15.
4. For articles and books capitalize only the first letter of the title and subtitle and proper nouns. However, as Figure 11-15 makes clear, use standard title capitalization for periodicals. Do not enclose article titles within quotation marks.
5. Arrange the entries in alphabetical order rather than in order of appearance in the text. You need enter each entry only once, doing away with the need for "Ibid." and other subsequent entry forms. Determine alphabetical order by the author's last name or, if no author is listed, by title (disregarding *the, a,* or *an*).

REFERENCES

Editors, Astronomy. 1979. Voyager approaches Jupiter. Astronomy 7, no. 4:14-15.

Groth, E.J. et al. 1977. The clustering of galaxies. Scientific American 237, no. 5:76-98.

Leaky, R.E., and R. Lewin. 1977. Origins: What new discoveries reveal about the emergence of our species and its possible future. New York: E.P. Dutton.

Schwartzenburg, D., and R. Burnham. 1979. The grand adventure. Astronomy 7, no. 3:7-15, 22-24.

The Viking Lander Imaging Team. 1978. The Martian landscape. NASA SP-425. Washington, D.C.: U.S. National Aeronautics and Space Administration.

Yeomans, Donald K. 1981. The comet Halley handbook. JPL 400-91 1/81. Pasadena, Calif.: Jet Propulsion Laboratory.

FIGURE 11-15 List of References

6. Omit page numbers from book entries. You will see the reason for this in a moment. In periodical entries, give inclusive page numbers for each article.

With a reference list, source citation becomes very simple. At a point of reference in the text, place a parenthetical citation like one of these: (Brock 1983) or (Asher 1978, 93) or (Eisen 1980, 93–96). The first (Brock 1983) refers the reader to the entry in the bibliography with author Brock and date 1983. You will use this form when you are not citing a specific page. The second form (Asher 1978, 93) refers the reader to page 93 of the reference authored by Asher. The third form (Eisen 1980, 93–96) refers the reader to pages 93 to 96 of the Eisen work. The second and third forms are used when you wish to cite a specific page in an entry. If the work has two or three authors, include the last name of each. For more than three authors, use et al.: (Ratti et al. 1979, 45–47).

Parenthetical references should be on the line with the text and not above the line. Their location is normally at the end of the material used and inside the end punctuation.

Example:

As early as 1973, experts were predicting that "by 1990 we will be using continuous-thrust spacecraft for all space exploration projects" (Whitburn 1973, 93). The development of continuous-thrust propulsion systems—which now appears to be a certainly—will greatly reduce flight times for all space missions (Ratti et al. 1979; Jones and Stein 1978).

Within the reference list or bibliography itself, if two or more publications are listed for the same author or authors, only the first entry gives the name. In immediately succeeding entries the author's name is replaced by a short line followed by a period:

Leakey, R. E., and R. Lewin. 1977. *Origins: What new discoveries reveal about the emergence of our species and its possible future.* New York: E. P. Dutton.

———. 1978. *People of the lake.* Garden City, N.Y.: Doubleday & Co., Anchor Press.

The parenthetical citation for the first entry would be (Leakey and Lewin 1977), for the second (Leakey and Lewin 1978).

Copyright

Fairly stringent copyright laws protect published work. When you are writing a student report that you do not intend to publish, you need not concern yourself with this problem. If, however, you intend to publish a report, you should become familiar with copyright law. You can find a good summary of it in *The Chicago Manual*, on pages 107–128.

EXERCISES

1. Reprinted in this exercise is the discussion section from the report titled *Mercury in Food*.[11] It was written for both food technologists and intelligent lay persons. Its purpose is to examine the possible dangers of mercury poisoning in the food we eat. It reaches conclusions but does not offer recommendations. For the purpose of this exercise, pretend that you have written the report as a class assignment. Provide the following elements:

   • Letter of transmittal to your writing teacher
   • Title page that includes a descriptive abstract
   • Introduction
   • Summary
   • Conclusions

MERCURY IN FOOD

Mercury and the Environment

During normal growth process, plants absorb mercury from the soil and air; in some instances, plants even concentrate it to small droplets of the metal. Some bacterial organisms, when exposed to inorganic mercury, can convert it to organic mercury compounds. These microscopic organisms and the material they produce, generally called alkyl mercury, may be consumed by fish or animals. Some animals and vegetables have the ability to convert organic forms of mercury back to inorganic compounds.

This constant cycling of mercury from one form to another has gone on for eons without any recognizable toxic effect on the food supply of the world. It is extremely doubtful that the concentration of mercury in the oceans has increased significantly as a result of man's use of the metal.

Awareness of Potential Dangerous Effects Increasing

Until the last two decades, man had only been vaguely aware of the problems arising from the misuse of mercury. A series of isolated incidents tragically demonstrated the potential dangers.

In 1953, a veritable epidemic hit the fishermen and their families in the villages on Minamata Bay, Japan. A number of people who were highly dependent on seafood showed signs of brain damage. Some of these cases were fatal. An intense investigation revealed that a local chemical plant was discharging a waste stream containing organic mercury into the bay. This was absorbed by the fish in the area and eventually passed on to the villagers.

After finding the cause of the problem, authorities were able to eliminate the source of pollution. Mercury levels in the bay returned to normal and once again the local fish was safe to eat.

Mercury showed its insidious effect in Sweden when naturalists noted that certain species of birds were diminishing. Here a study disclosed that these birds had been feeding on grain seeds that had been treated with mercury as a portection against fungus. A similar product was also a favorite pesticide. In 1966 the Swedish government banned or restricted these uses of mercury compounds. The bird life is reviving and the awareness of the potential problems is very much alive.

The practice of treating seeds with mercury compounds has also had its impact on humans. In 1968, a New Mexico farmer fed treated grain seed to his hogs and he and his family subsequently ate the meat from these animals. As a result, three of the farmer's children were crippled and a fourth child was born blind and retarded. Bags containing mercury treated seeds carry a label warning that the contents are poisonous to animals and humans, and such seeds are dyed a bright pink to differentiate them from untreated seeds. Unfortunately, however, these warnings were ignored or misunderstood by those involved. As a result of this, and similar incidents which occured in other parts of the world, the United States Department of Agriculture in March of 1969 banned the practice of treating seeds with organic mercury compounds.

Maximum Allowable Exposure Levels Set

Urine specimens indicate that man and other animals absorb some mercury from their food and water. People who work in industries using mercury may show urine levels ten or twenty times higher than those who are not directly involved with the metal. The United States Department of Labor has established maximum allowable exposure levels to protect workers. Studies have shown that it is possible for a worker to absorb approximately one milligram of mercury per day from industrial exposure when working within the established allowable limits. Periodic physical examinations, which emphasize special methods of evaluation for possible neurological damage, indicate that the allowable levels are safe for the employees' health.

Guidelines Established for Levels in Food

The United States Food and Drug Administration has conducted routine analysis of foods for mercury content for several years. Practically all foods tested showed levels of mercury concentration well within the norms for the natural environmental content of the element. Only fish and fishery products showed concentrations greater than could be considered to be normal. As a result of their investigations and augmented by the Japanese and Swedish experiences, the Food and Drug Administration in 1969 established a 0.5-parts-per-million (ppm) guideline as the maximum safe limit for mercury in fish.

Only Fish Products Exceed Limits

Swordfish and tuna are the only commercially popular fish that have shown a mercury content exceeding 0.5 ppm. These two species of fish accumulate organic mercury compounds as they grow larger because they consume vast quantities of smaller fish.

Since both tuna and swordfish are caught at sea, far from any possible sources of industrial pollution, the mercury in their systems must come from natural sources. Most probably man, for many years, has eaten tuna and swordfish with concentrations of mercury higher than the established limit without signs of any harmful effect. Analysis of museum specimens of tuna caught during the period 1879 to 1909 reveals that they contain levels of mercury as high as those in fish being caught today. Scientists must, therefore, conclude that mercury levels in tuna, and very probably swordfish, have not appreciably changed in the past 90 years.

Why Man Hasn't Suffered from Eating Fish

Recent studies at the University of Wisconsin have discovered that some fish, including tuna, have a built in mechanism for blocking and reducing the toxicity of mercury in their tissues. This research may answer the question of how man has safely eaten fish containing mercury concentrations higher than those allowed by the Food and Drug Administration. Most experts agree that 0.5 ppm level for fish has a considerable margin of safety built into it. It is heartening to recognize there have been no reported cases of mercury poisoning in the U.S.A. from eating fish.

2. Compile a list of difficult and unusual terms you may have to introduce in the final report of your own selected project. Will the vocabulary problem be great enough to require you to include a glossary?

3. Prepare at least a tentative table of contents for your proposed final report. Will you need a numbering system? If so, will the traditional system be adequate or should you use one of the more complicated systems such as the multiple-decimal system?

4. Decide what material, if any, you should place in the appendix of your final report. Why would you not include this material somewhere in the body of your report? If it is placed in an appendix, who would use the material and for what purpose?

5. How would you argue the point that reports generally profit if their format is conventional and the mechanical elements are given standard treatment?

6. In a published report or textbook locate as many of the mechanical elements discussed in this chapter as possible. What variations and departures do you find?

7. How does a reference list differ from a list of notes? What are the virtues and limitations of each system?

8. The following sources are given in the order in which they are cited in a report but without attention to proper format. List them in proper order and with proper format as they would appear in (a) a list of notes and (b) a scientific list of references.

Joseph M. Williams, Style, 1981. Scott, Foresman and Company. Glenview, Illinois, p.12.

Brown, James W., Richard B. Lewis, and Fred F. Harcleroad. AV Instruction, Media and Methods. McGraw-Hill Book Co. New York, 1969. Page 12.

Lalter Arno Wittich and Charles Francis Schuller, Audiovisual Materials, Their Nature and Uses. Harper & Row: New York, Evanston, London, 1964, page 44.

The Severity of an Earthquake, by the Geological Survey, U.S. Department of the Interior. U.S. Government Printing Office, Washington, D.C., 1979, page 16.

"The Invisible Threat: The Stifled Story of Electron Waves." By Susan

Schiefelbein. Saturday Review, September 15, 1979. Vol. 6, No. 18, inclusive pages 16–20, specific citation to page 18.

S. Michael Halloran, "Technical Writing and the Rhetoric of Science," Journal of Technical Writing and Communication, Vol. 3, Number 2, 1978. Inclusive pages 77–88, specific citation to page 80.

Stress in Community Groups by Jerry W. Robinson, Jr., and others. Cooperative Extension Service, University of Illinois at Urbana-Champaign, 1975. Pages 6–10.

The Development of Writing Abilities (11–18), by James Britton, Tony Burgess, Nancy Martin, Alex McLeod, and Harold Rosen. London, 1975, Macmillan Education, page 63.

Chapter 12

Graphical Elements of Reports

FROM YOUR READING of earlier chapters you should have concluded that report writing requires you to make many kinds of decisions. Some of these decisions, and highly important ones, involve the use of graphics. When should you use prose to convey your ideas, and when should you turn to graphics instead? The answers are not always clear, and sometimes you have to experiment with several means to find which will work best. But for your general guidance we can offer three principles:

- Prefer graphics if your ideas are primarily quantitative, structural, or pictorial.
- Estimate whether your readers are primarily "eye minded" or "word minded." Generally, technically minded people are also eye minded.
- Decide where your own skill lies—with words or with visual means.

Prose and graphics are usually closely related. You may need to explain some graphics in detail, while a brief mention will suffice for others. The amount of explanation will depend on the importance of the graphic to your exposition, the complexity of the graphic, and the technical expertise of your audience. Generally speaking, the more important and complex the graphic, and the less expert your audience, the more prose explanation you need. In any event, always at least mention the graphic to your reader. And mention it early in the discussion. Do not lead readers through a complicated prose explanation, and then refer to the graphic that simplifies the whole explanation. Send them to the graphic immediately, and they can cut back and forth between the prose and the graphic as necessary.

When a graphic is important to your explanation, locate it as close to the explanation as you can. If it is small, put it right on the text page. If the graphic itself takes up a page, place it facing the prose explanation or at least on an adjoining page. When graphics are less important to your explanation, they may be located further away, perhaps in an appendix.

Graphics are generally divided into two categories: tables and figures. In a table the data are organized into columns and rows. Necessary numbers, titles, explanations, and so forth are provided. A figure is any graphic that is not a table, a category that includes graphs, drawings, diagrams, and photographs. In a formal report, the titles of figures and tables are usually listed separately in a list of figures and a list of tables, both of which appear under the heading ILLUSTRATIONS. In the rest of this chapter we give you basic explanations of tables and figures.

TABLES

A table is any arrangement of data set up in vertical columns and horizontal rows, classified as either formal or informal.

Informal Tables

We "read" an informal table much as we would read normal prose conveying essentially the same information. Just as we usually read a paragraph only once, so do we scan through an informal table only once. It follows that informal tables must be relatively simple and immediately clear. As you may guess, the writer who inserts an informal table is actually exercising a choice between using prose and using the tabular display.

Informal tables have certain earmarks to distinguish them from the formal variety:

- They are brief and simple.
- They are not identified by table number and often have no title.
- So that they will be seen as a continuation of the text, they have no ruled frame around them, and they have only the necessary minimum of internal ruled lines.
- Lacking titles and table numbers, informal tables cannot be listed in the list of tables at the front of the report or in the index.

The example in Figure 12-1 shows a common variety of informal table. Portions of the surrounding text have been reproduced to illustrate how the sense of the table ties in with that of the prose preceding and following it.

Informal tables have certain virtues. Because they use minimal

The relative effectiveness of these four nematocides can be seen in a study done at The West Virginia University where each of the nematocides used brought the following percentages of reduction in nematode population in red pine seedling beds:

Methyl bromide	99.2
D-D	85.0
Mylone	68.0
Vapam	64.0

Therefore, in this study methyl bromide was the most satisfactory material for controlling the plant parasitic nematodes in the red pine seedbed. In other studies as well, methyl bromide proved most effective in the control of plant parasitic nematodes.

FIGURE 12-1   Informal Table

verbal machinery, they are space-savers, often conveying factual information that would require prose occupying several times the same space. They give readers relief by breaking up pages of solid prose. They signal readers that something different or special is being said. Readers can slow down and be sure of picking up the sense.

Formal Tables

Formal tables constitute a major problem in the use of graphics. For clarity of discussion here, we first identify in Figure 12-2 the standard parts of such tables by name, location and function.

You should notice several points in our skeleton table. First, the number and title of the table are traditionally placed above the table, either centered or at the left margin. Usually tables are numbered consecutively throughout a report with either arabic or roman numerals or occasionally capital letters. If a report is divided into chapters, the tables may be numbered consecutively throughout each chapter: I-A, I-B, II-A, II-B, etc.; or 1-1, 1-2; 2-1, 2-2, etc.

Every column and subcolumn and every line or group of lines must have a heading that clearly identifies the data. All headings should have parallel grammatical form (see page 461). Substantive headings consisting of nouns or noun phrases are usually most appropriate. The stub heading classifies the line headings only; it should not introduce the column headings. In other words, stub and column headings have vertical reference, whereas line headings have horizontal reference. Violations of these principles are shown in the very faulty Table A in Figure 12-3.

TABLE NUMBER
TITLE

Stub Heading[a]	Column Heading[b]	Column Heading	
		Subheading	Subheading
Line Heading	Individual "cells" for tabulated data		
Subheading			
Subheading			
Line Heading			

[a]Footnote

[b]Footnote

FIGURE 12-2 Typical Table Format

Table A

Model Number	100	101	102	103
Year of origin	1974	1978	1979	1982
Lens f-number	5.6	3.5	2.8	1.4
Shutter speed (max)	1/300	1/500	1/800	1/1000
Retail cost new	$120	$135	$200	$225

Table B

Descriptive Data	Model Numbers			
	100	101	102	103
Year of origin	1974	1978	1979	1982
Lens f-number	5.6	3.5	2.8	1.4
Shutter speed (maximum)	1/300	1/500	1/800	1/1000
Retail cost new (dollars)	120	135	200	225

Table C

Model Number	Year of Origin	Lens f-number	Shutter Speed (maximum in seconds)	Retail Cost New ($)
100	1974	5.6	1/300	120
101	1978	3.5	1/500	135
102	1979	2.8	1/800	200
103	1982	1.4	1/1000	225

FIGURE 12-3 Evolution of a Table

Note how a few simple revisions make Table B more logical and easier to read. However, even Table B's arrangement is not the best. Assume that the main intent of the table is to compare the different models. In such a case people clearly prefer Table C, in which the data directly compared are arranged in columns rather than rows.

Notice that in these sample tables we have eliminated the vertical lines of the skeleton table. The modern trend is to eliminate as much clutter from tables as possible. Therefore, vertical columns are usually separated simply by spacing rather than lines. Also, the horizontal double line between the headings and data is frequently eliminated and a single line substituted. Be careful with your spacing. Too little results in crowded, hard-to-read tables. Too much results in readers' having difficulty seeing the relationships you want them to see.

Here are the major conventions governing the construction of tables and their incorporation into the text of reports.

- Make every table as simple, clear, and logical as it can be made.
- Mention every table in the text, preferably before it appears.
- Make every reasonable effort to insert a table into the text where the reader is expected to refer to it.
- If it is necessary to refer to an earlier or later table, give its page number as well as table number.
- Keep titles clear and succinct. Avoid phrases like "A Summary of" and "Presentation of Data Concerning" as in "A Presentation of Data Concerning Energy Production and Consumption: 1960 to 1980." Simply say, "Energy Production and Consumption: 1960 to 1980."
- Arrange column headings and line headings in some rational order: alphabetical, geographical, temporal, or quantitative.
- Where applicable, include units of measure, such as miles, degrees, or percentages, in column and line headings.
- In columns, align whole numbers on the right-hand digits, fractional numbers on the decimal points.
- Indicate the source of data appearing in a table by acknowledgment in the text, by a footnote, or both.
- As indicated in our skeleton table on page 249, footnotes are internal to the table. Designate them by superscript lower-case letters or by symbols (* # §). Many stylebooks recommend against numbers as footnotes for fear they will be mistaken for data. However, you will see numbers used, notably in the *Statistical Abstract of the United States.*

Figures 12-4 and 12-5 illustrate typical tables. The table in Figure 12-4 from the 1981 *Statistical Abstract of the United States* illustrates several things besides table layout. Notice the use of a headnote directly beneath the title that gives information needed for interpreting the table. Notice also the *source* line at the bottom. When, as in Figure 12-4, ail the information in a table comes from one source, such a line is commonly used. When the information comes from various sources, as in Figure 12-5, footnotes are used.

FEDERAL OUTLAYS FOR EDUCATION AND RELATED ACTIVITIES: 1960 TO 1981

[In millions of dollars. Years ending **June 30** except, beginning **1977**, ending **Sept. 30**. See headnote, table 220]

| YEAR | OUTLAYS SUPPORTING EDUCTION IN EDUCATIONAL INSTITUTIONS | | | | | | Other outlays for education and related activities |
| | Total | Grants | | | | Loans, higher education | |
		Total	Elementary, secondary	Higher education	Vocational technical		
1960	1,734	1,493	490	830	173	240	2,267
1966	5,844	5,232	2,037	2,272	923	612	3,820
1970	9,236	8,728	3,206	3,911	1,611	508	3,329
1972	11,782	11,433	3,840	5,172	2,421	349	4,516
1973	12,689	12,343	4,033	5,965	2,344	346	4,706
1974	13,082	12,729	4,130	6,064	2,535	352	4,847
1975	17,604	17,124	4,884	7,992	4,248	480	5,863
1976	19,553	19,157	4,679	9,675	4,803	396	6,137
1977	18,780	18,459	4,965	8,899	4,594	322	7,568
1978	21,598	20,795	5,707	9,385	5,703	802	7,442
1979	24,383	22,995	6,714	9,322	6,958	1,388	8,390
1980, est.	26,753	24,952	7,355	9,904	7,692	1,802	9,230
1981, est.	26,771	25,275	7,737	9,828	7,710	1,496	9,264

Source: U.S. National Center for Education Statistics, *Digest of Education Statistics*, annual.

FIGURE 12-4 Sample Table Showing Source

Amino acid composition of processed wild rice, rice, oat groats, and wheat

Amino Acid	Wild rice[a] Minn.	Wild rice[a] Wisc.	Brown rice[b]	Polished rice[b]	Oat groats[b]	Whole wheat[b]
	------------------------------ g per 100 g recovered amino acids ------------------------------					
Alanine	6.2	5.9	6.3	5.4	5.0	3.5
(Ammonia)	–	2.0	2.4	2.8	2.7	3.9
Arginine	8.2	7.3	9.3	8.1	6.9	4.7
Asparatic acid	10.6	10.1	10.6	9.2	8.9	5.1
Cysteine	0.2	1.4	1.2	2.7	1.6	2.3
Glutamic acid	19.5	19.2	22.9	17.6	23.9	30.8
Glycine	4.9	5.0	5.2	4.5	4.9	4.0
Histidine	2.8	2.6	2.7	2.2	2.2	2.3
Isoleucine	4.4	4.4	3.8	4.6	3.9	3.9
Leucine	7.5	7.4	8.8	8.1	7.4	6.8
Lysine	4.5	4.4	4.0	3.4	4.2	2.6
Methionine	3.1	2.6	1.9	2.7	2.5	1.7
Phenylalanine	5.1	5.2	5.6	5.3	5.3	4.6
Proline	4.0	4.3	4.6	4.5	4.7	9.5
Serine	5.5	5.6	5.9	5.0	4.2	4.9
Threonine	3.3	3.6	3.8	3.5	3.3	3.0
Tyrosine	3.9	3.1	3.5	4.5	3.1	3.1
Valine	6.2	5.9	5.4	6.5	5.3	4.8
SLTM[c]	10.9	10.6	9.7	9.6	10.0	7.3
	------------------------------ percent ------------------------------					
Total protein	14.2	13.5	12.8	7.3	17.1	12.0

[a] Data from University of Minnesota, Agricultural Experiment Station and University of Wisconsin–Madison, Agricultural Experiment Station.

[b] Data from table in "1975 Report on Wild Rice Processors' Conference" by University of Wisconsin–Madison.

[c] Sum of lysine, threonine, and methionine.

FIGURE 12-5 Sample Table Showing Footnotes [From Ervin A. Oelke et al., *Wild Rice Production in Minnesota*, (St. Paul: University of Minnesota Agricultural Extension Service, 1982), 38.]

GRAPHS

In this section we describe some of the basic kinds of graphs you can construct to illustrate your reports. Before you read the ground rules of graphing that follow, look at Figure 12-6 to get a general idea of what distinguishes each kind of graph from the others. (Later we discuss the different graphs in detail, with more precise illustrations.)

In making graphs, follow these general rules:

Keep the graph simple. Provide no more detail than you absolutely need, nor more ideas than the reader can easily grasp, only essential facts. Keep in mind the purpose and audience of the graph.

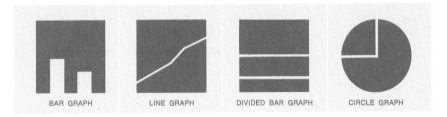

BAR GRAPH   LINE GRAPH   DIVIDED BAR GRAPH   CIRCLE GRAPH

FIGURE 12-6   Graph Types

- Place titles either above or below graphs, but be consistent with your placement. Generally, titles are aligned with the left margin. As with tables, keep titles clear and succinct.
- Make the graph big enough. The answer to what is "big enough" is subjective but also rather obvious. No squinting should be necessary.
- Keep the graph orderly. In making comparisons, present the quantities in some such order as from large to small or small to large or chronologically, according to your purpose.
- Make the illustration attractive. A graph should be well balanced and blend into the page design.
- Use any device possible to help the reader interpret your graph. Color, arrows, heavier lines, and annotation may be used to this end. Including a summary of the graph material either on the face of the graph or in a caption below the title will aid the reader's understanding.
- For most readers, avoid two- and three-dimensional presentation of data, as illustrated in Figure 12-7. Areas and volumes may be appropriate on occasion but usually fail to make comparisons as clearly as a graph using one dimension. Graphs that attempt to depict volume are very difficult to interpret correctly.
- Make the units agree. Design your graphs so that the reader cannot possibly compare dollars with tons, or miles with hours. Make sure that the graph will be read so as to compare dollars with dollars, tons with tons, and so on.

Bar Graphs

Bar graphs, illustrated in Figure 12-8, are easily and accurately read. The reader does not come away with conflicting visual and verbal impressions. In general, it makes no difference whether the bars run

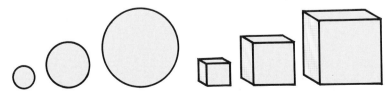

FIGURE 12-7   Two- and Three-Dimensional Presentation

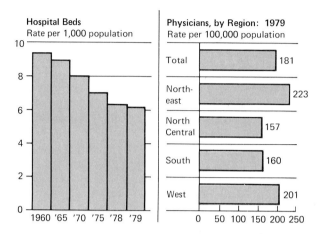

FIGURE 12-8 Typical Bar Graphs [From U.S.,
Bureau of the Census, *Statistical Abstract of the
United States*, 102nd ed. (Washington, D.C.: U.S.
Government Printing Office, 1981), 903.]

vertically or horizontally, but there are some exceptions. One factor
to be considered is page layout. Another is the nature of the data.
People expect certain kinds of comparisons to be read up and
down—altitudes for example—and others, like distances, to be read
from side to side. The degree of reading precision you desire will
determine whether you use a scale and, if you use it, how many lines
to include on it.

Avoid, when possible, using keys to graphs. Put information on the
bar itself or right next to it. When space or other limitations require
the use of keys, keep them simple.

Multiple Bars. By using multiple bars clumped together, as in
Figure 12-9, it's easy to include additional comparisons in a bar
chart. In Figure 12-9, we can compare each consumer group with
every other group and each year with every other year, thus getting
multiple comparisons.

Divided Bars. Through the use of divided bars, you can show
percentage breakdowns as in Figure 12-10. In Figure 12-10, notice
the use of lines to aid the reader interpret the data.

Line Graphs
Of all graphs used by scientists and engineers, the line graph is by
far the most popular and useful. Certainly, it lends itself the best to
purely technical information. One use for the line graph is to plot the

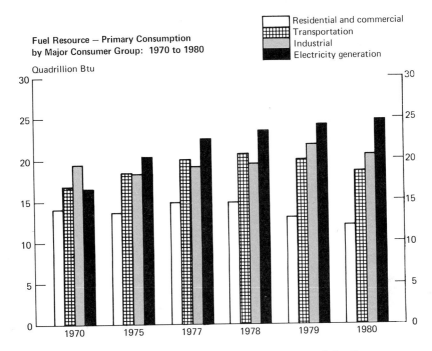

FIGURE 12-9   Multiple Bar Graph [From U.S., Bureau of the Census, *Statistical Abstract of the United States*, 102nd ed. (Washington, D.C.: U.S. Government Printing Office, 1981), 576.]

behavior of two variables: the independent variable and the dependent variable. The independent variable is normally plotted horizontally, that is, along the abscissa. The dependent variable is normally plotted vertically, that is, along the ordinate. The dependent variable is affected by changes in the independent variable.

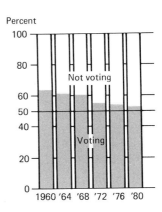

FIGURE 12-10   Divided Bar Graph, showing Voter Participation in Presidential Elections [From U.S., Bureau of the Census, *Statistical Abstract of the United States*, 102nd ed. (Washington, D.C.: U.S. Government Printing Office, 1981), 907.]

In order to avoid a cluttered appearance, values needed are often shown in hash marks on the abscissa and ordinate, as in Figure 12-11, rather than on a complete grid. Figure 12-11 also illustrates the use of the "suppressed zero." When the values plotted are so high that it would be impractical to start with zero, a scale starting at a point above zero is used. Whenever you must follow this practice be sure your reader understands clearly that the zero has been suppressed. One way to do this is to separate the scale, as in Figure 12-11.

When accuracy is of paramount importance, a complete grid must be used. Figure 12-12 illustrates a line graph plotted on graph paper with a fine grid that permits accuracy of construction and interpretation.

Line graphs are also excellent for showing and comparing trends. Figure 12-13 is a good example of such use. Notice how by shading and the use of a dotted line the four separate trends are kept distinct. Various line devices, such as dots (. . . .) and alternating dots and dashes ($\cdot-\cdot-\cdot-$), can also be used. When color is appropriate, you may be able to use it as well. However, if the graph will need to be photoreproduced, you will likely have to stay with black and white. Also, most journals will not print color because of the expense.

Sometimes, a line and a bar graph are combined into one graph,

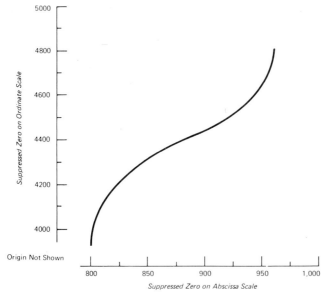

FIGURE 12-11 Line Graph with Hash Marks and Suppressed Zero

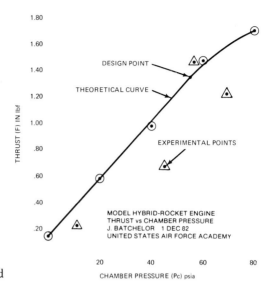

FIGURE 12-12 Line
Graph Requiring Fine Grid

Source: Chart prepared by U.S. Bureau of the Census. For data, see tables 873 and 876.

FIGURE 12-13 Line Graph Comparing Trends [From U.S., Bureau of the Census, *Statistical Abstract of the United States,* 102nd ed. (Washington, D.C.: U.S. Government Printing Office, 1981), 502.]

as in Figure 12-14. Such graphs can make quite complex comparisons possible, but because they are complex, reserve them for practiced graph readers. Lay people would likely have a tough time interpreting Figure 12-14. Notice also the use of explanatory notes, such as "Oil wells (left scale)," placed where they will do the most good, right on the lines themselves.

Circle Graphs

The circle or pie graph makes for ease and accuracy of reading. It shows proportions, the relative size of related quantities. The pie graph is commonly found in business and financial reports and is an excellent graph for executives. Figure 12-15 presents two typical pie graphs. In both graphs, note the orderly arrangement of the parts. As in Figure 12-15, it is good practice to begin the largest segment at twelve o'clock and then proceed clockwise in descending order of magnitude. (The exception to this practice in the 1970 graph is to provide consistency with the 1980 graph.) Notice that where space allows the percentage figures and identifications are placed inside the wedge. Whether they are inside or outside, percentage figures should be horizontal to the page and not to the lines separating the wedges.

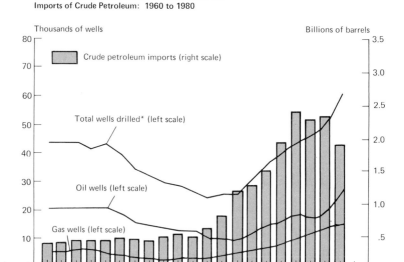

Crude Petroleum and Natural Gas—Wells Drilled and Imports of Crude Petroleum: 1960 to 1980

FIGURE 12-14 Bar and Line Graph Combined [From U.S., Bureau of the Census, *Statistical Abstract of the United States*, 102nd ed. (Washington, D.C.: U.S. Government Printing Office, 1981), 720.]

FIGURE 12-15   Pie Graphs
[From U.S., Bureau of the Census, *Statistical Abstract of the United States*, 102nd ed. (Washington, D.C.: U.S. Government Printing Office, 1981), 912.]

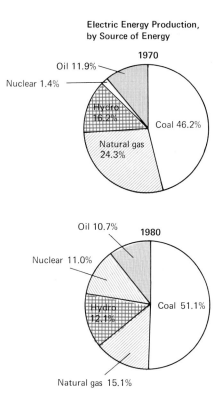

Electric Energy Production, by Source of Energy

Pictograms

A pictogram is a pictorial representation of data. In essence, it is usually a dressed-up bar graph. For example, the bars may be drawn to represent men or women or factories or whatever. In drawing a pictogram the problem is to make sure that the reader interprets the drawing as a pictogram—for example, in Figure 12-16, as a pile of coins rather than as the volume or area covered. If the design is drawn so that it is seen in terms of the desired units, it can be read quite accurately. A pictogram has the attention-getting value of novelty and is sometimes useful for a lay audience.

Table-Graph Relationship

Obviously, often you could present the same data in a table, in a bar graph, or in a line graph. Which method or combination of methods you choose will depend upon your intention and your audience.

In the table in Figure 12-17, you see the number of hours and minutes that an automobile water pump survived testing at various revolutions per minute. This table shows clearly how the life of the

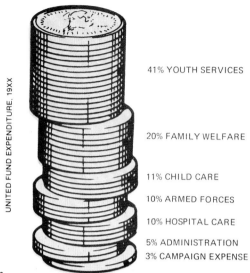

FIGURE 12-16 Pictogram

41% YOUTH SERVICES

20% FAMILY WELFARE

11% CHILD CARE

10% ARMED FORCES

10% HOSPITAL CARE

5% ADMINISTRATION
3% CAMPAIGN EXPENSE

UNITED FUND EXPENDITURE, 19XX

water pump decreases as its speed of revolution is increased. But the table does not show very clearly the shape of the relationship of rpm to time before failure. If you were writing this paper for a predominantly lay audience, you might choose to show this shape with a simple bar graph. The bar graph in Figure 12-17 shows clearly the relationship between rpm and the time before failure. Because it is customary to show time on the horizontal axis, the graph ignores the convention that the dependent variable is plotted on the ordinate.

An audience primarily composed of technically minded people would probably prefer a line graph such as the one in Figure 12-17. They would recognize, as nontechnical readers might not, that the increases in rpm actually constitute a continuous variable. For them, the slope of the curve would show the continuous relationship of rpm and time to failure. The line graph also reverts to the conventional and plots the dependent variable along the vertical. The zero is supressed on the abscissa, but the numbers on the scale make this fact quite clear.

DRAWINGS, PHOTOGRAPHS, AND DIAGRAMS

In this section we touch only briefly upon the use of drawings, photographs, and diagrams. The production of these graphics generally calls for the services of a specialist. However, students can use such graphics to a limited extent, and engineers, scientists, and technical writers should at least have some idea of how to use them.

TABLE IV
TESTING TIME BEFORE FAILURE

Test No.	RPM	Hours-Minutes
1	4000	4:08
2	5000	4:02
3	6000	3:55
4	7000	3:45
5	8000	3:26
6	9000	2:45
7	10,000	1:15

Bar Graph: Testing Time Before Failure

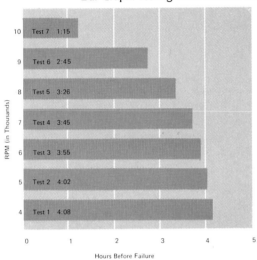

Line Graph: Testing Time Before Failure

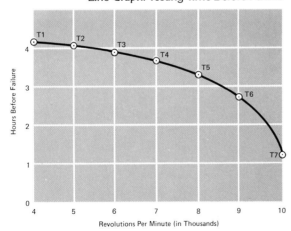

FIGURE 12–17 Table-Graph Relationship

261

Drawings and Photographs

Drawings have several advantages over photographs. For one thing the artist can include as much or as little detail as needed. For example, Figure 12-18 reproduces a page from a Honeywell technical manual, *Instructions for Guarded Null Director, Model 3990*. The artist's only intention is to portray the dimensions of the bench model of the 3990. A simple drawing, free from a photograph's realistic clutter, is best for this.

SECTION 2

INSTALLATION

2-1. GENERAL

The Model 3990 null detector is shipped completely assembled and is ready for operation following unpacking, inspection, and checkout. Exercise normal care in unpacking the instrument from the shipping carton. After unpacking, carefully inspect the case, controls, and connectors for visible evidence of shipping damage. IN CASE DAMAGE HAS BEEN INCURRED, IMMEDIATELY NOTIFY THE RESPONSIBLE CARRIER AND INITIATE CLAIM PROCEDURE.

2-2. INSTALLATION

For the bench model null detector, the requisites for placing the instrument in operation are a source of line (117 VAC ± 10% or 230 VAC ± 10%, 50-60Hz) or battery (11 VDC or 14 VDC) power and an operating area relatively free of electromagnetic field. Refer to Figure 2-1 and 2-2 for outline dimensions of the bench and rack models.

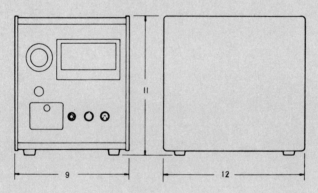

Figure 2-1. Outline Dimensions of the Null Detector, Bench Model

FIGURE 12-18 Drawing Showing Instrument Dimensions (Courtesy Penn Airborne Products Company.)

Another advantage of the drawing is that the artist can "cut away" barriers to sight and allow the reader to see inside objects as in the internal view of *Sea Lab III* in Figure 12-19. As Figure 12-19 also illustrates, drawings are easy to label.

Photographs, however, can be labeled as well and when labeled can be useful for maintenance technicians. Figure 12-20 reproduces a photograph of one of the circuit boards for the Honeywell Model 3990. The annotations on this photograph refer the reader to an adjoining table—part of which we have also reproduced—that gives maintenance people information about part numbers, spare parts, schematic locations, and so forth. Properly annotated photographs help readers find their way around in highly complex pieces of equipment. Photographs and electron micrographs are also sometimes the best way to show experimental results, particularly in the various biological sciences.

The companion illustrations of Halley's Comet in Figure 12-21 clearly depict what is often a major difference between a photograph and a drawing. The drawing provides information for the reader's mind. The photograph appeals to the mind's eye. Thus they are different not only in appearance but in their function and purpose. Both have their place in many reports.

Diagrams

Diagrams are, of course, drawings, but they rely more on symbolism than do the types of drawings we have been discussing. Like all graphic representation, the purpose of diagrams is to explain and clarify the material presented. The diagram in Figure 12-22 showing observers where to look for Halley's Comet is an excellent example of how a well-done and rather simple diagram can aid reader understanding.

You can use diagrams to show time as in Figure 12-23, physical process as in Figure 12-24, and human process as in Figure 12-25. Diagrams that show the *flow* of process are called flow charts. Those that show the progression of *time* are called, not unnaturally, time charts.

Diagrams can be as simple or as complicated as the writer wants them. Figure 12-26 shows a simplified block diagram of the Honeywell Model 3990. In this particular case the author's intention is to give the technician who will operate the instrument a functional description of it. We have reproduced the entire page so that you can see how the writer introduces the diagram. Compare the simplification evident in Figure 12-26 with the complication in Figure 12-27. Figure 12-27 reproduces merely a portion of a complete schematic diagram of the Model 3990, intended for the technician who must

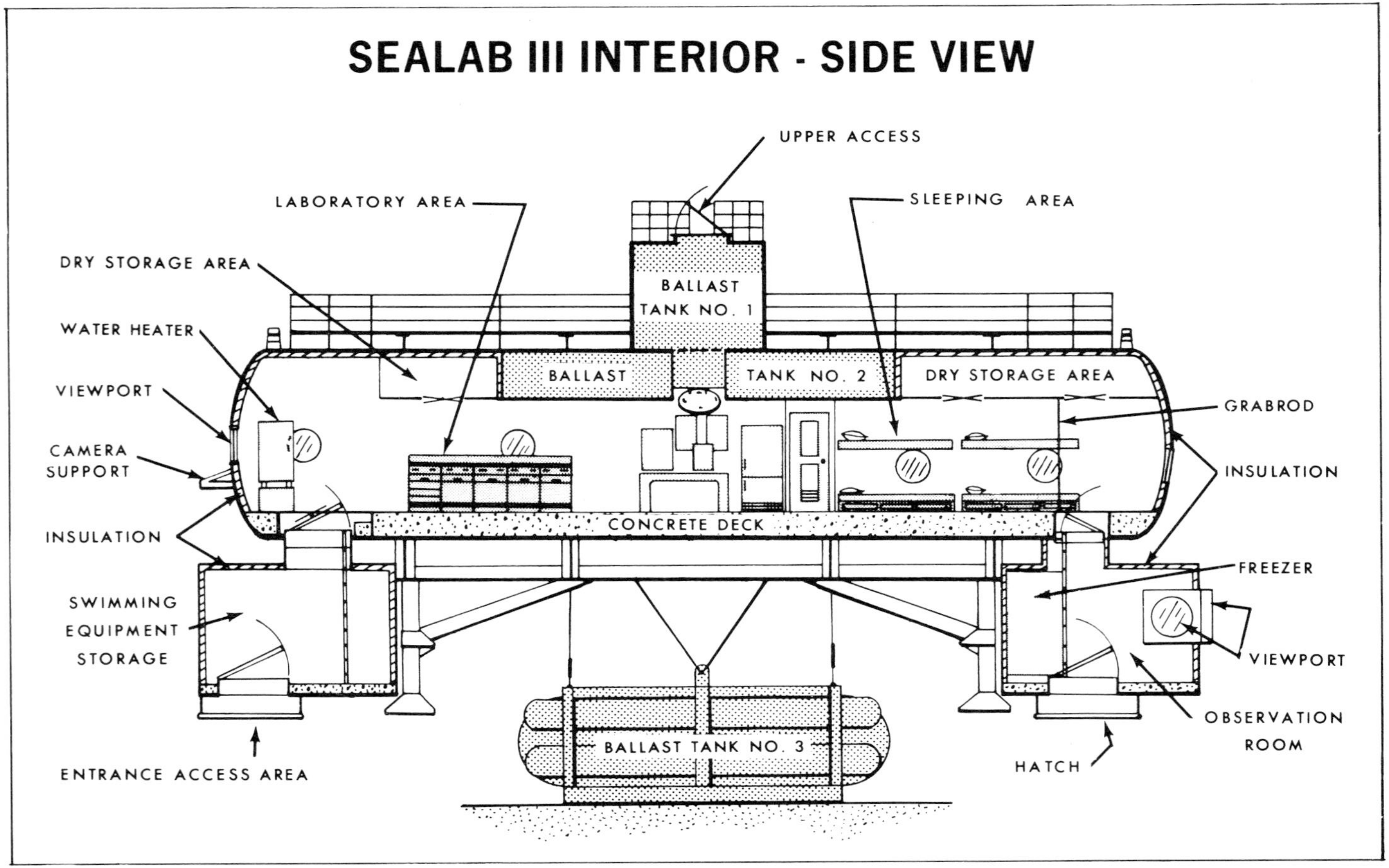

SEALAB III INTERIOR - SIDE VIEW

FIGURE 12-19 Drawing of Sealab III

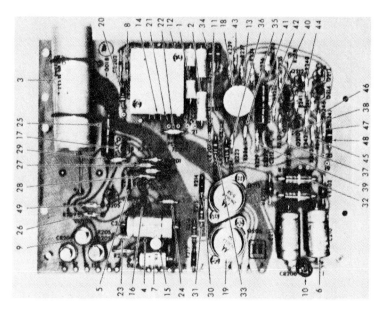

Figure 6-6. DC Amplifier Circuit Board

Fig	Index	Schem	Description	Code	Manufacturer's Part Number	Honeywell Part No.	Qty/Assy	Spares
6-4	10 11	CR206 CR207, CR208	Bridge, WO-2 Diode, 1N914	11711 00508	WO-2 1N914	18739755-002 18756865	1 2	2
	12	DB201, DB202	Lamp, Glow	03508	NE-2H	18737777	2	2
	13 14 15	Q201	Heatsink Plug, Mounting Transistor	86684	2N3053	18756892 18757774 18783169	1 4 1	
	16	Q202, Q203, Q209, Q211, Q213, Q214	Transistor	04713	2N3904	18782172	6	
	17	Q204, Q206, Q210, Q212, Q207, Q208	Transistor	04713	2N3906	18782173	4	
	18	Q207, Q208	Transistor Assy			18756814	2	2
	19	Q206, Q215	Transistor	04713	2N3614	18781655-004	2	
	20 21	R201, R202, R203	Resistor, 43 K, 1/2 W, ±5% Photocell	44655 03911	LEDFA43000-5 CL60UCL	18750076-589 18755808-004	1 1	2
	22 23	R204, R205, R206	Resistor, 15 K, 1/8 W, ±1% Resistor, 750Ω, 1/8 W, ±1%	07716 07716	CEA-TO-1%-15 K CEA-TO-1%-750Ω	18757185-318 18757185-185	1 2	
	24 25 26	R207, R208, R209, R233, R235	Resistor, 464Ω, 1/8 W, ±1% Resistor, 500Ω, 1/2 W, ±10% Resistor, 301Ω, 1/8 W, ±1%	07716 80294 07716	CEA-TO-1%-464Ω 308TP-1-501 CEA-TO-1%-301Ω	18757185-165 18750069-003 18757185-147	1 1 3	
	27 28	R210, R211, R212, R213	Resistor, 909Ω, 1/8 W, ±1% Resistor, 4.02 K, 1/8 W, ±1%	07716 07716	CEA-TO-1%-909Ω CEA-TO-1%-4.02 K	18757185-193 18757185-259	1 2	
	29	R214, R217	Resistor, 158Ω, 1/8 W, ±1%	07716	CEA-TO-1%-158Ω	18757185-120	1	
	30	R214, R217	Resistor, 620, 1/2 W, ±5%	44655	LEDFA68-5	18750076-520	2	
	31	R215, R216	Resistor, 9100, 1/2 W, ±10%	44655	LIDC0810-10	18750078-544	2	
	32	R218, R219	Resistor, 5100, 2 W, ±5%	44655	LIDGC510-5	18750078-044	2	
	33	R220, R224	Resistor, 6.8 K, 1/2 W, ±10%	44655	LIDL08100-10	18750078-843	2	
	34	R221, R222	Resistor, 2.2 K, 1/2 W, ±10%	44655	LIDL02200-10	18750078-836	2	
	35	R223, R225	Resistor, 24.9 K, 1/8 W, ±1%	07716	CEA-TO-1%-24.9 K	18757185-339	2	
	36	R226	Resistor, 49.9 K, 1/8 W, ±1%	07716	CEA-TO-1%-49.9 K	18757185-382	2	
6-4	37	R227, R228	Resistor, 1 K, 1/8 W, ±1%	07716	CEA-TO-1%-1 K	18757185-201	2	

FIGURE 12-20 Photograph and Corresponding Table (Courtesy Penn Airborne Products Company.)

Head of Halley's Comet as seen on May 8, 1910 (Mt. Wilson Observatory photograph).

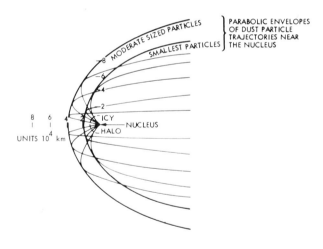

Schematic drawings of the principal gaseous and particulate features of a typical comet.

FIGURE 12-21   Drawing-Photograph Comparison [From Science Working Group, *International Halley Watch* (Washington, D.C.: National Aeronautics and Space Administration, 1980), vi and 15.]

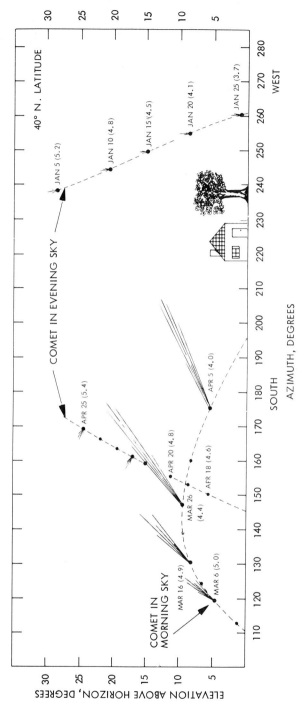

FIGURE 12-22 Comet Halley Diagram [From U.S., National Aeronautics and Space Administration, *The Comet Halley Handbook* (Pasadena, Calif.: Jet Propulsion Laboratory, California Institute of Technology, 1981), 10.]

Stages of Development	Days	Date	GDD
Germination	0		0
Emergence	12	May	138
Floating Leaf	29		468
Aerial Leaf	39	June	686
Early Tillering	49		928
Jointing	67		1394
Boot	75	July	1590
Early Flowering	83		1834
Mid Flowering	91		2078
Early Grain Formation	105	August	2466
Maturity	121		2940

Left side vertical labels: Vegetative Growth Phase / Reproductive Growth Phase

Wild rice plant development (K2 variety) in Aitkin County. GDD ≡ growing degree days using 40° F as the base temperature.

FIGURE 12-23 Time Chart [From Ervin A. Oelke et al., *Wild Rice Production in Minnesota*, (St. Paul: University of Minnesota Agricultural Extension Service, 1982), 19.]

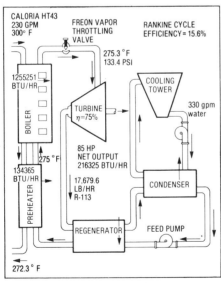

CALORIA HT43
230 GPM
300° F

FREON VAPOR THROTTLING VALVE

RANKINE CYCLE EFFICIENCY = 15.6%

275.3 °F
133.4 PSI

1255251 BTU/HR

BOILER

TURBINE η=75%

COOLING TOWER

330 gpm water

85 HP NET OUTPUT 216325 BTU/HR

275 °F

134365 BTU/HR

PREHEATER

17,679.6 LB/HR R-113

CONDENSER

REGENERATOR

FEED PUMP

272.3 ° F

THE RANKINE CYCLE IS A THERMODYNAMIC MEANS of obtaining shaft power from heat. The cycle is not new, but our application of it is. Here the engines generate shaft power by extracting heat from the secondary loop in a boiler and a preheater and use this heat to vaporize the refrigerant which is then expanded through a turbine.

FIGURE 12-24 Physical Process Flow Chart (From Vis Vidins, "The Honeywell Plaza Solar System," *Scientific Honeyweller*, June 1980, 6. Courtesy *Scientific Honeyweller*, all rights reserved.)

Application Route for Permission to Dredge Lake

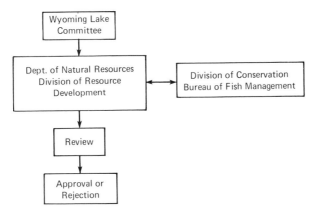

FIGURE 12-25   Human Process Flow Chart (Courtesy Barbara J. Buschatz.)

maintain the instrument. Maintenance people obviously need complete detail whereas operators do not. As we have stressed many times, analyze your audience and give them what they need.

A CLOSING WORD

In this chapter we have tried to show that the construction and use of graphics demand much thought, skill, and experience on the part of authors and their colleagues in the art department. We have done little more than hint at the many ways that graphics can make your reports more understandable and more lively. We urge you to follow up on the information in this chapter by reading more about graphics in the books on this subject in our bibliography. Also, if you intend to publish in a journal, read the stylebook that governs that journal for special instructions in the preparation of graphics. For example, the *Council of Biology Editors Style Manual* devotes 12 pages to the subject.

EXERCISES

1.   Figure 12-28 presents a table concerning United States energy supply and disposition. You are to make a graph based upon that table. The table contains more information than you would want to put on one graph. Therefore, you must select what information you choose to

is also made for safeguarding against accidentally introduced voltage over-
loads through the use of diodes in the input filter. Stability is provided by a
mechanical chopper modulator and a photoconductive demodulator. The modulator and demodulator are phase-synchronized by a common voltage source
that operates at 94 Hz to minimize line frequency pickup effects.

B.   Figure 4-1 shows that the null detector consists of an input filter, a mechanical
chopper feeding into an input transformer, an AC amplifier section feeding into
an output transformer, a demodulator, an output filter, and a DC amplifier.
Power for the modulator, demodulator, and amplifiers is furnished by an in-
verter (DC to DC converter) driven by either a conventional 120/230-volt power supply or a 12-volt lead acid battery. Three AC amplifiers are used in the
0.1 microvolt position. two amplifiers in the 1 microvolt and 10 microvolt positions. and a single amplifier in the remaining ranges. Range switch S301
connects AC amplifiers to operating circuits as shown in the following diagram

Switch Position	Amplifier(s) Affected
0.1 μv	All
1 μv	1st & 2nd
10 μv	1st & 3rd
Above 10 μv	1st only

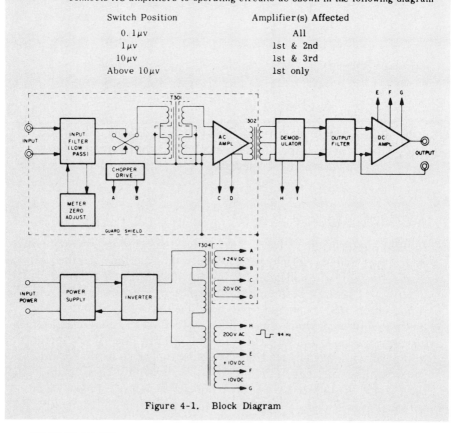

Figure 4-1. Block Diagram

FIGURE 12-26   Block Diagram (Courtesy Penn Airborne Products
Company.)

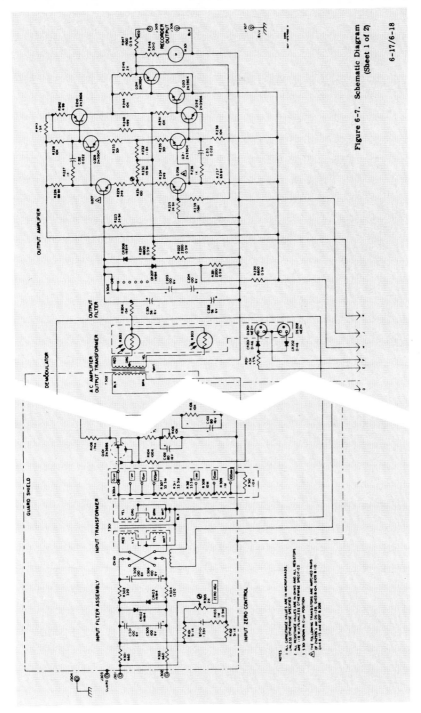

Figure 6-7. Schematic Diagram
(Sheet 1 of 2)

6-17/6-18

FIGURE 12-27 Schematic Diagram for Technician (Courtesy Penn Air-borne Products Company.)

ENERGY SUPPLY AND DISPOSITION, BY TYPE OF FUEL: 1960 TO 1980

[In quadrillion British thermal units, except percent. For Btu conversion factors, see text, p. 575]

TYPE OF FUEL	1960	1965	1970	1972	1973	1974	1975	1976	1977	1978	1979	1980, prel.
Total supply[1]	**45.6**	**54.9**	**69.5**	**73.8**	**76.7**	**75.0**	**73.1**	**76.7**	**78.4**	**80.1**	**81.9**	**80.0**
Production	41.8	49.7	62.5	62.8	62.4	61.2	60.1	60.1	60.3	61.2	63.9	64.8
Crude oil[2]	14.9	16.5	20.4	20.0	19.5	18.6	17.7	17.3	17.5	18.4	18.1	18.3
Natural gas liquids	1.5	1.9	2.5	2.6	2.6	2.5	2.4	2.3	2.3	2.3	2.3	2.3
Natural gas[3]	12.7	15.8	21.7	22.2	22.2	21.2	19.6	19.5	19.6	19.5	20.1	19.7
Coal	11.1	13.4	15.1	14.5	14.4	14.5	15.2	15.9	15.8	15.0	17.7	18.9
Nuclear power	(Z)	(Z)	.2	.6	.9	1.3	1.9	2.1	2.7	3.0	2.8	2.7
Hydropower	1.6	2.1	2.6	2.9	2.9	3.2	3.2	3.0	2.3	3.0	3.0	2.9
Geothermal and other	—	—	(Z)	(Z)	.1	.1	.1	.1	.1	.1	.1	.1
Imports[4]	4.2	5.9	8.4	11.5	14.7	14.4	14.1	16.8	20.1	19.3	19.6	15.8
Percent of supply	9.2	10.7	12.1	15.6	19.2	19.2	19.3	21.9	25.6	24.1	23.9	19.8
Crude oil	2.2	2.7	2.8	4.7	6.9	7.4	8.7	11.2	14.0	13.5	13.8	11.1
Refined petroleum products[5]	1.8	2.8	4.7	5.6	6.6	5.7	4.2	4.4	4.7	4.4	4.1	3.4
Natural gas	.2	.5	.9	1.1	1.1	1.0	1.0	1.0	1.0	1.0	1.3	1.0
Total disposition	**45.6**	**54.9**	**69.5**	**73.8**	**76.7**	**75.0**	**73.1**	**76.7**	**78.4**	**80.1**	**81.9**	**80.0**
Consumption	44.1	53.0	66.8	71.6	74.6	72.8	70.7	74.5	76.3	78.2	79.0	76.3
Refined petroleum products	19.9	23.3	29.5	33.0	34.8	33.5	32.7	35.2	37.1	38.0	37.1	34.3
Natural gas[3]	12.4	15.8	21.8	22.7	22.5	21.7	20.0	20.4	19.9	20.0	20.7	20.4
Coal	10.1	11.9	12.7	12.5	13.3	12.9	12.8	13.7	14.0	13.9	15.1	15.7
Nuclear power	(Z)	(Z)	.2	.6	.9	1.3	1.9	2.1	2.7	3.0	2.7	2.7
Hydropower[6]	1.7	2.1	2.7	2.9	3.0	3.3	3.2	3.1	2.5	3.2	3.2	3.1
Geothermal and other	—	—	(Z)	(Z)	.1	.1	.1	.1	.1	.1	.1	.1
Exports	1.5	1.9	2.7	2.1	2.1	2.2	2.4	2.2	2.1	2.0	2.9	3.8
Coal	1.0	1.4	1.9	1.5	1.5	1.6	1.8	1.6	1.5	1.1	1.8	2.5

— Represents zero. Z Less than 50 trillion. [1] Includes adjustments for stock changes, losses, gains, and miscellaneous changes, not shown separately. [2] Includes lease condensate. [3] Dry marketed gas. [4] Includes other types, not shown separately. [5] Includes imports of unfinished oils and natural gas liquids. [6] Includes industrial generation and net electricity imports.

Source: U.S. Energy Information Administration, *Annual Report to Congress*, vol. II.

FIGURE 12-28 Information for Exercise 1. [From U.S., Bureau of the Census, *Statistical Abstract of the United States*, 102 ed. (Washington D.C.: U.S. Government Printing Office, 1981), 578.]

portray. As in writing, base your selection on purpose and audience. Accompany your completed graph with a paragraph that explains how your purpose and audience guided you to the finished graph.

2. Look at the diagrams in Figures 12-23, 12-24, and 12-25. All three show the flow of either time or process. Construct a chart that shows either time or process or some combination of the two. For example, you could chart the flow of your college education or a major writing project you are currently working on. Or you could chart the process by which film is developed or a check proceeds through the banking system. There are, indeed, many processes you are learning about that could be charted in order to aid some reader to understand them. Accompany your chart with a paragraph that explains how purpose and audience guided you in constructing your chart.

3. Prepare a list of the illustrations (figures and tables) you intend to include in your project report.

Part 2 Applications

In Part 2, we discuss how to apply the basic techniques of Part 1 in various types of technical reports and in speaking. Here we move from basic matters into professional work, covering correspondence, oral reports, and five kinds of technical reports: instructions, proposals, progress reports, feasibility reports, and empirical research reports.

If you are still a student, we ask you to project yourself into the future. Imagine, as you write the assignments your instructor gives, that you are already practicing in your chosen profession. Care about your work now as you will in the future. Do not be satisfied with anything less than a professional piece of work.

Chapter **13**

Correspondence

ON THE JOB, you will often be responsible for a share of your organization's correspondence. Remember that to the receiver any letter from an organization represents that organization. If you write letters that are misleading, dull, or rude, you present an image of an organization that is misleading, dull, or rude.

We do not attempt in this chapter to give you a course in business correspondence. There are many fine books on the subject, some of them listed in our bibliography. However, we do cover those aspects of correspondence that will be most useful to you. We talk briefly about the tone and format of business letters. Then we give you specific instructions in how to write letters of inquiry, replies to letters of inquiry, letters of complaint and adjustment, the correspondence of the job hunt, and memorandums.

STYLE AND TONE

Everything we say about style in Chapter 9 goes doubly when you are writing letters. Letters must be clear, so clear that the readers cannot possibly misunderstand them. Use short paragraphs, lists, clear sentence structure, and specific words. Above all, avoid pomposity and the passive voice. The passive voice gives a cold formality to a letter's tone. Compare "It is requested that you send . . ." with "I would appreciate your sending. . . ." Fill your letters with *I* and *you*, particularly *you*.

Do not make yourself a repository for all the clichés that writers before you have used. *Tell* people, do not *advise* them. Do not say, "This is to acknowledge" or "I would like to thank you." (Simply say, "Thank you.") Do not open letters with "Re the" or "Referring to." Do not tell people "We are in receipt of your letter of the 10th and have noted its contents." They know that, or will know it when you answer their questions, grant their request, or whatever. If you want to pinpoint the date that a letter was written to you, do it in a courteous or an unobtrusive way. Say, "Thank you for your letter of November 10" or "We don't blame you for being angry with us; your letter of November 10 made our mistake very clear." Perhaps the best solution is a subject or reference line that includes such file data. (See Figures 13-1 and 13-2.)

Many people have stressed the need for brevity in letters, and certainly it is a good thing to be concise. But do not get carried away with the notion. Letters are not telegrams. Too brief a letter gives an impression of brusqueness, even rudeness. Often a longer letter gives a better impression. The reader feels that you took more time to consider the problem. Particularly avoid brevity when you must refuse someone something. People appreciate your taking the time to explain a refusal.

Remember that a letter substitutes for conversation. Be natural. Consider the human being reading your letter when you write. Develop what has been labeled the *you-attitude*.

To some extent, we suppose, the you-attitude means that your letter contains a higher percentage of *you*'s than *I*'s. But it goes beyond this mechanical usage of certain pronouns.

With the you-attitude you see things from your reader's point of view. You think about what the letter will mean to the reader, not what it means to you. We can illustrate simply. Suppose you have an interview scheduled with a prospective employer and, unavoidably, you must change dates. You could write as follows:

```
Dear Ms. Moody:

A change in final examination schedules here at school
makes it impossible for me to keep our appointment on
June 10.

I am really disappointed.  I was looking forward to
coming to Los Angeles.  I feel inconvenienced, but I
hope we can work out an appointment.  Please let me
know on what date we can arrange a new meeting.

Sincerely yours,
```

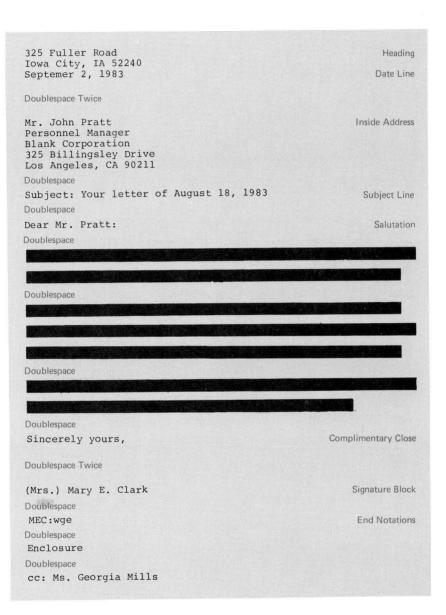

325 Fuller Road   Heading
Iowa City, IA 52240
Septemer 2, 1983   Date Line

Doublespace Twice

Mr. John Pratt   Inside Address
Personnel Manager
Blank Corporation
325 Billingsley Drive
Los Angeles, CA 90211
Doublespace
Subject: Your letter of August 18, 1983   Subject Line
Doublespace
Dear Mr. Pratt: Salutation
Doublespace

Doublespace

Doublespace

Doublespace
Sincerely yours,   Complimentary Close

Doublespace Twice

(Mrs.) Mary E. Clark   Signature Block
Doublespace
MEC:wge   End Notations
Doublespace
Enclosure
Doublespace
cc: Ms. Georgia Mills

FIGURE 13-1   Block Letter

```
                                        325 Fuller Road
                                        Iowa City, IA 52240
                                        July 27, 1983
```

Doublespace Twice

```
Mr. John R. Galt, Manager
Blank Corporation
325 Billingsley Drive
Los Angeles, CA 90211
```

Doublespace

```
Re: Your letter of July 7, 1983
```

Doublespace

```
Dear Mr. Galt:
```

Doublespace

▬▬▬▬▬▬▬▬▬▬▬▬▬▬▬▬▬▬▬▬▬▬▬▬▬▬▬▬
▬▬▬▬▬▬▬▬▬▬▬▬▬▬▬▬▬▬▬▬▬▬▬▬▬▬▬▬

Doublespace

▬▬▬▬▬▬▬▬▬▬▬▬▬▬▬▬▬▬▬▬▬▬▬▬▬▬▬▬
▬▬▬▬▬▬▬▬▬▬▬▬▬▬▬▬▬▬▬▬▬▬▬▬▬▬▬▬
▬▬▬▬▬▬▬▬▬▬▬▬▬▬▬▬▬▬

Doublespace

▬▬▬▬▬▬▬▬▬▬▬▬▬▬▬▬▬▬▬▬▬▬▬▬▬▬▬▬
▬▬▬▬▬▬▬▬▬▬▬▬▬▬▬▬▬▬▬▬▬▬▬▬▬▬▬▬

Doublespace

```
                                     Sincerely yours,
```

Doublespace Twice

```
                                     William G. Clark
```

Doublespace

```
cc: Mr. John Muller
    Mr. Eugene Wright
```

FIGURE 13-2 Semi-Block Letter

Now this letter is clear enough, and it may even get a new appointment for its writer. But it has an I-attitude, not the you-attitude. The writer thinks only of himself, and the reader may be vaguely or even greatly annoyed. This next version will please a reader far more:

Dear Ms. Moody:

A change in final examination schedules here at school makes it impossible for me to keep our appointment on June 10. When I should have been talking to you, I'll be taking an exam in chemistry.

I hope the change has not seriously inconvenienced you, Ms. Moody. Please accept my appologies.

Will you be able to work me in at a new time? Final exam week ends on June 12. Please choose any date after that convenient for you.

Sincerely yours,

Mechanically, the second letter contains more *you*'s, but more to the point, it considers the inconvenience caused the reader, not the sender. And it makes it easy for the reader to set up a new date.

With the you-attitude, you consider both your reader's feelings and needs. Here is the way a camera company politely corrected an amateur photographer's film loading procedure:

Dear Mr. Bayless:

We are sending you a roll of film to replace the one overexposed when you used your new Film-X camera for the first time. And we'll certainly replace the camera if it proves to be defective.

Trouble such as you have experienced is often caused by the photographer's inadvertently forgetting to set the proper film speed while loading the camera. The ASA dial on the front of the camera must be set to match the ASA film speed number of the film you are working with. Failure to do so will result in either under-or overexposed film such as you experienced.

If you'll check Instruction 5 on page 3 of your Film-X Manual, you'll find complete information on how to set the ASA dial.

We hope that your problem was nothing more than an improperly set ASA. If not, however, follow the warranty instructions that came with the camera, and we'll replace your Film-X.

Sincerely yours,

With this letter, the writer begins by offering a gift. Next he puts the reader's mind at ease by telling him that the company will replace the camera if need be. Then he attempts to correct the reader's film loading procedure, the real reason, in the writer's opinion, for the trouble.

Suppose the letter writer had taken an attitude like the one expressed in this letter:

```
Dear Mr. Bayless:

The type of film failure you wrote to us about stems from
faulty loading of the Film-X.  Your method and not the
camera is defective.

Review the instructions that came with the camera.  Pay
particular attention to Instruction 5.

Sincerely yours,
```

The second letter will surely put the reader's back up. Do people really write foolish, rude letters like this one? Unfotunately, perhaps because they are pressed for time, or because of a sense of superiority, they do—and worse. When people make a mistake, correct the consequences without making them feel like fools. We all make mistakes, and we appreciate gentle correction, not the lash of sarcasm.

Our advice about style and tone can be summed up with what we call the four C's of correspondence: Correspondence should be *Clear, Concise, Complete,* and *Courteous.* Sometimes, *Concise* and *Complete* may be in conflict. If in doubt, opt for completeness. Always remember the importance of being *Courteous.* Taking time for the you-attitude is the good and human way to act. Luckily for us all, it's also very good business.

FORMAT

Almost any organization you join will have rules about its letter formats. You will either have a secretary to type your correspondence, or you will have to learn the rules for yourself. In this section we give you only enough rules and illustrations so that you can turn out a good-looking, correct, and acceptable business letter on your own. If for no other reason, you will find this a necessary skill when you go job hunting.

Figures 13-1 through 13-4 on pages 279, 280, 283, and 284 illustrate four widely accepted business styles; the block, the semiblock, an alternative block, and the simplified. Figure 13-5 on page 285 illustrates the continuation page of a block letter. We have indicated in these samples the spacing, margins, and punctuation you should use. In the text that follows we discuss briefly the different styles and then

```
December 18, 1983

Doublespace Twice

Mr. John R. Galt, Manager
Blank Corporation
325 Billingsley Drive
Los Angeles, CA 90211
Doublespace
Subject: Claim for elevator accident
Doublespace
Dear Mr. Galt:
Doublespace
```

██
██

Doublespace

██
██
████████████████████████████████

Doublespace

██
███████████████████████

```
Doublespace
Sincerely yours,

Doublespace Twice
_____

William G. Clark
325 Fuller Road
Iowa City, IA 52240
Doublespace
Encl: Copy of accident report
      Filled-in claim
Doublespace
cc: Mrs. Nina Baldachi
```

FIGURE 13-3   Alternative Block Letter

give you some of the basic rules you should know about the parts of a letter. Before continuing with the text, look at Figures 13-1 through 13-5.

For most business letters you may have to write, any of the four styles shown would be acceptable. In letters of inquiry or complaint, where you probably do not have anyone specific in a company to address, we suggest the simplified style. For letters of application we

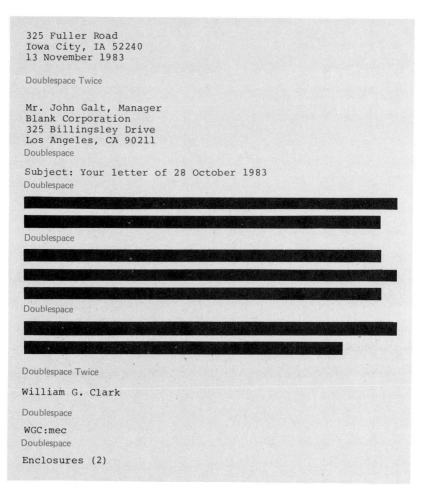

325 Fuller Road
Iowa City, IA 52240
13 November 1983

Doublespace Twice

Mr. John Galt, Manager
Blank Corporation
325 Billingsley Drive
Los Angeles, CA 90211
Doublespace

Subject: Your letter of 28 October 1983
Doublespace

Doublespace

Doublespace

Doublespace Twice

William G. Clark

Doublespace

WGC:mec
Doublespace

Enclosures (2)

FIGURE 13-4 Simplified Letter

suggest the block or semiblock style without the use of a subject line. Some people still find the simplified letter without the conventional salutation and complimentary close a bit too brusque. Unless you know for certain that the company you are applying to prefers the simplified form, do not take a chance with it.

If you are an amateur typist doing your own typing, we suggest the block style as the best all-around style. All the conventional parts of a letter are included, but everything is lined up along the left-hand margin. You do not have to bother with tab settings and other complications. Some people feel that a block letter looks a bit lopsided, but it is a common style that no one will object to. The alternative block style is also an acceptable choice. No matter which style you choose, leave generous margins, at least an inch all around, and bal-

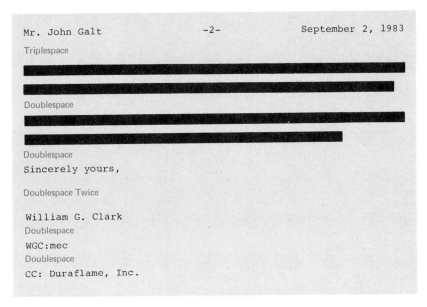

Mr. John Galt   -2-   September 2, 1983

Triplespace

Doublespace

Doublespace
Sincerely yours,

Doublespace Twice

William G. Clark
Doublespace
WGC:mec
Doublespace
CC: Duraflame, Inc.

FIGURE 13-5 Continuation Page of Block Letter

ance the letter vertically on the page. Because letters should look good with lots of white space, you should seldom allow paragraphs to run more than seven or eight lines. And don't be afraid of writing paragraphs as short as one or two lines, particularly at the end of a letter.

Heading

We have assumed in our models that you do not have letterhead stationery of your own and have shown you how to type your heading. In the semiblock style, the heading is approximately flush right. In the other formats shown, the heading is flush left. Do not abbreviate words like street or road. Write them out in full. You may abbreviate your state. Figure 13-6 gives you the state abbreviations preferred by the U.S. Postal Service. If you do have business letterheads made up, have them printed on good quality white bond in a simple style.

Date Line

In a letter without a printed letterhead, the date line is part of your heading in the block, semiblock, and simplified styles. In the alternative block style, it's flush left and four spaces above the inside address. When you do have a printed letterhead, the date line is flush left in the block and simplified styles and approximately flush right in semiblock. Place it two spaces below a printed letterhead or in a

Alabama	AL*	Montana	MT
Alaska	AK	Nebraska	NE
Arizona	AZ	Nevada	NV
Arkansas	AR	New Hampshire	NH
California	CA	New Jersey	NJ
Colorado	CO	New Mexico	NM
Connecticut	CT	New York	NY
Delaware	DE	North Carolina	NC
District of		North Dakota	ND
Columbia	DC	Ohio	OH
Florida	FL	Oklahoma	OK
Georgia	GA	Oregon	OR
Guam	GU	Pennsylvania	PA
Hawaii	HI	Puerto Rico	PR
Idaho	ID	Rhode Island	RI
Illinois	IL	South Carolina	SC
Indiana	IN	South Dakota	SD
Iowa	IA	Tennessee	TN
Kansas	KS	Texas	TX
Kentucky	KY	Utah	UT
Louisiana	LA	Vermont	VT
Maine	ME	Virginia	VA
Maryland	MD	Virgin Islands	VI
Massachusetts	MA	Washington	WA
Michigan	MI	West Virginia	WV
Minnesota	MN	Wisconsin	WI
Mississippi	MS	Wyoming	WY
Missouri	MO		

*Notice that both letters of the abbreviation are capitalized and that no period is used.

FIGURE 13-6 State Abbreviations Recommended by the U.S. Postal Service

manner to help balance the letter on the page. Write the date out fully either as June 3, 19XX or 3 June 19XX. Do not abbreviate the month or add *st*, *nd*, etc. (e.g., 1st, 2nd), to the number of the day.

Inside Address

The inside address is placed flush left in all the formats shown. Make sure the inside address is complete. Follow exactly the form used by the person or company you are writing to. If your correspondent abbreviates *Company* as *Co.*, you should also. Use *S. Edward Smith* rather than *Samuel E. Smith* if that is the way Smith wants it. Do use courtesy titles such as *Mr.*, *Dr.*, and *Colonel* before the name. The usual abbreviations used are *Mr.*, *Mrs.*, *Ms.*, and *Dr.* Place one-word titles such as *Manager* or *Superintendent* immediately after the name. When a title is longer than one word, place it on the next line by itself. Do not put a title after the name that means the same thing as a courtesy title. For example, don't write *Dr. Samuel E. Smith, Ph.D.*

Subject Line

Place the subject line flush left in all the styles. It is usually preceded by the word *Subject* or *Re*, though sometimes you will see it with nothing before it. In this line you can give the date of the letter being answered or a substantive heading such as "complaint adjustment."

Salutation

Place the salutation flush left. Convention still calls for the use of *Dear*. Always use a name in the salutation when one is available to use. When you use a name, be sure it is in the inside address as well. Also, use the same courtesy title as in the inside address, such as *Dear Dr. Sibley* or *Dear Ms. McCarthy*. You may use a first name salutation, *Dear Sarah*, when you are on friendly terms with the recipient. Follow the salutation with a colon.

What do you do when you are writing a company blindly and have no specific name to address? In the past, convention has called for the use of *Dear Sir* or *Gentlemen*. But because many people now feel uncomfortable assuming the recipient is always male, this convention is losing ground. Some people now use a *Dear* followed by the name of the department being written to, such as *Dear Customer Relations Department*. Still others substitute a *Hello* or *Good Day* for the traditional salutation. In other words, people are experimenting a bit instead of automatically following tradition. One solution, perhaps the best, is to choose the simplified style where no salutation is used. Whatever you do, we strongly urge you not to begin letters with *Dear Person* or *Dear People*. These salutations are distasteful to many.

Body

In the average length letter, single-space between the lines and double-space between the paragraphs. In a particularly short letter, double-space between the lines and triple-space between the paragraphs. Avoid splitting words between lines. Never split a date or a person's name between two lines. If you are going to have a continuation page, the last paragraph on the preceding page should contain at least two lines.

Complimentary Close

In the block and alternative block style, place the complimentary close flush left. In the semiblock style, align the close with the heading (or with the date line in a letterhead letter). Settle for a simple close, such as *Sincerely yours* or *Very truly yours.* Capitalize only the first letter of the close and place a comma after the close.

Signature Block

Type your name four spaces below the complimentary close. If you have a title, type it below your name. Sign your name immediately above your typed name. Avoid fancy flourishes and unintelligible signatures until you are president of the company.

If you are a man, sign your letters with your first name, middle initial, and last name. If your first name could be mistaken for a woman's such as *Marion,* you might want to use your first two initials or use your first initial and middle name, as in *M. David Smith.* Your signature and your typed name should agree.

If you are a woman, sign your letters with first name, middle initial, and last name. If married, use your own first name, not your husband's. Again, the typed name below should agree with the signature. Before the typed name, women without an honorific title such as *Doctor* or *Professor* may choose to include, in parentheses, a preferred title: *Miss, Mrs.,* or *Ms.* (see Figure 13-1). The form *Ms.* (pronounced *miz* or by saying each of its two letters, *m-s*) is gaining popularity as a replacement for *Miss* or *Mrs.* It is a polite way to address a woman when you do not know her marital status, and it is a practical form for women who feel that marital status need not be identified in a professional setting.

End Notations

Various end notations may be placed at the bottom of a letter, always flush left. The most common ones indicate identification, enclosure, and carbon copy.

Identification.   The notation for identification is composed of the letter writer's initials in capital letters and the typist's initials in lower case:

DHC:lnh

Enclosure.   The enclosure line indicates that additional material has been enclosed with the basic letter. You may use several forms:

Enclosure
Enclosures (2)
Encl: Employment application blank

Carbon Copy.   Both courtesy and ethics require that you inform the recipient of a letter when you have sent a copy of the letter to someone else. In form, the copy notation looks like this:

cc: Ms. Elaine Mills

See Figures 13-1 to 13-5 for proper spacing and sequence of these three notations.

Continuation Page

Do not use letterhead stationery for a continuation page. Use plain white bond of the same quality as the first page. Try to begin the continuation page with a new paragraph. If you cannot, be sure the continued paragraph runs at least two lines. As shown in Figure 13-5, the continuation page is headed by three items: name of addressee, page number, and date.

LETTERS OF INQUIRY

As a student, business person, or simply a private person, you will often have occasion to write letters of inquiry. Students often overlook the rich sources of information for reports that they can tap with a few well-placed, courteous letters. Such letters can bring brochures, photographs, samples, and even very quotable answers from experts in the field the report deals with.

Sometimes companies solicit inquiries about their products through their advertisements and catalogs. In such cases your letter of inquiry can be short and to the point:

```
Your advertisement on page 89 of the January 1984
Scientific American invited inquiries about your new
Film-X developing process.

I am a college student and president of a 20-member
campus photography club.  The members of the club and
I would appreciate any information about this new
process that you can send us.

We are specifically interested in modernizing the
film-developing facilities of the club.
```

As in the above letter, you should include three important steps:

1. Identify the advertisement that solicited your inquiry.
2. Identify yourself and establish your need for the information.
3. Request the information. Specify the precise area in which you are interested. Obviously, in this step you also identify the area in which the com-

pany may expect to make a sale to you. You thus, in a subtle way, point out to the company why it is in their best interest to answer your inquiry promptly and fully: a good example of the you-attitude in action.

An unsolicited letter of inquiry cannot be quite so short. After all in an unsolicited letter you are asking a favor, and you must avoid the risk of appearing brusque or discourteous. In an unsolicited letter you include five steps:

1. Identify yourself.
2. State clearly and specifically the information or materials that you want.
3. Establish your need for the requested information.
4. Tell the recipient why you have chosen him or her as a source for this information.
5. Close graciously, but do not say "thank you in advance."

The first four steps may be presented in various combinations rather than as distinct steps, or in different order, but none of the steps should be overlooked.

Identification

In an unsolicited letter, identify yourself. We mean more here than merely using a title in your signature block. Rather, you should identify yourself in terms of the information sought. That is, you are not merely a student, you are a student in a dietetics class seeking information for a paper about iron enrichment of flour. Or you are a member of a committee investigating child abuse in your town. Or you are an engineer for a state department of transportation seeking information for a study of noise walls on city freeways. Certainly, the more prestigious your identification of yourself, the more likely you are to get the information you want. But do not misrepresent yourself in any way. Misrepresentation will usually lead to embarrassment.

Some years ago, one of the authors of this text naively wrote for information on bookshelves, using his college letterhead instead of his private stationery. He was deluged with brochures that were followed up by a long-distance telephone call from a sales manager asking how many shelves the college was going to need: "Is the college expanding its library facilities?"

"No," the lame answer came, "I just wanted to put up a few shelves in my living room."

Do not misrepresent yourself. But do honestly represent yourself in the best light you can. Most companies are quite good about answering student inquiries. They recognize in students the buyers and the employees of the future and are eager to court their goodwill.

Inquiry

State clearly what you want. Be very specific about what you want and do not ask for too much. Avoid the shotgun approach of asking for "all available information" in some wide area.

Particularly, do not expect other people to do your work for you. Do not, for example, write to Bell Telephone asking for references to articles on wave guides. A little time spent by you in the indexing and abstracting publications will produce the same information. On the other hand, it would be quite appropriate to write to a Bell engineer or scientist asking for a clarification or amplification of some point in a recent article that he or she had written. Science thrives on the latter kind of correspondence.

If you do need involved information, put your questions in an easy-to-answer questionnaire form. If you have only a few questions, include them in your letter. Indent the questions and number them. Be sure to keep a copy of your letter. The reply you receive may answer your questions by number without restating them.

For a larger group of questions, or when you are sending the questionnaire to many people, make the questionnaire a separate attachment. Be sure in your letter to refer the reader to the attachment. Questionnaires are tricky. Unless they are presented properly and carefully made up they will probably be ignored. Do not ask too many questions. If possible, phrase the questions for yes or no answers or provide multiple-choice options. Sometimes, meaningful questions just cannot be asked in this objective way. In such cases, do ask questions that require the respondent to write an answer. But try to phrase your questions so that answers to them can be short. Provide sufficient space for the answer expected. If you are asking for confidential information, stress that you will keep it confidential. The questionnaire should be typed, printed, well mimeographed, or photocopied. Perhaps the best solution for multiple copies of anything these days is to use one of the many inexpensive photocopying services available. Look under *Photocopying* in the Yellow Pages of your telephone book. For further information about putting together a questionnaire, see pages 84–89.

Need

Tell the recipient why you need the information. Perhaps you are writing a report, conducting a survey, buying a camera, or simply satisfying a healthy, scientific curiosity. Whatever your reason, do not be complicated or devious here. Simply state your need clearly and honestly. Often this step can best be combined with the identification step. If there is some deadline by which you must have an answer, say what it is.

Choice of Recipient

Tell the recipient of the letter why you have chosen him or her or the recipient's company as a source for the information. Perhaps you are writing a paper about stereo equipment and you consider this company to be one of the foremost manufacturers of FM tuners. Perhaps you have read the recipient's recent article on space medicine. You specifically want an amplification of some point.

Obviously, this section is a good place to pay the recipient a sincere compliment. But do not flatter or be sickeningly sweet. Phoniness will get you nowhere.

Point out any benefit the recipient may gain by answering your request. For example, if you are conducting a survey among many companies, promise the recipient a tabulation and analysis of results when the survey is completed.

Gracious Close

Close by expressing your appreciation for any help the receiver may give you. But do not use the tired, old formula, "Thanking you in advance, I remain. . . ." Do write a thank-you note even if all you get is a refusal. Who knows? The second letter may cause a change of heart.

Sample Letters

A complete letter of inquiry might look like this one:

```
Dear Mr. Hanson:

I am a second-year student at Florida Technological
Institute.  For a course in technical writing that
I am currently taking, I am writing a paper on the
proper way to educate Americans in the use of the
metric system.  The paper is due on 3 March.

The specific question that concerns me is whether
metric measurements should be taught in relation to
present standard measurements, such as the foot and
pound, or whether they should be taught independently
of other measurements.  I see both methods in use.

In the journals where I have been researching the subject,
your are frequently mentioned as a major authority in
the field.  Would you be kind enough to give me your
opinion about how metrics should be taught?

Any help you can give me will be greatly appreciated,
and I will, of course, cite you in my paper.

Sincerely yours,
```

We created the preceding letter to illustrate in a somewhat mechanical way all the steps in a good letter of inquiry. Reproduced below is a real letter received by one of us. The writer of the letter covers all the required steps in a relaxed, nonmechanical way. The letter also illustrates the use of a questionnaire approach within a letter. Notice, too, how the writer makes legitimate use of a person—Mary Schaeffer, past president of the Society for Technical Communication—known to both him and the letter's recipient.

```
Dear Professor Pearsall:

At the joint meeting of the STC Board and the Phoenix
Chapter, Mary Schaeffer mentioned that you were preparing
a survey of technical communication programs.  We are
interested in developing a technical communication degree
program at our university.  But, before we prepare a
proposal, we want to study the programs at other schools.

Briefly, we would like to answer two questions:

     (1)  Would such a degree program be of real value
          to both students and industry?

     (2)  Would an interdisciplinary approach be best?
          (We have excellent schools of printing,
          journalism, engineering, etc.)

Of course, we are not asking you to answer these questions
for us, but we would like to have your personal opinion and
any information on proposed and existing technical communica-
tion programs that you can send us.

This is a particularly critical moment for us because the
next catalog will cover two years.  As you can see, we
must move rapidly and surely. We will greatly appreciate
any assistance you can give us.

Sincerely,
```

REPLIES TO LETTERS OF INQUIRY

In replying to a letter of inquiry, be as complete as you can. Probably you will be trying to avoid a second exchange of letters of questions and answers. If you can, answer the questions in the order in which they appear in the inquiry. If the inquirer represents an organization and has written on letterhead stationery, you may safely assume that a file copy of the original letter has been retained. In that case, you don't have to repeat the original questions—just answer them. If the inquirer writes as a private person, you cannot assume that a copy of the letter has been kept. Therefore, you will need to repeat enough of the original question or questions to remind the inquirer of what has been asked.

You might answer the earlier letter about metric education in the following fashion:

Dear Ms. Montez:

The question of how to educate people about the metric system concerns a great many educators. I suppose it's inevitable that those familiar with the present system will be tempted to convert metric measurements to ones they already know. For example, people will say "A kilogram, that's about two pounds."

In my opinion, however, such conversion is not the best way to teach metrics. Rather, people should be taught to think in terms of what the metric measurement really measures, to associate it with familiar things. Here are some examples:

  --A paper clip is about a centimeter wide.
  --A dollar bill weighs about a gram.
  --A comfortable room is 20 degrees C.
  --Water freezes at 0 degrees C.
  --At a normal walking pace, we can go about five kilometers
    an hour.

These are the kinds of associations we have made all our lives with the present system. We need to do the same for metric measurements. As in learning a foreign language, we really learn it only when we stop translating in our heads and begin to think in the new language.

I hope this answers your question. You can get a valuable little book about metrics by writing to the Office of Public Information, Federal Reserve Bank of Minneapolis, Minneapolis, Minnesota 55480. Ask for The United States and the Metric System.

Sincerely,

Because the writer of the first letter wrote as a private person, she is, in this answer, reminded of the original question. The tone of the letter is friendly. The question is answered succinctly but completely. An additional source of information is mentioned, an excellent idea that should be followed when possible.

When you can, include as enclosures to your letter previously prepared materials that provide adequate answers to the questions asked. If you do so, provide whatever explanation of these materials is needed. Reproduced here is the answer to the letter of inquiry on page 293 that asks for information about technical communication programs:

```
Dear Professor Garden:

The answer to both your questions is yes.  Rather than
take time to support my answer in this letter, I've
enclosed several documents that should do the job for me.

The first is a proposal I submitted for a technical
communication program at this university.  My personal
views on the value and nature of a program in TC are in
the proposal.  The program, by the way, was approved
and is well under way.

The second enclosure is an article by John Walter that
expresses his views on these subjects.

I've also enclosed for you a brochure explaining our
program, a catalog in which our program is described
(p.44), and a list of other schools that have TC programs.

I hope all this helps you in your efforts to begin a TC
program at your school.  Write again if I can help in
any way.

Sincerely yours,
```

LETTERS OF COMPLAINT AND ADJUSTMENT

Mistakes and failures happen in business as they do everywhere else. Promised deliveries don't arrive on time or at all. Expensive equipment fails at a critical moment. If you're on the receiving end of such problems, you'll probably want to register a complaint with someone. Chances are good that you'll want an adjustment—some compensation—for your loss or inconvenience. To seek such adjustment, you'll write a letter of complaint or, as it is also called, a claim letter.

Your attitude in a letter of complaint should be firm but fair. There is no reason for you to be discourteous. You'll do better in most instances if you write the letter with the attitude that the offending company will want to make a proper adjustment once they know the problem. Don't bluster and threaten to withdraw your trade on a first offense. Of course, after repeated offenses you will seek another firm to deal with, and that should be made clear at the appropriate time.

Be very specific about what is wrong and detail any inconvenience caused you. Be sure to give any necessary product identification such as serial numbers. At the end of your letter, motivate the receiver to make a fair adjustment. If you know exactly what adjustment you want, spell it out. If not, allow the suggested adjustment to come from the company you are dealing with. Figure 13-7 shows you what

```
26 Shady Woods Road
White Bear Lake MN 55101
28 October 1983


Customer Relations Department
Chapman Products, Inc.
1925 Jerome Street
Brooklyn, NY 11205

Subject:  Broken sanding belts

This past September, I purchased two Chapman sanding belts,
medium grade, #85610, at the Fitler Lumber Company in St.
Paul, Minnesota.  I paid a premium price for the Chapman
belts because of your reputation.  However, the belts have
proved to be unsatisfactory, and I am returning them to you
in a separate package (at a cost to me of $2.36).

The belt I have labeled #1 was used only 10 minutes before it
broke.  The belt labeled #2 was used only 5 minutes when the
glue failed and it broke.

I attempted to return the belts to Fitler's for a refund.
The manager refused me a refund and said I'd have to write
to you.

I am disturbed on two counts.  First, I paid a premium price
for your belts.  I did not expect an inferior product.
Second, does the retailer have no responsibility for the
Chapman products he sells?  Do I have to write to you every
time I have a problem with one of your products?

I'm sure that Chapman is proud of its reputation and will
want to adjust this matter fairly.

John Griffin
John Griffin
```

FIGURE 13-7 Letter of Complaint

a complete letter of complaint looks like. Note that it is in simplified format and addressed to the Customer Relations Department. You can safely assume that most companies have an office or an employee specifically responsible for complaints; in any case, if the letter is addressed in this manner it will reach the appropriate person much more rapidly.

What happens at the other end of the line? You have received a letter of complaint and must write the adjustment letter. What should be your attitude? Oddly enough, most organizations welcome letters of complaint. At least they prefer customers who complain rather than customers who think "never again" when a mistake occurs and take their trade elsewhere. Most organizations will go out of their way to satisfy complaints that seem at all fair. A skillful writer will attempt to use the adjustment letter as a means of promoting future business.

Letters of adjustment fall into two categories—granting of the adjustment requested or a refusal. The first is easy to write, the second more difficult.

When you are granting an adjustment, be cheerful about it. Remember, your main goal is to build goodwill and future business. Begin by expressing regret about the problem or stating that you are pleased to hear from the customer—or both. Our earlier comments about the you-attitude (pages 278–282) should be much in your mind while writing an adjustment letter. Explain the circumstances that caused the problem. State specifically what the adjustment will be. Handle any special problems that may have accompanied the complaint and close the letter. Figure 13-8 shows you such a letter.

A letter refusing an adjustment is obviously more difficult to write. You want, if possible, to keep the customer's goodwill. You want at the very least to forestall future complaints. In stating your refusal, you must exercise great tact.

Begin with a friendly opener. Try to find some common ground with the complainant. Express regret about the situation. Even though you may think the complaint is totally unfair, don't be discourteous. Incidentally, not everyone writes a letter as courteous as the one in Figure 13-7. Sometimes, in fact, people are downright abusive. If so, attempt to shrug it off. Just as you would not pour gasoline on a fire, don't answer abuse with abuse. Pour on some cooling words instead.

Second, explain the reason for the refusal. Be very specific here and answer at some length. The very length of your reply will help to convince the reader that you have considered the problem seriously. Third, at the end of your explanation, state your refusal. Sometimes, however, your explanation can be so complete that your refusal is obvious and does not need to be stated. If so, the best strategy is probably not to state the refusal directly. If, however, the complaint is serious enough that legal action may result, you must state your refusal quite clearly to avoid future misunderstandings. In any case do not state the refusal until you have thoroughly presented the reasons for it.

Fourth, if you can, offer a partial or substitute adjustment. Finally, close your letter in a friendly way.

Companies selling products are not the only organizations that

```
                        CHAPMAN PRODUCTS, INC.
                         1925 Jerome Street
                         Brooklyn, NY 11205


November 11, 1983


Mr. John Griffin
26 Shady Woods Road
White Bear Lake, MN 55101

Dear Mr. Griffin:

Thank you for your letter of October 28.  We're sorry that
you had a problem with a Chapman product.  But we are happy
that you wrote to us about your dissatisfaction.  We need
to hear from our customers if we are to provide them satis-
factory products.

The numbers on the belts you returned indicate that they
were manufactured in 1973.  Sanding belts, like many other
products, have a "shelf life," and the belts you purchased
had exceeded theirs.  Age, heat, and humidity had weakened
them.

Mr. Griffin, we stand behind our products.  Although we are
sure that the belts you purchased were not defective when we
shipped them, we wish to replace them for you.  You are being
shipped a box of 10 belts, medium grade.  We're sure that
these belts will live up to the Chapman name.

We also suggest that you look in the yellow pages of your
telephone directory under "Hardware-Retail" for authorized
Chapman dealers.  We can only suggest to independent dealers
how they should shelve and sell our products.  We can exert
more quality control with our own dealers.  We know that you
can find a Chapman dealer who will give you excellent service.

Sincerely,

*Theresa R. Brummer*
Theresa R. Brummer
Customer Service

TB/ay

cc:  Fitler Lumber Company
```

FIGURE 13-8 Letter of Adjustment

receive letters of complaint. Public service organizations do also. The letter that follows illustrates a refusal from such an organization—a state department of transportation. In this case, a citizen had written stating that a curved section of highway near her home was dangerous. She requested that the curve be rebuilt and straightened. The reply uses the strategy that we have outlined:

1. It begins in a friendly way.
2. It details why the department of transportation cannot rebuild the road immediately.
3. The refusal is so obvious that it is not stated explicitly.
4. A substitute is offered.
5. It closes in a friendly way.

In the case of this letter, the goal is not to keep a paying customer but to keep a taxpayer friendly. In either case, the strategy is to offer an honest, detailed, factual explanation in a cheerful way.

```
Dear Mrs. Ferguson:

Thank you for your letter concerning the section of
Trunk Highway (TH) 50 near your home.  The Department
of Transportation shares your concern about the safety
of TH 50, particularly the section between Prestonburg
and Pikeville, near which your home is located.

TH 50, as you mentioned, has a goodly number of hills
and curves and, because of the terrain, some steep
embankments.  However, its accident rate—3.58 accidents
per million vehicle miles—is far from the worst in the
state.  In fact, there are 64 other highways with worse
safety records.  We do have studies under way that will
result ultimately in the relocation of TH 50 to terrain
that will allow safer construction.  These things take
time, as I'm sure you know.  We have to coordinate our
plans with county and town authorities and the federal
government.  Money is assigned to these projects by
priority, and judged by its accident rate, TH 50 does
not have top priority.

Accidents along TH 50 are not concentrated at any one
curve.  They are spread out over the entire highway.
Reconstruction at any one location would cause little
change in the overall accident record.

However, we are currently evaluating the need for guard-
rails along the entire length of TH 50.  Within a year
we will likely construct guardrails at a number of
locations. Most certainly, we will place a guardrail at
the curve that concerns you.  This should correct the
situation to some extent.

We appreciate your concern.  Please write to us again
if we can be of further help.

Sincerely yours,
```

THE CORRESPONDENCE OF THE JOB HUNT

In many cases, the first knowledge prospective employers will have of you is the letter of application and resume that you send to them. A good letter of application and resume will not guarantee that you get a job, but bad ones will probably guarantee that you do not. In this section we tell you about the letter of application, the resume, and several follow-up letters needed during the job hunt.

The Letter of Application

The good letter of application begins before you ever sit down to your typewriter. Find out all you can about the company you are applying to. Talk to friends who may know something about the company. Read the company's advertisements. If they have a company magazine, get it and read it. Send away for copies of their brochures and annual reports. (See "Letters of Inquiry," page 289.) Investigate the company in sources such as the *College Placement Annual, Standard Statistics, Standard Corporation Reports,* and Dun and Bradstreet's *Middle Market Directory, Million Dollar Directory,* and *Reference Book of Manufacturers.* For information about the occupation you plan to enter look into the *Occupational Outlook Handbook,* published by the U.S. Bureau of Labor Statistics, or the *Career Index.* Consult magazines and newspapers that regularly carry business news such as the *Harvard Business Review,* the *Wall Street Journal,* and *Business Week.* To see what has recently appeared in the business press about the company, see the *F & S Index of Corporations and Industries* and the *Business Periodicals Index.* For general coverage see the *New York Times Index* and the *Reader's Guide to Periodical Literature.* Analyze the company to see what achievements it is most proud of and to determine its goals.

For information about jobs with the federal government and how to get them, look into *Working for the USA* and the *Federal Career Directory.* The latter is published yearly. For jobs at the state level, ask the state department of personnel for information.

We have listed only some of the most basic sources where you can find information to prepare for the job hunt. Your librarian can help you find many more, some perhaps quite specifically oriented to the occupation you are most interested in. We cannot emphasize enough how important this preparation is. Listen to what one interviewer, Michael Shaughnessy of General Motors, has to say on the subject:

> It's really impressive to a recruiter when a job candidate knows about the company. If you're a national recruiter, and you've been on the road

for days and days, you have no idea how pleasant it is to have a student say, "I know your company is doing such and such, and has plants here and here, and I'd like to work on this particular project." Otherwise I have to go into my standard spiel, and God knows I've certainly heard myself give that often enough.[1]

Analyze yourself as well as potential employers. What are your strengths? What are your weaknesses? How well have you performed in past jobs? Have you shown initiative? Have you improved procedures? Have you accepted responsibility? Have you been promoted or given a merit raise? How can you present yourself most attractively? What skills do you possess that relate directly to what the employer seems to need? Are you really qualified for the job you want? Do not apply for a job if you lack the necessary qualifications. You may fake your way to the interview but not beyond.

Plan the mechanics of your letter of application carefully. Buy the best quality white bond paper. This is not a time to skimp. Plan to type your letter, of course, or have it typed. Use a standard typeface such as pica or elite. Do not use italic. With pica you get ten letters to the inch, and with elite you get twelve. This can be useful information when you need either to expand or contract the space your information occupies in a letter or resume. Make sure your letter is mechanically perfect, free from erasures and grammatical errors. Be brief but not telegraphic. Keep the letter to one page. Don't send a carbon or a letter duplicated in any way. Accompany each letter, however, with a duplicated resume. We have more to say about this feature later.

Pay attention to the style of the letter and the resume that accompanies it. The tone you want in your letter is one of self-confidence. You must avoid both arrogance and humility. You must sound interested and somewhat eager, but not fawning. Do not give the impression that you must have the job. Nor do you want to seem uncaring about whether you do get the job.

When describing your accomplishments in the letter and resume, use action verbs. They help to give your writing brevity, specificity, and force. For example, don't just say that you worked as a sales clerk. Rather, tell how you maintained inventories, sold merchandise, prepared displays, implemented new procedures, and supervised and trained new clerks. Here's a sampling of such words:

administer	cut	edit
analyze	design	evaluate
conduct	develop	exhibit
create	direct	expand

improve	oversee	reorganize
manage	plan	support
operate	produce	was promoted
organize	reduce costs	write

You cannot avoid the use of *I* in a letter of application. But take the you-attitude as much as you can. Think about what you can do for the prospective employer. The letter of application is not the place to be worried about salary and pension plans. Above all, be mature and dignified. Forget about tricky and flashy approaches. Write a well-organized, informative letter that highlights those skills your analysis of the company shows it desires most. We will discuss the application letter in terms of a beginning, a body, and an ending.

The Beginning. Beginnings are tough. Do not begin aggressively or cutely. Beginnings such as "WANTED: an alert, aggressive employer who will recognize an alert, aggressive young forester" will usually send your letter wastebasket-bound. A good beginning, if it is available to you, is a bit of legitimate name dropping. Use this beginning only if you have permission and if the name dropped will mean something to the prospective employer. If you qualify on both counts begin with an opener like this:

```
Dear Ms. Marchand:

Professor John J. Jones of State University's Food
Science faculty has suggested that I apply for the post
of food supervisor that you have open.  In June I will
receive my Bachelor of Science degree in Food Science
from State University.  Also, I have spent the last
two summers working in food preparation for Memorial
Hospital in Melbourne.
```

Remember that you are trying to arouse immediate interest about yourself in the potential employer. Another way to do this is to refer to something about the company that interests you. Such a reference establishes that you have done your homework. Then try to show how some preparation on your part ties you into this special interest. See Figure 13-9 for an example of such an opener.

Sometimes the best approach is a simple statement about the job you seek accompanied by something in your experience that fits you for the job, as in this example:

> Your opening for a food supervisor has come to my attention. In June of this year, I will graduate from State University with a Bachelor of Science degree in Food Science. I have spent the last two summers working in food preparation for Memorial Hospital in Melbourne. I believe that both my education and work experience qualify me to be a food supervisor on your staff.

Be specific about the job you want. As the vice-president of one firm told us, "We have all the people we need who can do anything. We want people who can do something." Quite often, if the job you want is not open, the employer may offer you an alternative one. But employers are not impressed with vague statements such as, "I'm willing and able to work at any job you may have open in research, production, or sales."

The Body. In the body of your letter you select items from your education and experience that show your qualifications for the job you seek. Remember always, you are trying to show the employer how well you will fit into the job and the organization.

In selecting your items, it pays to know what things employers value the most. One thorough piece of research[2] shows that for recent college graduates employers give priority as follows:

First priority	Major field of study
	Academic performance
	Work experience
	Plant or home-office interview
	Campus interview
Second priority	Extracurricular activities
	Recommendations of former employer
	Academic activities and awards
Third priority	Type of college or university attended
	Recommendations from faculty or school official
Fourth priority	Standardized test scores
	In-house test scores
	Military rank

Try to include information from the areas that employers seem to value the most, but put emphasis on those areas in which you come off best. If your grades are good, mention them prominently. If you stand low in your class—in the lowest quarter, perhaps—maintain a discreet silence. Speak to the employer's interests and at the same time highlight your own accomplishments. Show how it would be to the employer's advantage to hire you. The following paragraph, an excellent example of the you-attitude in action, does all these things:

```
I understand that the research team of which I might be a
part works as a single unit in the measurement and collection
of some data.  Because of this, team members need a general
knowledge and ability in fishery techniques as well as a
specific job skill.  Therefore, I would like to point out
that last summer I worked for the Department of Natural
Resources on a fish population study.  On that job I gained
electro-fishing and seining experience and also learned
how to collect and identify aquatic invertebrates.
```

By being specific about your accomplishments, you avoid the appearance of bragging. It is much better to say, "I was president of my senior class," than to say, "I am a natural leader."

One tip about job experience: Mention job experience even if it does not relate to the job you seek. Employers feel a student who has worked is more apt to be mature than one who has not.

Do not forget hobbies that relate to the job. You are trying to establish that you are interested in, as well as qualified for, the job.

Do not mention salary unless you are answering an advertisement that specifically requests you to. Keep the you-attitude. Do not worry about pension plans, vacations, and coffee breaks at this stage of the game. Keep the prospective employer's interests in the foreground. Your self-interest is taken for granted.

If you are already working and not a student, you construct the body in much the same fashion. The significant difference is that you will put more emphasis on work experience than on college experience. Do not complain about your present employer. Such complaints will lead the prospective employer to mistrust you.

In the last paragraph of the body refer the employer to your enclosed resume. Mention your willingness to supply additional information such as letters concerning your work, research reports, and college transcripts.

The Ending. The goal of the letter of application is an interview with the prospective employer. In your ending you request this interview. Your request should be neither humble nor overaggressive. Simply indicate that you are available for an interview at the employer's convenience and give any special instructions needed for reaching you. If the prospective employer is in a distant city, indicate (if you can) some convenient time and place where you might meet a representative of the company, such as the convention of a professional society. If the employer is really interested, you may be invited to visit the company at their expense. But you cannot ask for this expense-paid interview. If you really want the job, be prepared to pay your own way.

The Complete Letter.   Figure 13-9 shows the complete letter of application. Take a minute to read it now. The beginning of the letter shows that the writer has been interested enough in the company to investigate it. The desired job is specifically mentioned. The middle portion highlights the course work and work experience that relate directly to the job sought. The close makes an interview convenient for the employer to arrange.

The Resume

With your letter of application you enclose a resume that provides your prospective employer with a convenient summary of your education and experience. To whom should you send letters and resumes? When answering an advertisement, you should follow whatever instructions are given there. When operating on your own, send them, if at all possible, to the person within the organization for whom you would be working; that is, the person who directly supervises the position. This person normally has the power to hire for the position. Your research into the company may turn up the name you need. If not, don't be afraid of calling the company switchboard and asking directly for the name and title you need. Write to personnel directors only as a last resort. Whatever you do, write to *someone* by name. Don't send "To Whom It May Concern" letters on your job hunt. It's wasted effort. Sometimes, of course, you may gain an interview without having sent a letter of application—for example, when recruiters come to your campus. When you do, bring a resume with you and give it to the interviewer at the start of the interview. He or she will appreciate this help tremendously.

Recent research indicates that, as in the letter of application, neatness and brevity—ideally only one page—are of major importance in your resume.[3] A tip here: If you have wide experience to report you will find that the smaller elite type will help you fit it on one page. With really extensive experience, you might want to go to a printing process that reduces the print in size. In any case, you must invest a little money and have the resume printed or photocopied. Never send carbons or mimeographed sheets. Make the resume good-looking. Leave generous margins and white space. Use distinctive headings and subheadings. The sample in Figure 13-10 provides you with a good example of what a resume should look like. Take a look at it now before you read the following comments.

635 Shuflin Road
Watertown, CA 90233
March 23, 1984

Mr. Morrell R. Solem
Director of Research
Price Industries, Inc.
2163 Airport Drive
St. Louis, MO 63136

Dear Mr. Solem:

I read in the January issue of Metal Age that Dr. Charles E.
Gore of your company is conducting extensive research into
the application of X-ray diffraction to the solutions of
problems in physical metallurgy. I have conducted experiments
at Watertown College in the same area under the guidance of
Professor John J. O'Brien. I would like to become a research
assistant with your firm and, if possible, work for Dr. Gore.

In June, I will graduate from Watertown with a Bachelor of
Science degree in Metallurgical Engineering. At present, I
am in the upper 25 percent of my class. In addition to my
work with Professor O'Brien, I have taken as many courses
relating to metal inspection problems as I could.

For the past two summers, I have worked for Watertown Concrete
Test Services where I have qualified as a laboratory technician
for hardened concrete testing. I know how to find and apply
the specifications of the American Society for Testing and
Materials. This experience has taught me a good deal about
modern inspection techniques. Because this practical experience
supplements the theory learned at school, I could fit into a
research laboratory with a minimum of training.

You will find more detailed information about my education and
work experience in the resume enclosed with this letter. I
can supply job descriptions concerning past employment and my
report of my X-ray diffraction research.

In April, I will attend the annual meeting of the American
Institute of Metallurgical Engineers in Detroit. Would it
be possible for me to talk with some member of Price Industries
at that time?

Sincerely yours,

Jane E. Lucas
Jane E. Lucas

Enclosure

FIGURE 13-9 Letter of Application

RESUME OF JANE E. LUCAS

65 Shuflin Road
Watertown, California 90233
Phone: (213) 596-4236

Education
1982-1984

WATERTOWN COLLEGE   WATERTOWN, CALIFORNIA

Candidate for Bachelor of Science degree in
Metallurgical Engineering in June 1984. In
upper 25% of class with GPA of 3.2 on 4.0
scale. Have been yearbook photographer for
two years. Member of Outing Club, elected
president in senior year. Elected to Student
Intermediary Board in senior year. Oversaw
promoting and allocating funds for student
activities. Wrote a report on peer advising
that resulted in a change in college policy.
Earned 75% of college expenses.

1980-1982

SAN DIEGO COMMUNITY COLLEGE SAN DIEGO, CALIFORNIA

Received Associate of Arts degree in General
Studies in June 1982. Made Dean's List three
of four semesters. Member of debate team.
Participated in dramatics and intramural
athletics.

Business
Experience
1982, 1983

summers

WATERTOWN CONCRETE TEST SERVICES
                         WATERTOWN, CALIFORNIA
Qualified as laboratory technician for hardened
concrete testing under specification E329 of
American Society for Testing and Materials
(ASTM). Conducted following ASTM tests: Load
Test in Core Samples (ASTM C39), Penetration
Probe (ASTM C803-75T), and the Transverse
Resonant Frequency Determination (ASTM C666-73).
Implemented new reporting system for laboratory
results.

1982-1984

WATERTOWN ICE SKATING ARENA
                         WATERTOWN, CALIFORNIA
During academic year, work 15 hours a week as
ice monitor. Supervise skating and administer
first aid.

1978-1982

Summer and part-time jobs included newspaper
carrier, supermarket stock clerk, and salesperson for large department store.

Personal
Background

Grew up in San Diego, California. Travels
include Mexico and the Eastern United States.
Can converse in Spanish. Interests include
reading, backpacking, photography, and sports
(tennis, skiing, and running). Willing to
relocate.

References

Personal references available upon request.

February 1984

FIGURE 13-10 Resume Modeled After the One Taught Students at the
Harvard University Graduate School of Business Administration.

Address.   Use the address to which you want your mail sent. Give you phone number, and don't forget the area code.

Education.   List the colleges or universities you have attended in reverse chronological order; in other words, list the school you attended most recently first, the one before second, and so on. Do not list your high school. Give your major and date or expected date of graduation. Do not list courses, but list anything that is out of the ordinary, such as honors, special projects, and emphases in addition to the major. Extracurricular activities also go here. For most students, educational information should be placed first in the resume. People with extensive work experience, however, may choose to put that first.

Business Experience.   As you did with your educational experience, put your business experience in reverse chronological order. To save space and to avoid the repetition of *I* throughout the resume, use phrases rather than complete sentences. The style of the sample resume makes this technique clear. As we advise you to do in the letter of application, emphasize the experiences that show you in the best light for the kinds of jobs you seek. Use active verbs in your descriptions. Do not neglect less important jobs of the sort you may have had in high school, but use even more of a summary approach for them. You would probably put college internships and work-study programs here, though you might rather choose to put them under education. If you have military experience, put it here. Give highest rank held, list service schools attended, and describe duties. Make a special effort to show how military experience relates to the civilian work you seek.

Personal Background.   Provide some personal information about yourself, so that the company can see you as a whole human being. Also, personal information can in a subtle way point out desirable qualities you possess. Recent travels indicate a broadening of knowledge and probably a willingness to travel. Hobbies listed may relate to the work sought. Participation in sports, drama, or community activities indicates a liking for working with people. Cultural activities indicate you are not a person of narrow interests.

If you indicate you are married, you might want to say that you are willing to relocate. Don't say anything about health unless you can describe it as excellent. Because of state and federal laws concerning fair employment practices, certain information should not appear in your resume. Do not mention handicaps and do not include a photograph. You may, if you wish, choose to omit this entire section.

References. Do not use up precious space listing references. Simply indicate that they are available. You should have at least three. Be sure to get permission from people before you send their names to anyone. It's a smart idea to send a copy of your resume to people you use as references. This will help them to discuss you in a specific—and positive—way.

Dateline. At the bottom of the resume, place the date—month and year—in which you completed the resume.

Job Objective. We have not shown a job objective entry on our sample resume. Some research indicates that you should include one. But we rather agree with the Harvard Graduate School of Business Administration, which says,

> We recommend the use of a job objective in a resume only if you really feel sure of what you want to do. Sometimes students who do not feel such a sense of certainty attempt a vague catchall job objective. It usually shows. Even if successful, such a bluff is unwise. A job objective is a two-edged instrument. It will heighten interest in some employers, but will diminish it in others.[4]

If you omit the job objective in your resume, you can, of course, mention it in your letter of application. If you do decide to include this entry, it should look something like this:

```
Job Objective
     Desire work in food service management.
```

Place the job objective entry immediately after the address and align it with the rest of the entries.

Follow-Up Letters

Write follow-up letters (1) when you have had no answer to your letter of application in two weeks; (2) after an interview; (3) when a company refuses you a job; and (4) to accept or refuse a job.

When a company has not answered your original letter of application, write again. Be gracious, not complaining. Say something like this:

> Dear Mr. Souther:
>
> On 12 April I applied for a post with your company. I have not heard from you, so perhaps my original letter has been misplaced. I enclose a copy of it.
>
> If you have already reached some decision concerning my application, I would appreciate your letting me know.
>
> I look forward to hearing from you.
>
> Sincerely yours,

After an interview, be smart and within a week's time follow up your interview with a letter. Such a letter draws favorable attention to yourself as someone who understands business courtesy and good communication practice. Express appreciation for the interview. Draw attention to any of your qualifications that seemed to be important to the interviewer. Express your willingness to live with any special conditions of employment such as relocation. Make clear your hope for the next stage in the process, perhaps a further interview. If you include a specific question in your letter, it may hasten a reply. Your letter might look like this one:

> Dear Ms. Marchand:
>
> Thank you for speaking with me last Tuesday about the food supervisor position you have open.
>
> Working in a hospital food service relates well to my experience and interests. A feasibility study I am currently writing as a senior project deals with a hospital food service's ability to provide more varied diets to people with restricted dietary requirements. Would you like me to send you a copy?
>
> I understand that my work with you would include alternating weekly night shifts with weekly day shifts. This requirement presents no difficulty for me.
>
> Tuesdays and Thursdays are best for me for any future interviews you may wish. But I can arrange a time at your convenience.
>
> Sincerely yours,

When a company refuses you a job, good tactics dictate that you acknowledge the refusal. Express regret that no opening exists at the present time and express the hope that they may consider you in the future. You never know; they might.

Writing an acceptance letter presents few problems. Be brief. Thank the employer for the job offer and accept the job. Settle when you will report for work and express pleasure at the prospect of working for the organization. A good letter of acceptance might read as follows:

Dear Mr. Solem:

Thank you for offering me a job as research assistant with your firm. I happily accept. I can easily be at work by 1 July as you have requested.

I look forward to working with Price Industries and particularly to the opportunity of doing research with Dr. Gore.

Sincerely yours,

Writing a letter of refusal can be difficult. Be as gracious as possible. Be brief but not so brief as to suggest rudeness or indifference. Make it clear that you appreciate the offer. If you can, give a reason for your refusal. The employer who has spent time and money in interviewing you and corresponding with you deserves these courtesies. And, of course, your own self-interest is involved. Some day you may wish to reapply to an organization that for the moment you must turn down. A good letter of refusal might look like this one:

Dear Ms. White:

I enjoyed my visit to the research department of your company. I would very much have liked to work with the people I met there. I thank you for offering me the opportunity to do so.

However, after much serious thought, I have decided that the research opportunities offered me in another job are closer to the interests I developed at the University. Therefore, I have accepted the other job and regret that I cannot accept yours.

I appreciate the courtesy and thoughtfulness that you and your associates have extended me.

Sincerely yours,

LETTER AND MEMORANDUM REPORTS

Often a short report will be written as a letter or memorandum (usually called memo). Such reports seldom go over several pages. Consider memos as letters that stay within the organization, with a few differences. In a memo, more often than in a letter, you will be writing to someone you know quite well. The recipient may also be familiar with most of the background information. Therefore, more so than in letters, you can and should come directly to the subject and purpose of your message. Avoid irrelevant chitchat, but do remember that memos go from one human being to another. Avoid being coldly formal. Do not be concise to the point of brusqueness. Use a relaxed, conversational style. Use *I* and *you,* particularly *you.*

Perhaps the chief difference between memos and letters is format. Figure 13-11 shows a typical memo format. Frequently, organizations have printed memo forms. The printed form saves secretarial time and reminds the writer of all the parts of a memo. Figure 13-12 illustrates such a printed memo form. (Incidentally, the memo in Figure 13-12 is a real one. It illustrates both a well-written memo and the interest that organizations have in good communication.) When continuation pages are needed in memos, use the same format shown at the top of the continuation page illustrated in Figure 13-5, on page 285.

MEMORANDUM

Doublespace Twice

Date: 21 January 1984 cc: William Martin
Herbert Johnson
Gail McClure

Doublespace

To: Keith Wharton

Doublespace

From: James Tammen

Doublespace

Subject: Reallocation procedures

Doublespace

Doublespace

FIGURE 13-11 Memorandum Format

OFFICE MEMO

Date   15 December 1978

To   All Departments
<small>DEPARTMENT OR OFFICE</small>

Attention Department Heads

From   Robert Lauer, Employee Relations
<small>NAME OF PERSON, OFFICE OR AGENCY (INCLUDE CITY AND STATE)</small>

Subject Report Writing Class Update

The following is a list of objectives for a Report Writing Class pilot program. This list is drawn from your comments on the needs analysis done December 5, 1978. Each objective is a statement of what the trainees will be able to do if training is successful. These needs will be communicated to the consultant selected to run our pilot.

Please review the list to determine relevance to your department's needs. Call me by December 29, 1978 if you have any suggestions for improving the list.

Report Writing Trainees should be able to

1. Identify the target audience(s) of each of their reports by functional area and job level.

2. Write a one sentence statement detailing the purpose of the report, that is, what the information is to be used for.

3. List the significant findings of the report in descending order of importance to the target audience.

4. Construct a framework (outline) for a report that is appropriate to both the target audience and stated purpose.

5. Compose (from the outline) a report that is

    A. Informative & concise

    B. Logically organized

    C. Free of departmental jargon (foreign to target audience)

    D. Free of mechanical errors

    E. Readable

    F. An accurate depiction, where appropriate, of statistical data.

WHEN A REPLY IS REQUIRED, PLEASE RETURN THIS MEMO WITH YOUR HANDWRITTEN COMMENTS IF THAT IS CONVENIENT.

<small>6696 Rev. 3-72 Printed in U.S.A.</small>

FIGURE 13-12 Company Memorandum (Courtesy of Robert Lauer, St. Paul Fire and Marine Insurance Company.)

How to write a business letter

Some thoughts from Malcolm Forbes
President and Editor-in-Chief of Forbes Magazine

International Paper asked Malcolm Forbes to share some things he's learned about writing a good business letter. One rule, "Be crystal clear."

A good business letter can get you a job interview.

Get you off the hook.

Or get you money.

It's totally asinine to blow your chances of getting *whatever* you want–with a business letter that turns people off instead of turning them on.

The best place to learn to write is in school. If you're still there, pick your teachers' brains.

If not, big deal. I learned to ride a motorcycle at 50 and fly balloons at 52. It's never too late to learn.

Over 10,000 business letters come across my desk every year. They seem to fall into three categories: stultifying if not stupid, mundane (most of them), and first rate (rare). Here's the approach

I've found that separates the winners from the losers (most of it's just good common sense)–it starts *before* you write your letter:

Know what you want

If you don't, write it down–in one sentence. "I want to get an interview within the next two weeks." That simple.

List the major points you want to get across–it'll keep you on course.

If you're *answering* a letter, check the points that need answering and keep the letter in front of you while you write. This way you won't forget anything–*that* would cause another round of letters.

And for goodness' sake, answer promptly if you're going to answer at all. Don't sit on a letter–*that* invites the person on the other end to sit on whatever you want from *him*.

Plunge right in

Call him by name–not "Dear Sir, Madam, or Ms." "Dear Mr. Chrisanthopoulos"–and be sure to spell it right. That'll get him (thus, you) off to a good start.

(Usually, you can get his name just by phoning his company–or from a business **d**irectory in your nearest library.)

Tell what your letter is about in the first paragraph. One or two sentences. Don't keep your reader guessing or he might file your letter away–even before he finishes it.

In the round file.

If you're answering a letter, refer to the date

it was written. So the reader won't waste time hunting for it.

People who read business letters are as human as thee and me. Reading a letter shouldn't be a chore–*reward* the reader for the time he gives you.

Write so he'll enjoy it

Write the entire letter from his point of view–what's in it for *him?* Beat him to the draw–surprise him by answering the questions and objections he might have.

Be positive–he'll be more receptive to what you have to say.

Be nice. Contrary to the cliché, genuinely nice guys most often finish first or very near it. I admit it's not easy when you've got a gripe. To be agreeable while disagreeing– that's an art.

Be natural–write the way you talk. Imagine him sitting in front of you–what would you *say* to him? Business jargon too often is cold, stiff, unnatural.

Suppose I came up to you and said, "I acknowledge receipt of your letter and I beg to thank you." You'd think, "Huh? You're putting me on."

The acid test–read your letter *out loud* when you're done. You

"Be natural. Imagine him sitting in front of you–what would you say to him?"

might get a shock–but you'll know for sure if it sounds natural.

Don't be cute or flippant. The reader won't take you seriously. This doesn't mean you've got to be dull. You prefer your letter to knock 'em dead rather than bore 'em to death.

Three points to remember:

Have a sense of humor. That's refreshing anywhere–a nice surprise

FIGURE 13-13 How to Write a Business Letter (Courtesy International Paper Company.)

in a business letter.

Be specific. If I tell you there's a new fuel that could save gasoline, you might not believe me. But suppose I tell you this:

"Gasohol" – 10% alcohol, 90% gasoline – works as well as straight gasoline. Since you can make alcohol from grain or corn stalks, wood or wood waste, coal – even garbage, it's worth some real follow-through.

Now you've got something to sink your teeth into.

Lean heavier on nouns and verbs, lighter on adjectives. Use the active voice instead of the passive. Your writing will have more guts.

Which of these is stronger? Active voice: "I kicked out my money manager." Or, passive voice: "My money manager was kicked out by me." (By the way, neither is true. My son, Malcolm Jr., manages most Forbes money – he's a brilliant moneyman.)

"I learned to ride a motorcycle at 50 and fly balloons at 52. It's never too late to learn anything."

Give it the best you've got

When you don't want something enough to make the effort, making an effort is a waste.

Make your letter look appetizing – or you'll strike out before you even get to bat. Type it – on good-quality 8½" x 11" stationery. Keep it neat. And use paragraphing that makes it easier to read.

Keep your letter short – to one page, if possible. Keep your paragraphs short. After all, who's going to benefit if your letter is quick and easy to read?

You.

For emphasis, underline impor-

tant words. And sometimes indent sentences as well as paragraphs.

Like this. See how well it works? (But save it for something special.)

Make it perfect. No typos, no misspellings, no factual errors. If you're sloppy and let mistakes slip by, the person reading your letter will think you don't know better or don't care. Do you?

Be crystal clear. You won't get what you're after if your reader doesn't get the message.

Use good English. If you're still in school, take all the English and writing courses you can. The way you write and speak can really help – or hurt.

If you're not in school (even if you are), get the little 71-page gem by Strunk & White, *Elements of Style*. It's in paperback. It's fun to read and loaded with tips on good English and good writing.

Don't put on airs. Pretense invariably impresses only the pretender.

Don't exaggerate. Even once. Your reader will suspect everything else you write.

Distinguish opinions from facts. Your opinions may be the best in the world. But they're not gospel. You owe it to your reader to let him know which is which. He'll appreciate it and he'll admire you. The dumbest people I know are those who Know It All.

Be honest. It'll get you further in the long run. If you're not, you won't rest easy until you're

found out. (The latter, not speaking from experience.)

Edit ruthlessly. Somebody ~~has~~ said that words are ~~a lot~~ like inflated money – the more ~~of them that~~ you use, the less each one ~~of them~~ is worth. ~~Right on.~~ Go through your entire letter ~~just~~ as many times as it takes. ~~Search out and~~ **A**nnihilate all unnecessary words, ~~and~~ sentences – even ~~entire~~ *paragraphs*.

"Don't exaggerate. Even once. Your reader will suspect everything else you write."

Sum it up and get out

The last paragraph should tell the reader exactly what you want *him* to do – or what *you're* going to do. Short and sweet. "May I have an appointment? Next Monday, the 16th, I'll call your secretary to see when it'll be most convenient for you."

Close with something simple like, "Sincerely." And for heaven's sake sign legibly. The biggest ego trip I know is a completely illegible signature.

Good luck.

I hope you get what you're after.

Sincerely,

Malcolm S. Forbes

With these differences in mind, then, we can treat memos and letter reports the same. A common plan for both is as follows:

1. Begin by telling the reader the subject and purpose. Perhaps you are reporting on an inspection tour or summarizing the agreements reached in a consultation between you and the recipient. You may be reporting the results of a research project, or the beginning of one. Or you may be writing a progress report on a project that is under way but not completed. Whatever the subject and purpose may be, state them clearly. If someone has requested the report, name the requester.

2. What step two should be depends to some extent on company preference and upon the situation. Many companies and organizations prefer that any conclusions, decisions, and recommendations reached should be given at this point in the report. In a neutral situation, where the reader is not likely to feel hostile or upset because of the conclusions, we recommend this plan. But when you anticipate a hostile reader reaction, you would be safer if you gave your facts first and then your conclusions.

3. Develop your subject. Use the same techniques and rhetorical principles that you would use in a longer report. Quite often, however, because of the brevity desired in a letter or memo, your style may approach that of the informative abstract. (See pages 214–216.) You may also find listing a useful technique in letters and memos. (See page 171.) Remember to consider your audience precisely as you do in longer reports. If you are writing to an executive, do not fill your letter or memo with jargon and technical terms. If you must report a mass of statistics, try to round them off. If absolute accuracy is necessary, perhaps you can give the figures in an informal table. (See pages 247–248.) You may use headings in a letter or memo just as you would in a longer report. Many a business communication would profit by a few well-placed headings to provide better transition between points. (See pages 228–229.)

4. Frequently, you will conclude by telling the reader you will be happy to follow up with additional information, personal consultations, or other appropriate action.

All of the reports that we discuss in Chapters 14 through 17 can be and often are written as memos or letters. All of the rhetorical modes discussed in Chapters 6 and 7, used singly or in combination, are found in memos and letters.

However, when a report gets beyond four or five pages long, it's probably too long to be presented as a memo or letter. At this point, you must consider changing to a format that allows you to use such report elements as descriptive abstracts, title pages, and prefaces. See Chapters 8 and 11, particularly pages 161–165 and 200–217, for more on this matter.

HOW TO WRITE A BUSINESS LETTER

In *Business Week* magazine we came upon an advertisement, "How to Write a Business Letter," written for the International Paper Company by Malcolm Forbes, publisher of *Forbes* magazine. Forbes' thoughts about successful business letter writing agree so well with ours that, by way of a conclusion to this chapter, we reproduce the ad for you in Figure 13-13. Like Robert Lauer's memo in Figure 13-12, it illustrates that business people take good communication seriously.

EXERCISES

1. Write an unsolicited letter of inquiry to some company asking for sample materials or information. If you really need the information or material, mail the letter, but do not mail it as an exercise.

2. Write a letter to some organization applying for full- or part-time work. Accompany your letter with a resume. It may well be that you are seeking work and can write your letter with a specific organization in mind.

3. Imagine that you are working for a firm that provides a service or manufactures a product you know something about. Someone has written the firm a letter of inquiry asking about the service or product. Your task is to answer the letter.

4. Think about some service or product that has recently caused you dissatisfaction. Find out the appropriate person or organization to write to and write that person or organization a letter of complaint.

5. Swap your letter of complaint written for Exercise 4 for the letter of complaint written by another member of the class. Your assignment is to answer the other class member's letter. You may have to do a little research to get the data you will need for your answer.

6. Write a memo to some college official or to an executive at your place of work. Many of the papers you have written earlier in your writing course are probably suitable for a memorandum format. Or you could choose some procedure, such as college registration, and suggest a new and better procedure. Perhaps your memo could be to an instructor suggesting course changes. A look at Figure 3-3 on page 34 could also suggest a wide range of topics and approaches that you could use in a memo.

Chapter 14

Instructions for Performing a Process

YOU WILL FIND, once on the job, that instructing others to follow some procedure is a common task. Sometimes the instructions are given orally. However, when the procedure is done by many people or is done repeatedly, written instructions are a better choice. Instructions may be quite simple—perhaps comprising half a typed page—or exceedingly complex—a bookshelf full of manuals. They may be highly technical, dealing with operating machinery or programming computers, for example. Or they may be executive or business oriented, explaining how to create a certain kind of brochure or how to route memorandums through a company. The task is not to be taken lightly. A Shakespearean scholar who had also served in the British Army wrote the following:

> The most effective elementary training [in writing] I ever received was not from masters at school but in composing daily orders and instructions as staff captain in charge of the administration of seventy-two miscellaneous military units. It is far easier to discuss Hamlet's complexes than to write orders which ensure that five working parties from five different units arrive at the right place at the right time equipped with the proper tools for the job. One soon learns that the most seemingly simple statement can bear two meanings and that when instructions are misunderstood the fault usually lies with the original order.[1]

Sets of instructions may contain six possible sections:

- Introduction
- Theory or Principles of Operation

- List of Equipment and Materials Needed
- Description of the Mechanism
- Performance Instructions
- Troubleshooting Procedures

We do not present this list as a rigid format. For example, you may find that you do not need a theory section, or you may include it as part of your introduction. You may want to vary the order of the sections. You may want to describe or list equipment as it is needed while performing the process rather than in a separate section. Often nothing more is provided than the performance instructions. We describe all six sections so that you'll know how to write all of them. Take the information we give you and adjust it to suit your needs.

THE INTRODUCTION

Introductions to sets of instructions are usually short and to the point. Normally, they tell what the instructions cover and for whom they are intended. They may be as short as this introduction for a brochure on tree planting:

> Most home owners are interested in having healthy, attractive trees on their property. If you pay little attention to planting, your trees may grow slowly, continually lack vigor, or even die. However, your trees are more likely to thrive if they are properly planted and cared for. This bulletin describes the proper methods of handling and planting to give your landscape trees the best chance of success.[2]

After reading this introduction you know the audience—home owners. You know the subject matter is tree planting and care. You know the purpose is to show the reader the procedures to follow for healthy trees. Included in the scope of the bulletin are the proper handling and planting of landscape trees.

Look at Figure 14-1. There the introduction for a government consumer bulletin on paint and painting clearly lays out audience, subject, and purpose. The scope and plan of organization are shown by printing the table of contents on the same page as the introduction—an imaginative breakaway from the conventional method of dealing with these two items separately.

Both the planting and painting bulletins attempt to persuade the reader of the desirability of attending carefully to the instructions. Such motivation is common. The use of motivation is even more

Introduction

A good paint job not only adds beauty to your home, both interior and exterior; it also protects your investment. The right paint, properly applied to a surface carefully prepared, is an excellent barrier to weathering and decay.

But therein lies a problem. Surface preparation is an exacting task. The paint must be of good quality, and of a type designed for the particular job you want done. And it must be applied as the manufacturer intended or it may not cover well or dry properly to provide a satisfactory appearance.

PAINT AND PAINTING has been written for the nonprofessional painter. Its simple, clearcut instructions and illustrations will help you to do a good job of painting the interior and exterior of your home with a minimum of trouble and expense.

Because there are so many types of modern paints, however, it is essential that you READ LABELS CAREFULLY AND FOLLOW INSTRUCTIONS EXACTLY AS RECOMMENDED BY THE MANUFACTURER. The guidelines in this booklet are general in nature; the instructions from the manufacturer are specific to the particular product you are using. For the best results, use both of them.

Contents

FIGURE 14-1 Sample Introduction [From U.S., General Services Administration, *Paint and Painting* (Washington, D.C.: U.S. Government Printing Office, 1977).]

obvious in the following introduction to a short course on "capabilities books":

> Whether new company or old, large or small, industrial or consumer, almost every firm whose management expects to stay in business needs an updated "capabilities" book.
> Type of product sold, or kind of service rendered, makes no difference: Capability books are designed to convince prospects that this company must be given serious consideration.
> There are many reasons why the need today is almost universal. Most companies have a completely different profile from that of just a few years ago.
>
> *Diversification* is one reason.
>
> *Change in ownership* is another.
>
> *Space-age technologies and materials* have created new products for new markets.
>
> *Computerization* has resulted in other changes.
>
> Unless the company takes the trouble to tell its story of these changes, and to provide a broad explanation of what it can do now, important new users of its services and its products will be lost.
> In essence, the successful capability book is a door opener into new markets. It also provides significant side benefits.
> This fourth volume in the Mead Short Course examines the new capabilities book, and reproduces current examples from a number of different kinds of companies.[3]

Good introductions to instructions, then, are not much different from the introductions we describe for you in Chapter 11, "Formal Elements of Reports." (See pages 218–221.) They usually state subject, purpose, scope, and plan of organization. Frequently, they also attempt to motivate the reader.

When introductions are longer than the ones we have shown you, it is usually because the writers have chosen to include theory or principles of operation in the introduction. This is an accepted practice, but we tell you how to give such information in the next section.

THEORY OR PRINCIPLES OF OPERATION

Many sets of instructions contain a section that deals with the theory or principles of operation that underlie the procedures explained. Sometimes historical background is also included. These sections may be called "Theory" or "Principles of Operation," or they may have substantive titles like "Color Do's and Don'ts," "Purpose and Use of Conditioners," and "Basic Forage Blower Operation." Information about theory is presented for several reasons. People have a natural curiosity about the principles behind a procedure. Also, they want to know the purpose and use of the procedure. The good TV repair technician wants to know why turning the "vertical control" steadies the picture. Understanding the purposes behind simple adjustments enables the technician to investigate complex problems. What if nothing happens when the "vertical control" is turned? With a background in theory, the technician will know more readily where to look in the TV set to find a malfunction.

Such sections can be quite simple. In the following excerpt labeled "Color Do's and Don'ts," some basic color design theory is presented in easy-to-understand language:

DO Use light colors in a small room to make it seem larger.

DO Aim for a continuing color flow through your home—from room to room using harmonious colors in adjoining areas.

DO Paint the ceiling of a room in a deeper color than walls, if you want it to appear lower; paint it in a lighter shade for the opposite effect.

DO Study color swatches in both daylight and nightlight. Colors often change under artificial lighting.

DON'T Paint woodwork and trim of a small room in a color which is different from the background color or the room will appear cluttered and smaller.

DON'T Paint radiators, pipes and similar projections in a color which contrasts with walls or they will be emphasized.

DON'T Choose neutral or negative colors just because they are safe, or the result will be dull and uninteresting.

DON'T Use glossy paints on walls or ceilings of living areas since the shining surface creates glare.[4]

As simple as this excerpt is, it presents color principles. Readers are not only told to use light colors in a small room, but why—"to make it seem larger."

Theory sections can also be lengthy and complex. Here is an excerpt about hay baling from a John Deere manual:

PURPOSE AND USE

Baling is essentially a "packaging" operation. The range of materials that can be packaged with a baler is wide—from high-quality hays to crop residues—especially small-grain straw.

The popularity of baling is greater than any other hay-packaging method. Such wide acceptance has come because farmers like the size, shape, and density of bales. Bales are small enough to be man-handled for stacking and feeding, but dense enough for efficient inside storage. Loose hay may take twice as much storage space as an equal weight of baled hay. This high density (10 to 14 pounds per cubic foot) is extremely important to commercial hay growers because it makes long-distance transportation of baled hay economically feasible.

Baling used to be primarily a custom operation. Now, most hay farmers own balers. Small, lower-priced PTO balers match the needs of many smaller operations. Owning a baler lets a farmer time his own haying operations, which is essential for high-quality hay. Some very-small-acreage hay growers cannot economically justify ownership of a baler, but custom baling is readily available in most areas.[5]

In baling hay, a farmer must pay a good deal of attention to having the hay at the proper moisture content. Why is this so? John Deere explains:

MOISTURE CONTENT

Moisture content of the crop at baling time also influences the effectiveness of a baler. Hay baled at the proper moisture content yields more nutrition and forms well-shaped, solid bales. Such bales stack and store efficiently. Also, hay baled at the proper moisture content maintains high density throughout the storage period, and the bales are easier to handle and feed than poorly shaped, loose bales.

When hay is baled too dry, leaves shatter easily. Excessive leaf loss may result in a drastic reduction in quality. Hay baled too dry will not compact tightly; result: loose, poorly formed bales. In addition, the windrow pickup is not efficient in lifting dry hay.

Wet hay can be baled, but the baler operates under excessive strain. Wet hay is difficult to push through the bale chamber. This increases the load on the baler plunger and other baler parts. Also wet bales are normally heavy, even with the bale tension bars under reduced tension. Wet bales may dry and shrink in storage and become loose and sloppy. Or, bales of wet hay may heat and mold. Bales that have heated and molded in storage have lower quality and may even cause livestock to become sick.

Proper hay moisture content for baling depends on type and maturity of crop, weather conditions, and desired storage life. Effects of each factor must be carefully evaluated. Good managers and operators learn to estimate the best baling moisture content for their local conditions.[6]

Theory sections are not found only in instructions that deal with technical subjects such as hay baling. They can be found in any instructional document. In Mead Paper's short course on capabilities booklets, the writer tells the reader the objectives of the booklets:

Today's successful capabilities booklet is a well-designed, carefully planned, broad-brushed presentation of a company's various abilities.

It is not a parade of products and prices.

It is not a financial report.

It is not a bare-bones listing of equipment.

It is not a history or anniversary recital of corporate milestones.

The capabilities book educates and informs. It tells the reader in general terms how the company can help him, and why. It presents convincing photography with factual deductive writing. It is prepared specifically as a marketing tool—but its advantages range far beyond that goal. *It is a long-term sales builder.*

As previously mentioned, the successful capabilities book is prepared with a primary objective of expanding customers by generating sales in new markets.

Important *secondary uses* often can be anticipated. Here are a few:

Salesmen have a powerful reference tool to present.

Inquiries to the company can be answered with a short covering note accompanying the new book.

Employee recruiting is strengthened—at college and other levels.

Internal distribution helps educate employees who rarely are in a position to comprehend the corporate scope.

Financial people get a broad overview to supplement the company's financial reporting.

Community leaders and the local press get a new, accurate picture of the company's importance.[7]

As our excerpts illustrate, many diverse items of information can be placed in a theory or principles section. Remember, however, that the major purpose of the section is to emphasize the principles that underlie the actions later described in the performance instructions. The emphasis is on objectives, purpose, and use. In this section you're telling your readers why. Later you'll tell them how. You tell your readers what they need to know. Often, you should also tell them what, out of curiosity, they want to know. And you use language suitable to their level of knowledge in the subject matter. Obviously, good audience analysis is a must for any instruction writer.

LIST OF EQUIPMENT AND MATERIALS NEEDED

In a list of equipment and materials, you tell your readers what they will need to accomplish the process. A simple example would be a list of cooking utensils and ingredients that precedes a recipe. Sometimes in straightforward processes, or with knowledgeable audiences, the list of equipment is not used. Instead the instructions tell the readers what equipment they need as they need it: "Take a rubber mallet and tap the hubcap to be sure it's secure."

When a list is used, frequently each item is simply listed by name, as in the following list of painters' equipment. In this list the writer felt that the audience needed little further explanation and provided it in only three places:

CHECKLIST FOR PAINTERS

Before starting to paint, make sure you have all the tools needed to do the job right. Following are some of the items usually required:

Paint brushes	***Paint tray
Dust brush	Patching plaster
Stiff bristle brush	Putty or glazing compound
Wire brush	Putty knife
Caulking gun	Rags
*Cans	Rollers
Drop cloths	Roller extension poles
**Emery cloth	Sandpaper or production
Hammer	paper
Ladder	Spackling compound
Masking tape	Steel wool
Mixing paddle	Turpentine or other
Paint	solvents
Paint bucket	Wire combs
Paint scraper	
Paint strainer, wire mesh,	
or cheesecloth	

*For cleaning brushes with solvents.

**For cleaning and polishing metal surfaces.

***For painting with rollers.[8]

Sometimes, however, your audience analysis may indicate that more information is needed. You may want, for instance, to explain the properties and uses of the items used—as in the partially reproduced annotated list in Figure 14-2.

Often a table is useful in a materials or equipment list. Figure 14-3 is an example of a well-made table. It brings together paint products with their uses. Notice, also, the classification scheme in the table, based upon the substance being painted.

If the equipment cannot be easily obtained, you'll do your readers a service by telling them where they can find the hard-to-get items.

DESCRIPTION OF THE MECHANISM

Instructions devoted to the operation and maintenance of a specific mechanism include a section describing the mechanism. Also, when it is central in some process, the mechanism is frequently described.

Paint Types, Properties, and Uses

Types	Properties	Typical Uses
Oil base primers	Good adhesion and sealing; resistant to cracking and flaking when applied to unprimed wood; good brushing and leveling; controlled penetration; and low sheen. Unsuitable as a top coat and should be covered with finish paint within a week or two after application.	As primer on unpainted woodwork or surfaces previously coated with house paint.
Anti-rust primers	Prevent corrosion on iron and steel surfaces. Slow-drying type provides protection through good penetration into cracks and crevices. Fast-drying types are used only on smooth, clean surfaces, and those which are water resistant are effective where surfaces are subject to severe humidity conditions or fresh water immersion.	Priming of steel and other ferrous metal surfaces when good resistance to corrosion is required.
Galvanizing primers	High percentage of zinc dust provides good anti-rust protection and adhesion. Galvanizing/zinc dust primers give excellent coverage, one coat usually being sufficient on new surfaces. Two coats are ample for surfaces exposed to high humidity.	Priming of new or old galvanized metal and steel surfaces. Satisfactory as finish coat if color (metallic gray) is not objectionable.
House paints (oil or oil alkyd base)	Made with drying oils or drying oil combined with alkyd resin. Excellent brushing and penetrating properties. Provides good adhesion, elasticity, durability, and resistance to blistering on wood and other porous surfaces. Often modified with alkyd resins to speed drying time. Apply with brush to obtain strong bond, especially on old painted surfaces.	General exterior use on properly primed or previously painted wood or metal surfaces.

FIGURE 14-2 Annotated List [From U.S., General Services Administration, *Paint and Painting* (Washington, D.C.: U.S. Government Printing Office, 1977), 19.]

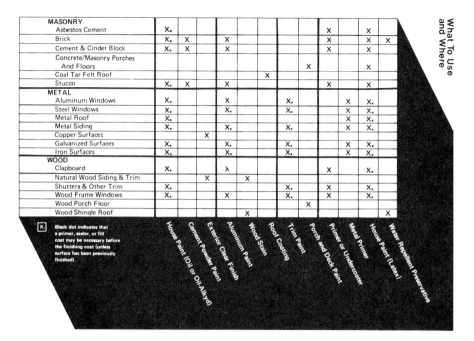

	House Paint (Oil or Oil-Alkyd)	Cement Powder Paint	Exterior Clear Finish	Aluminum Paint	Wood Stain	Roof Coating	Trim Paint	Porch and Deck Paint	Primer or Undercoater	Metal Primer	House Paint (Latex)	Water Repellent Preservative
MASONRY												
Asbestos Cement	X.								X		X	
Brick	X.	X		X					X		X	X
Cement & Cinder Block	X.	X		X					X		X	
Concrete/Masonry Porches And Floors								X			X	
Coal Tar Felt Roof						X						
Stucco	X.	X		X					X		X	
METAL												
Aluminum Windows	X.			X			X.			X	X.	
Steel Windows	X.			X.			X.			X	X.	
Metal Roof	X.									X	X.	
Metal Siding	X.			X.			X.			X	X.	
Copper Surfaces			X									
Galvanized Surfaces	X.			X.			X.			X	X.	
Iron Surfaces	X.			X.			X.			X	X.	
WOOD												
Clapboard	X.			λ					X		X.	
Natural Wood Siding & Trim			X		X							
Shutters & Other Trim	X.						X.		X		X.	
Wood Frame Windows	X.			X			X.		X		X.	
Wood Porch Floor								X				
Wood Shingle Roof					X							X

X. Black dot indicates that a primer, sealer, or fill coat may be necessary before the finishing coat (unless surface has been previously finished).

FIGURE 14-3 Sample Table [From U.S., General Services Administration, *Paint and Painting*. (Washington, D.C.: U.S. Government Printing Office, 1977), 18.]

In such sections, follow the principles for technical description given on pages 132–141. Break the mechanism into its component parts and describe how they function. The following excerpt is typical. (We have not reproduced all the figures referred to. For the ones we have reproduced, see accompanying Figure 14-4.)

The primary components of a baler are as follows (see Figs. 12 and 13):

- Pickup
- Auger and/or feed tines
- PTO shaft
- Feeder teeth
- Plunger
- Hay dogs
- Bale chamber
- Bale-measuring wheel
- Tying mechanism
- Needles
- Wire twister or twine knotter
- Bale chute

COMPONENT FUNCTIONS

The *pickup* lifts hay from the windrow and conveys it to the auger or feed rake (Fig. 14). *Hay compressors* on the pickup hold hay down for

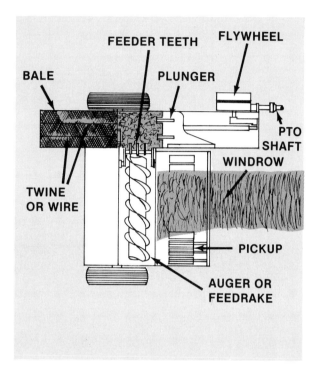

Fig. 15—Hay Movement into the Baling Chamber

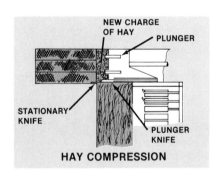

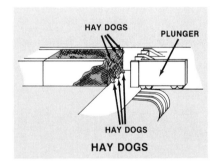

Fig. 17—Hay Is Compressed by the Plunger   Fig. 18—Hay Dogs Hold Compressed Hay

FIGURE 14-4 Illustrations to Accompany a Description of a Mechanism [From *Fundamentals of Machine Operation: Hay and Forage Harvesting* (Moline, Ill.: Deere and Company, 1976), 93, 95. Copyright © 1976 by Deere and Company. All rights reserved.]

uniform feeding, and keep strong winds from blowing hay out of the pickup.

Depending on the baler design, an *auger* or *feed rake* or *tines* deliver hay to the edge of the *bale chamber. Feeder teeth* then deliver hay into the baling chamber. The feeder teeth are timed to enter the baling chamber when the plunger is retracted (in the forward position) (Fig. 15).

The *plunger* is driven by a crank arm and pitman, and moves back and forth in the baling chamber about 80 strokes a minute. As the plunger moves into the baling chamber, the plunger knife (Fig. 16) moves past a stationary knife to slice off hay not totally within the chamber. The plunger then compresses the new charge of hay into the bale chamber (Fig. 17).

To keep the partly formed bale compressed in the bale chamber as the plunger retracts, *hay dogs* engage the bale (Fig. 18).

The process of feeding hay into the bale chamber and compressing it with the plunger is repeated until the bale is formed. The density of the bale is determined by adjusting the spring-loaded upper and lower tension bars on the bale chamber (Fig. 19). On some models, tension bars are controlled hydraulically.

The *bale-measuring wheel* rotates as the bale progresses through the bale chamber (Fig. 20). When the wheel has completed a preselected cycle, the *tying mechanism* is tripped. Bale length may be changed by adjusting the bale-measuring cycle.

Operation of the tying mechanism is timed to synchronize with the plunger movement. When the plunger is at the rearwardmost position and hay is fully compressed, *needles* deliver the wire or twine to the tying mechanism (Fig. 21). As the wire or twine is grasped by the tying mechanism, the needles retract, and the bale is tied.

The entire process is repeated as each completed bale passes through the bale chamber. The bale is finally forced out onto the ground (Fig. 22) or loaded as described earlier.[9]

As in the John Deere example, the description is generally accompanied by numerous photographs and illustrations, many of them annotated. Sometimes exploded views of the mechanism are provided, as in Figure 14-5. We hasten to add that in this context "exploded" means the mechanism is drawn in such a way that its component parts are separated and are thus easier to identify. Figure 14-5 makes the concept quite clear.

Notice that the author of the baler description made certain assumptions about the audience and used without definition terms such as *windrow* (a row of raked hay) and *pitman* (a connecting rod). Again, as always, audience analysis is the key. To reach the knowledgeable farmers the manual is intended for, the writer had only to define those terms that would be new to the audience or used in an unusual way.

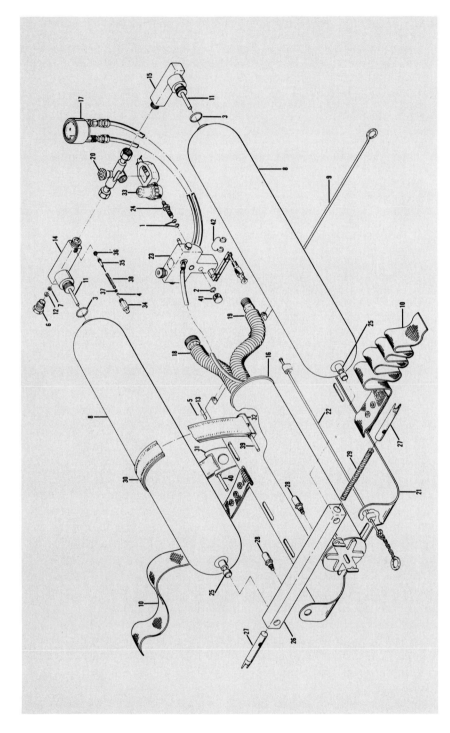

FIGURE 14-5 Exploded View of Equipment Assembly

PERFORMANCE INSTRUCTIONS

The actual instructions on how to perform the process obviously lie at the heart of any set of instructions. For a sample, let's look at a small portion of John Deere's instructions concerning balers—in this instance instructions on how to thread the needles of a baler with the wire that will eventually bind the hay bales:

> For typical wire needle threading (Fig. 40) [See our Figure 14-6]:
> - Make sure all wire pulleys turn freely.
> - Thread the wire from the right-hand coil through the guide (1).
> - Continue threading the wire around the front left-hand wire pulley (2).
> - With the needles in home position, thread the wire under the left-hand center wire pulley (3) and over the left-hand needle pulley.
> - Pull the wire back, loop it around the needle frame, and secure it with a twist (4).
> - Thread the left-hand wire through the guide (5), then repeat Steps 2, 3, and 4 through the right-hand pulleys and needle.[10]

See also the instructions for cleaning a paint brush in Figure 14-7. The same general principles apply to all performance instructions. Much of what we now tell you about writing them is well illustrated by the preceding excerpt and the instructions in Figure 14-7.

Style

When writing performance instructions, one of your major concerns is to use a clear, understandable style. To achieve this, write your instructions in the active voice and imperative mood: *Pull the wire back, loop it around the needle frame, and secure it with a twist.* The imperative mood is normal and acceptable in instructions. It's clear and precise and will not offend the reader. As an aid to simplicity, keep your sentences short. Ten- to twenty-word sentences are about right.

Notice also in both sets of instructions that a paragraph usually contains only one instruction and at the most two or three closely related instructions. Keep each step in a series clear and distinct from every other step.

Use familiar, direct language and avoid jargon. Tell your readers to *check* things or to *look them over*. Don't tell them to *conduct an investigation*. Tell your readers to *use* a wrench, not to *utilize* one. Fill your instructions with readily recognized verbs like *adjust, attach, bend, cap, center, close, drain, install, lock, replace, spin, turn,* and *wrap*. See also Chapter 9, "Achieving a Clear Style."

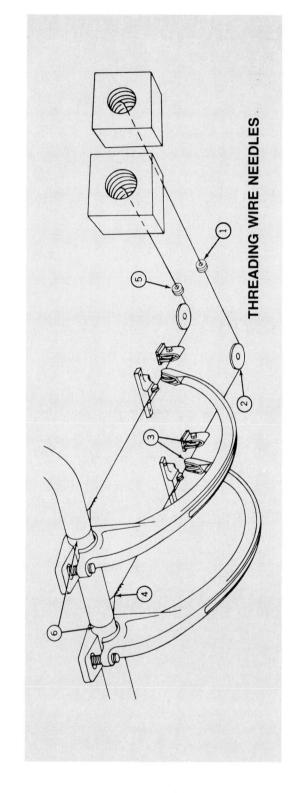

THREADING WIRE NEEDLES

FIGURE 14-6 Threading Needles of a Typical Wire-Tie Baler [From *Fundamentals of Machine Operation: Hay and Forage Harvesting* (Moline, Ill.: Deere and Company, 1976), 102. Copyright © by Deere and Company. All rights reserved.]

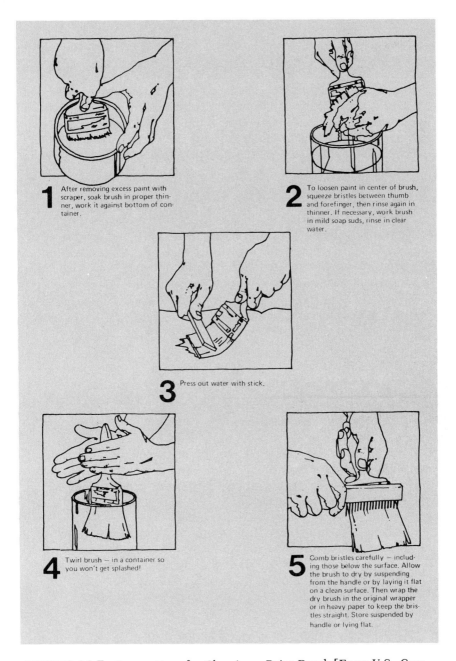

FIGURE 14-7 Instructions for Cleaning a Paint Brush [From U.S., General Services Administration, *Paint and Painting* (Washington, D.C.: U.S. Government Printing Office, 1977), 3.]

Graphics

Be generous with graphics. Word descriptions and graphics often complement each other. The words tell *what* action is to be done. The graphics show *where* it is to be done and often also show the *how*. Our samples demonstrate well the relationship between words and graphics. Note that graphics are often annotated with numbers or words to allow for easy reference to them. As Figure 14-7 demonstrates, graphics can be used to show the worker, or at least the worker's hands, actually performing the job. Figure 14-8 shows the close integration between words and graphics that the writers of manuals often achieve.

Organization

When writing performance instructions, organize the process being described into as many major routines and subroutines as needed. For example, a set of instructions for the overhaul and repair of a piece of machinery might be broken down as follows:

- Disassembly of major components
- Disassembly of components
- Cleaning
- Inspection
- Lubrication
- Repair
- Reassembly of components
- Testing of components
- Reassembly of major components

Notice that in this case the steps are in chronological order. Our samples, such as the one in Figure 14-7, also demonstrate chronological order. This matter of chronological order can be important in the extreme. Don't tell the reader,

> Place the mouthpiece in the mouth, squeeze the inhalation hose closed, and attempt to inhale through the mouthpiece. If it is possible to inhale with the inhalation hose closed off, the check valve is missing or defective.

And then add,

> Place the mouthpiece shut-off valve in the DIVING position before the test.

With the shut-off valve in the wrong position, the test will not come out right. Remember that your reader may be following you step by step.

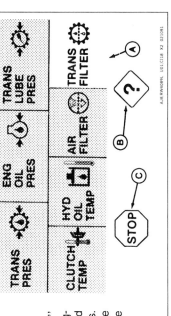

WATCH AIR FILTER RESTRICTION INDICATOR

Amber AIR FILTER indicator (A) and "Service Alert" (B) lamps glow to indicate a restriction to intake air flow through air filter. Service air filter elements as soon as possible. See Service Section.

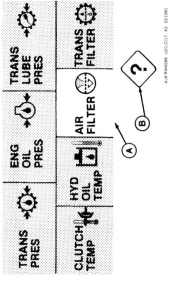

WATCH TRANSMISSION FILTER RESTRICTION INDICATOR

Amber TRANS FILTER indicator (A) and "Service Alert" lamp (B) glow if transmission filter becomes partially plugged. If filter is not replaced and restriction increases, red "Stop Engine" lamp (C) flashes and warning horn sounds. Stop tractor immediately and replace filter. See Maintenance Section. If filter replacement does not solve problem, see your John Deere dealer.

FIGURE 14-8 Integrated Text and Graphics [From *Operator's Manual: 8850 Tractor* (Moline, Ill.: Deere and Company, n.d.), 14.]

If not following a step in sequence might damage the equipment or injure the operator, word the step in the form of a warning:

WARNING

Avoid all contact with oil and grease. Oil coming in contact with high-pressure connections may result in an explosion. USE NO OIL.

Note that in our baler excerpt on page 331 the writer saved space and repetition by telling the reader to "repeat Steps 2, 3, and 4 through the right-hand pulleys and needle." This practice is legitimate, but use it with great care. Visualize your reader. Maybe he or she will be perched atop a shaky ladder, your instructions in one hand, a tool in the other. Under such circumstances the reader will not want to be flipping pages around to find the instructions that need to be repeated. You will be wiser and kinder to print once again all the instructions of the sequence. But if the reader will be working in a comfortable place, you would probably be safe enough to say, "Repeat steps. . . ."

This reader and situation analysis can help you make many similar decisions. Suppose, for example, that your readers are not expert technicians, and the process calls for them to use simple test equipment. In such a situation, you should include the instructions for operating the test equipment as part of the routine you are describing. On the other hand, suppose your readers are experienced technicians following your instructions at a comfortable workbench with a well-stocked library of manuals nearby. Then you can assume that they know how to operate any needed test equipment, or you can refer them to another manual that describes how to operate the test equipment.

Must your performance instructions always be in chronological order? When sequence is important in the proper performance of the process, then the answer is yes. But sometimes you will find that sequence is not crucial or perhaps can't even really be determined. In that case some sort of logical topical arrangement is called for (see pages 100–101). Look at the following instructions for driving a tractor on the road:

 CAUTION: Observe the following precautions when operating tractor on a road.

1. Be sure brakes are evenly adjusted, and couple pedals together before driving on a road. Avoid hard applications of brakes. A towed load of more than twice the weight of the tractor should have brakes. If not, drive slowly and avoid hills.
2. Be sure SMV emblem and warning lights are clean and visible. If

towed or rear-mounted equipment obstructs these safety devices, install SMV emblem and warning lights on equipment. See your John Deere dealer.

3. Turn light switch to "H" position. Never use flood lights or any lights which could blind or confuse other drivers. Always dim lights when meeting another vehicle.

4. If equipped use turn signals when turning. Be sure to return lever to center position after turning.

5. Drive slowly enough to maintain safe control at all times. Slow down for hillsides, rough ground, and sharp turns, especially when transporting heavy, rear-mounted equipment.

6. Before descending a hill, shift to a gear low enough to control speed without using brakes.[11]

These instructions are really a series of cautionary notes and sequence is not vital; therefore, the writer chose a topical arrangement.

Often a combination of chronological and topical instructions is used. For example, in Mead Paper's publication on capabilities booklets, the major routines of the performance instructions are in chronological order: preparation, production, and distribution. But look at the organization of the distribution section:

> Key to the ultimate success of the capabilities book is selecting the right people to receive it. In this case, building the list is far more difficult than it is for a product catalog or an annual report.
>
> Great care is required because in many cases it will be new prospects in new markets who will prove the most fruitful of all recipients. You won't want to miss important names, nor should you select poor prospects. Here are a few of the groups to be considered—either by mail or personal distribution:
>
> *New prospects*—executives and purchasing agents in companies in every industry where you should now be doing business but are not.
>
> *Existing customers*—especially executives in other divisions or areas not using your services and who possibly are not even aware of your new profile.
>
> *Financial community*—possibly the same list used for the annual report: financial analysts and business editors of newspapers and periodicals.
>
> *Shareholders*—depending on the nature of your company and the number of its shareholders. If the booklet will help strengthen stockholder relations—or help sell them as "customers" for company products—then it probably should be distributed to them.
>
> *Prospective employees*—for recruiting at campuses and other levels.
>
> *Suppliers*—sources of many indirect sales, suppliers are often neglected in the distribution of corporate literature. They can and should be loyal supporters, but are often in woeful need of "education."
>
> *Answering inquiries*—of a general nature directed to the company. In the case of big corporations, this can save a great deal of expensive office time.
>
> *Coupon returns*—companies which regularly advertise in print fre-

quently include a coupon return for requesting new literature such as the capabilities book.

Salesmen—certainly almost every sales representative should carry an adequate supply of the capability book, and hand it out wherever feasible.

Employees—probably not all of them, but certainly all those in key positions who should be aware of the company's new capabilities but for one reason or another are not.[12]

No particular sequence is possible, so a topical organization is used.

Ordinarily, then, you will use a chronological order in performance instructions. But sometimes you will find that a topical plan or a combination chronological-topical plan is the proper approach.

Format

Modern practice in format tends toward simplicity in numbering systems and generous spacing. Instruction writers do not want their readers making mistakes because they read cramped and crowded lines incorrectly. Figure 14-9 is a good example of today's style. It also illustrates rather well how to handle the computer-human being interaction. In side-by-side columns the instructions tell you what the computer does or displays and what the operator's proper response is.

Sometimes no numbering is used at all. The writer depends upon white space alone to keep the steps distinct or perhaps uses bullets as on page 334. However, in a long set of instructions you may wish to use a numbering system to keep the sequence straight or to allow for cross-reference, such as "See Section 4.2c." Often the conventional outline system of roman numerals, arabic numerals, capital letters, and lowercase letters will suffice (see page 230). Or you may want to use the decimal system (see page 231).

TROUBLESHOOTING

Many sets of instructions include a troubleshooting section designed to aid the technician. They usually take the form of a three-column chart with headings such as "Problem," "Possible Cause," and "Possible Remedy." Figure 14-10 shows two typical examples. Notice that the top chart in the figure uses graphics to illustrate the problem, an excellent technique that should be used where possible. Notice also that the remedies are given as instructions in the active voice, imperative mood.

The second example at the bottom of Figure 14-10 gives page references when appropriate to guide the reader to additional information.

ZEDIT for customer summary file

- Edits the customer summary file
- Prints a listing of those items in the customer summary file which have errors
- Activates customer summary file for sales analysis

What you need to run this procedure . . .

SALP30

Before you begin . . .

- Any errors in the customer summary file must be corrected before the file can be activated for sales analysis.
- You will be prompted to activate the files for sales analysis. Once you have activated the customer summary file, you will not be prompted to activate it again.

	What the system does or displays	What you do
1		Ready the system.
2	←READY	If the last procedure you ran loaded SALP30, go to Step 3. If not, insert diskette SALP30, then type in: ZLIBRARY SALP30
	Loads diskette.	
	←READY	After ←READY is displayed, remove diskette SALP30, and go to Step 3.
3	←READY	Type in: ZEDIT CUSTSUM
4	ACTIVATE FILE FOR SALES ANALYSIS? (YES/NO)	Type in: YES to activate sales analysis. Otherwise, type in: NO
5	Prints an edit listing. ←READY	File the listing for audit.

End of ZEDIT for customer summary file

FIGURE 14-9 Sample Format [From *System/32 Manufacturing Management Accounting System Sales Analysis Run Book* (Atlanta, Ga.: IBM Corporation, 1975). Reprinted by permission. Designed by Gerry Cohen.]

TROUBLE-SHOOTING

Most baler operating problems are caused by improper adjustment or delayed service. This chart is designed to help you when a problem develops, by suggesting a probable cause and the recommended solution.

Apply these suggested remedies carefully. Make certain the source of the trouble is not some place other than where the problem exists. A thorough understanding of the baler is a must if operating problems are to be corrected satisfactorily. Refer to the operator's manual for detailed repair procedures.

TROUBLE-SHOOTING CHART

PROBLEM	POSSIBLE CAUSE	POSSIBLE REMEDY
Knotter Difficulties — Twine Baler		
 KNOT IN TWINE OVER BALE	Tucker fingers did not pick up needle twine or move it into tying position properly.	Adjust tucker fingers. Adjust needles or twine disk. Check twine disk and twine-box-tension. Install plungerhead extensions.
	Hay dogs do not hold end of bale.	Free frozen hay dogs. Replace broken hay-dog springs. Reduce feeding rate. Install plungerhead extensions.
TWINE BROKEN IN KNOT	Extreme tension on twine around billhook during tying cycle causes twine to shear or pull apart.	Loosen twine-disk-holder spring. Smooth off all rough surfaces and edges on billhook.

ELECTRICAL SYSTEM

PROBLEM	POSSIBLE CAUSE	POSSIBLE REMEDY	PAGE REFERENCE
Battery will not charge	Loose or corroded connections.	Clean and tighten connections.	66
	Sulfated or worn-out battery.	Check electrolyte level and specific gravity.	67
	Loose or defective alternator belt.	Adjust belt tension or replace belt.	52
"CHG" indicator glows with engine running	Low engine speed.	Increase speed.	
	Defective battery.	Check electrolyte level and specific gravity.	67
	Defective alternator.	Have your John Deere dealer check alternator.	

FIGURE 14-10 Sample Troubleshooting Charts [Top sample from *Fundamentals of Machine Operation: Hay and Forage Harvesting* (Moline, Ill.: Deere and Company, 1976), 111. Copyright © by Deere and Company. All rights reserved. Bottom sample from *Operator's Manual: 850 and 950 Tractors* (Moline, Ill.: Deere and Company, n.d.), 76.]

PUTTING IT ALL TOGETHER

In the preceding sections we have discussed the possible parts of a set of instructions. To put it all together for you, we now present and discuss a short instructional document—a fact sheet called *Driving a Wellpoint.** In content and style, it represents a paper well within the reach of a student.

DRIVING A WELLPOINT
ROGER E. MACHMEIER

A wellpoint or drive point is a pipe with wall openings large enough to allow water to enter and small enough to keep the water-bearing formation in place. Wellpoints suitable for hand driving are available in sizes from 1¼ to 2 inches in diameter and from 18 inches to 3 feet long. The size of openings in the wellpoint is determined by the relative grain size of the material in the water-bearing formation. Some of the finest grains adjacent to the wellpoint should be removed by pumping to make the well more productive (see figure 1). However, a wellpoint should not be expected to yield large quantities of water.

Comments

The first two paragraphs combine an introduction and a theory section. Key terms are defined, and the use of wellpoints and shallow wells are explained. The third paragraph contains a warning. The author has wisely placed the warning early in the paper. Basically, the introduction is good. It could be improved, however. The first and second paragraphs would actually be better reversed. Also, some mention of the intended audience would have been helpful.

The fact sheet contains three excellent graphics. Figures 1 and 3 clarify the concept of a shallow well and the procedure for driving a wellpoint. Figure 2 supplements and amplifies the written definition of a wellpoint. Notice that the text refers to the figures whenever such reference is useful. All three figures are well annotated.

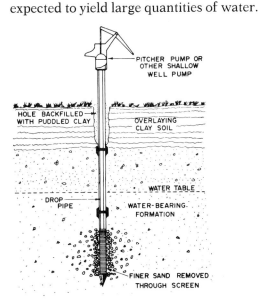

FIGURE 1 Installed wellpoint.

*Roger E. Machmeier, *Driving a Wellpoint* (St. Paul, Minn.: University of Minnesota, Agricultural Extension Service, 1972).

Why Wellpoints are Used

It may be desirable to develop a water supply for sprinkling lawns, gardens, etc. An economical supply can often be obtained from a shallow aquifer (water-bearing formation) through a wellpoint. The water table should be high, preferably within 10 feet and no farther than 15 feet underground. The wellpoint must be driven deep enough to penetrate a water-bearing formation below the water table, but it should not exceed 25 feet in depth.

Shallow water tables are susceptible to pollution. Drain fields, dry wells, animal wastes, heavy fertilizer applications, etc. can contaminate a shallow water table. Recharge is usually from rainfall, falling directly above and percolating downward to the water table. As the water moves downward it may carry contaminants. Thus, extreme care and periodic testing are necessary if the water is used for drinking.

Materials Needed

The following materials are needed: wellpoint, riser pipe (in 5- or 6-foot lengths, with 6-inch nipple), couplings, drive cap, and pipe thread compound.

Openings in the wellpoint should be large enough to permit the finer particles of the water-bearing formation to enter the wellpoint while keeping the coarser particles out. With proper sized openings, development of the well removes the finer particles and forms an envelope of porous and permeable material around the screen (see figure 1).

A local well contractor or hardware dealer may have valuable suggestions on the most suitable opening size for the wellpoint.

Several types of wellpoints are constructed of different materials. A few

The second major section, "Materials Needed," details the materials needed. (Note the use of headings to separate the sections.) The first paragraph under this heading lists the materials. The following paragraphs present details to help the reader select and use the materials wisely. The writer's audience analysis is evident in this and the next section, "Tools Needed." Because the writer goes into so much detail about the materials, we can deduce that he sees his audience as people inexperienced in the specific task, driving a

types are shown in figure 2. For mesh-covered wellpoints, the size of openings is designated by the mesh size. Common sizes, from larger to smaller openings, are 40-, 50-, 60-, 70-, and 80-mesh. For wellpoints with slot type openings, common slot sizes are 18, 12, 10, 8 and 7 slot. Slot sizes are the opening width in thousandths of an inch. No. 18 slot is 0.018-inch wide, No. 12 is 0.012-inch wide, etc.

wellpoint. But in the next section, "Tools Needed," he lists tools without comment, indicating that the writer sees his audience as being skilled tool users. The writer probably visualizes his audience as farmers or home owners used to doing their own routine maintenance.

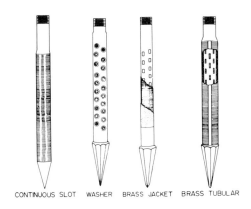

CONTINUOUS SLOT   WASHER   BRASS JACKET   BRASS TUBULAR

FIGURE 2   Types of wellpoints.

The riser pipe should be galvanized pipe in 5- or 6-foot lengths for convenient hand driving. A standard 21-foot length of pipe cut into four pieces normally is adequate for a driven well.

Special drive pipe couplings, which allow the pipe pieces to butt together, are desirable. The impact of driving is then transmitted through the pipe and not the pipe threads. Under severe driving conditions, standard pipe couplings may cause thread cracking or failure.

The drive cap is placed on the top of the pipe section being driven. The cap transmits the blow to the pipe and protects the threads.

All pipe joints should be screwed tightly after threads are carefully cleaned and oiled. White lead or pipe thread compound should be used to improve airtightness.

Tools Needed

A posthole digger or soil auger, sledge hammer, carpenter's level or plumb bob, and pipe wrenches are needed for hand-driving a wellpoint.

Installation Method

The well site should be at least 75 feet from sources of contamination such as a septic tank, cesspool, dry well, drain field, foundation drain, downspout, marshy area, or animal yard.

Dig or drill a vertical hole 1–2 inches larger than the wellpoint and as deep as possible with the posthole digger or auger (figure 3). The hole usually cannot be drilled more than a foot below the water table.

Rub a bar of soap over all the openings of the wellpoint to help prevent clay and sand from entering and to reduce friction during driving.

Attach a length of riser pipe to the open end of the wellpoint. Clean the threads, add the pipe thread compound to the outside threads, and make a tight connection with the pipe wrenches. Insert the wellpoint into the hole. Attach the nipple and drive cap to the top of the riser pipe. Do not use pipe thread compound on the drive cap.

Make sure that the pipe and wellpoint assembly are vertical by using the carpenter's level or the plumb bob. Use the sledge or driver to strike the drive cap with square, solid blows (figure 3).

When the drive cap is about 4 inches above the ground, unscrew the drive cap and nipple and place the unit on another section of pipe.

The "Installation Method" is the performance instructions. Notice that it begins with a wisely placed cautionary note about well location. The writer uses short sentences and paragraphs to keep his instructions clear and separate. He uses the active voice and imperative mood in most of this section. He assumes an experienced tool user who does not have to be told, for example, how to use a plumb bob or carpenter's level. But he does take the time to explain the use of the soap, a point even an experienced tool user might be curious about.

The writer follows chronological order throughout the performance instructions. He

Clean the threads on the new section of pipe and on the pipe in the ground and add pipe thread compound to the outside threads. Tighten the joint using two pipe wrenches working in opposite directions to avoid twisting the assembly in the ground, and resume driving.

frequently gives the reader helpful tips to make the job easier or to prevent complications.

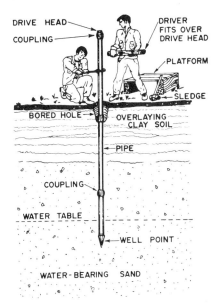

DRIVE HEAD

COUPLING

DRIVER FITS OVER DRIVE HEAD

PLATFORM

SLEDGE

BORED HOLE

OVERLAYING CLAY SOIL

PIPE

COUPLING

WATER TABLE

WELL POINT

WATER-BEARING SAND

FIGURE 3 Driving a wellpoint.

Continue the above procedure, adding sections of riser pipe as needed. To keep the threaded joints tight during driving, give the riser pipe an occasional half turn with a wrench. Use the wrench only to take up slack in the threaded joints and take care not to twist the pipe severely.

Pour water into the top of the pipe at regular intervals. This makes driving easier and will determine when a water-bearing formation is reached. When a gallon of water disappears into the formation within 2 minutes after being poured into the well, it is unnecessary to drive any deeper.

Enough of the riser pipe should extend above ground so the desired pump can be attached conveniently. For example, a hand pitcher pump should be about 3 feet above the ground. For an electric pump, the top of the riser should be a foot or less above ground. The pipe may have to be cut and threaded to obtain the proper height, or it may be possible to drive the pipe to the proper height.

To prevent pumped water and other surface water from moving downward along the pipe to the water-bearing formation, backfill the drilled hole with clay soil (figure 1). The clay should be extremely wet so it will puddle or flow like heavy grease. Fill the hole somewhat higher than ground level. Cement can be poured in a base around the riser pipe. However, using cement makes it extremely difficult to pull the well at a later date.

There are several ways to clean a new well by surging and pumping. One simple method is to take a wooden rod or closed-end pipe and—simulating piston action—rapidly work it up and down for about 5 minutes just below the water level in the well. This surging effect will draw fine, loose sand and silt into the well, leaving the coarser and more permeable material outside the wellpoint. Remove the fine sand from the well with a pitcher pump or other pump capable of handling sand.

Another means of cleaning out the sand is to jet water into the well with a garden hose inserted to the bottom. The sand and silt particles will wash out around the hose. Repeat until no more sand is obtained by pumping.

Attach the pump to be used following the manufacturer's instructions. To use the well water for domestic purposes, the

The format is simple. No numbering system is used. The writer puts only single instructions or very closely related instructions into each separate paragraph.

pump should be connected to a drop pipe installed inside the well (figure 1). A drop pipe is used so that the underground pipe would be under suction if the pump were connected directly to the well.

The well should be disinfected before the well water is used for drinking. For a 1½- to 2-inch well, add 1 cup of 5.25 percent hypochlorite laundry bleach to 3 quarts of water and pour the solution into the well. After 12 hours, pump the well until the chlorine odor is no longer objectionable. For additional information on disinfection procedures, see Agricultural Extension Service M-156, *Chlorination of Private Water Supplies.*

Forty-eight hours after pumping, the water can be sampled for bacteriological examination. Write to the Minnesota Department of Health, Minneapolis 55440 for a sampling kit that includes instructions and a sterile container. The water will be tested at no charge for bacteria, nitrates, and detergents. If a shallow well is used for domestic purposes, the water should be tested annually or semiannually.

It is usually a sound policy to have a reliable well driller develop a sanitary water supply for a home. Water purity and personal health should be the major consideration in developing a domestic water supply.

As is sometimes done, this set of instructions refers the reader to other instructions—*Chlorination of Private Water Supplies*—for additional information. However, the writer includes enough information on disinfection for the reader to accomplish the task. Never refer readers to documents that are not easily obtainable.

The concluding paragraph presents once again the warning that shallow wells are not the best source for drinking water. The writer obviously wants this point taken seriously.

The fact sheet you have just read is typical of the thousands of sets of instructions that are turned out yearly to help people carry out various procedures. It contains most of the sections we have described in this chapter. It does lack a section for "Description of the Mechanism." But the wellpoint, the major mechanism involved, is described in the introduction and again in the "Materials Needed" section. Only a troubleshooting section is completely missing. The paper illustrates well our advice to combine the sections as needed and to omit those sections not needed.

Notice that the paper has no particular conclusion, unless the final warning could be considered as such. For the most part, sets of instructions have no conclusions. They simply end with the last instruction. On occasion, particularly when writing for lay people, you might wish to close with a summary of the chief steps of the process or, perhaps, a complimentary close (see page 225). However, such endings are not general practice.

A FINAL WORD

Let us close this chapter on instructions with an important but often overlooked bit of advice. When you are writing instructions, check frequently with the people who are going to use them. Show them a sample of your theory section. Discuss it with them. See if they understand it. Does it contain too much theory or too little? Submit your performance instructions to the acid test. Let members of the audience for whom the instructions are intended—but who are not familiar with the process—attempt to perform the process following your instructions. Encourage them to tell you where your instructions are confusing. If they cannot follow your instructions, don't blame them. Examine your instructions to see where you have failed. Probably you will find you have left out some vital link in the process or assumed knowledge on the part of your readers that they do not possess.

EXERCISES

1. Writing instructions offers a wide range of possible papers. Short papers might consist of nothing more than an introduction and a set of performance instructions. Examples—good and bad—of such short instructions can be found in hobby kits and accompanying such things as toys, tents, and furniture that must be put together. Textbook laboratory procedures frequently are examples of a short set of instructions. Write a short set of instructions. Here are some suggested ideas:

   Developing film
   Making corrections on a typewriter
   Drawing a blood sample
   Applying fertilizer
   Using a lawn mower
   Setting a bicycle gear
   Building a dog house
   Replacing a broken window
   Cleaning a carpet
   Balancing a checkbook

2. The fact sheet presented on pages 341–347 is a good example of a longer set of instructions that uses most of the possible sections described in this chapter. Write a similar set of instructions that includes an introduction, a theory section, a list of equipment and materials needed, and performance instructions. If needed, also provide a description of a mechanism and troubleshooting procedures. Here are some suggested topics:

Laying a concrete patio
Tuning a radio transmitter
Writing (or following) a computer program
Setting up an accounting procedure for a small business
Conducting an agronomy field test
Checking blood pressure
Painting an automobile
Baking bread

Chapter 15

Proposals

ALL PROJECTS HAVE to begin somewhere and with someone. Initiative is required to attack the endless amount of work to be done in the world. Think of the projects and programs that have been started in your own community this past year. For example, the citizens of your community may have desired a public swimming pool for many years, but nothing happened until some individual took up the question and presented it to the town council as a proposal. The individual may have been a private citizen, a member of the council, or the editor of the local newspaper.

In business, industry, and government, proposals are an everyday necessity. As in community affairs, the original suggestion may come from the person, company, or agency that would like someone else to do a job that needs doing. On the other hand, a person, company, or agency that desires additional work may offer to render service to another. In either case, an oral or written proposal is required.

The flow chart (Figure 15-1) depicts the typical routes of a proposal with its inception through award of contract.

Block 1 indicates the "unsolicited situation," in which one offers, unasked, to perform work for another. Block 1a indicates the opposite or "solicited situation," in which someone requests the services of someone else. Scanning across the flow chart, you see the various stages that ensue until either a work contract is awarded or the proposed project dies out before any productive work is done.

Referring to block 1a of Figure 15-1, you see that companies, institutions, and agencies solicit work to be done by outsiders. Informally,

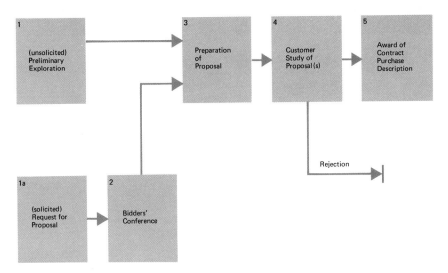

FIGURE 15-1 Flow Chart of Activities from Proposal to Award of Contract

they spread the news by word of mouth. Officially and formally, they issue a Request for Proposal (RFP) and distribute it to likely prospects. The RFP may be a one-page letter or it may be a bound document of a hundred pages or more.

The recipients who are interested in the project outlined in the RFP respond with a proposal (that is, they bid on it) in the hope they will win out against the competition. The interim is a time of anxious waiting. During times of normal prosperity the qualfied competitor stands, perhaps, one chance in ten of winning out in the competition. Happy is the contractor who is awarded the contract. Happier still is the person who can live with the contract and gain a profit from it.

Refer again to Figure 15-1, specifically, to block 1, at which point the course of the unsolicited proposal begins.

The term *unsolicited* as applied to proposals may be misleading. There is a great deal of exchange of information within the business community. The word gets around that X Company might be interested in having someone else conduct a research project for it. Those who are privileged to hear the word take it back to their own Y Company. Heads get together, inquiries are discreetly made, and a "feeler" letter may go to X Company. As a result, a proposal may be prepared on the basis of slim information. Again, chances are low that this kind of unsolicited proposal will be accepted, but the chance is worth taking if Y Company has research personnel to spare.

The customary procedure for unsolicited proposals is as follows:

1. An individual or group desiring to do a task for pay prepares an unsolicited proposal and sends it to an appropriate sponsor—a company, a private foundation, or a government agency.
2. The sponsor, who will pay for the task if approved, considers the merits of the proposal, the qualifications of the people involved, and the benefits that might result if the task has a productive outcome.
3. If the benefits seem worth the money, the proposal is funded.

When and if you reach the stage that an unsolicited proposal seems to be in order, go to the library to dig up all the supporting data you can find. Identify one or more sponsors who might be interested. Consult with knowledgeable persons in your organization. Write a letter to the proposed sponsor to obtain its specifications for proposals. Then settle down for weeks of work to create an attractive and persuasive proposal. The chances may be worth your investment of time and energy.

Because the "unsolicited proposal" brings all proposal functions into play, we limit the rest of our discussion to it. Recognize, however, that where our advice on proposal preparation necessarily has to be somewhat general, the directions in an RFP may be quite specific. When such is the case, be sure to follow the RFP directions to the letter. Failure to do so will usually result in a rejected proposal. Whether the proposal is solicited or unsolicited, we can safely assert that a proposal is designed to discharge two sales functions:

- To get a proposal accepted.
- To get you (or your company) accepted to perform work.

Every item of content, every point of diction, every matter of display and format must be designed to accomplish these dual objectives of all proposals.

BASIC PLAN OF PROPOSALS

Short proposals of several pages will probably be presented as a letter or memo (see pages 277–289 and 311–316). Longer proposals may be presented in a formal report format and be accompanied by a letter of transmittal (see pages 202–203). Remember that a proposal reaches your prospective client's desk as a "package." Once the wrappings are removed, the content will trigger a reaction. The initial reaction will be visual and tactile. In larger reports, a strong binding, an attractive cover, clean printing, and standard format will place the reader in a receptive mood. You must avoid "brochure-manship"—conspicuous overdressing, such as hand laid paper, half-calf covers, and deckled edges. But there is a happy medium between too much and too little. Neatness, good layout, commercial quality printing, and careful workmanship may tip the scale of judgment in your favor. Whether large or small, the report should be organized for selective reading through the use of headings (see pages 228–229). In one common format, the parts of a proposal consist of the following:

Project Summary
Project Description
    Introduction
    Rationale and Significance
    Plan of Work
    Facilities and Equipment
Personnel Qualifications
Budget
Appendixes

This basic plan is illustrated in Figure 15-2 on pages 363–365. The plan can accommodate proposals that run in length from a page or two to mammoth proposals, for millions and billions of dollars, that may fill a five-foot shelf.

PROJECT SUMMARY

Remember that proposals are read by busy people. Whether they are scientists or business people, when reading a proposal they are carrying out an executive, decision-making role: Should the proposal be rejected or accepted? Therefore, treat them like executives and put a summary of the key factors of your proposal up front (see pages 222–227). Include in your summary the objectives of your proposed work. Point out why the work is relevant to the reader or the subject matter area. Clearly and succinctly summarize the plan of work. Because you can't write a good summary until you have completed the rest of the proposal, the summary should be the last thing you write.

PROJECT DESCRIPTION

The reviewer of your proposal will judge your proposal largely by the material in your project description. The project description itself breaks down into four parts: introduction, rationale and significance, plan of work, and facilities and equipment.

Introduction

In the introduction you set down everything needed to inform your readers about the objectives of your proposed work. Make the subject and purpose—the *what* and the *why*—of your work completely clear. When proposal readers cannot tell immediately where your work is heading or what it will accomplish they lose interest rapidly. The introduction should point out, briefly, the *so-what* as well as the *why*. That is, in addition to telling what you hope to accomplish with the work, explain the implications of the work. In so doing, you are telling the readers how the work will be valuable and, therefore, why it is worth paying for. You will cover this same point much more thoroughly in the rationale and significance section.

Rationale and Significance

In your rationale and significance section you bring your sales ability to play. It's at this section that the readers look to see if they are going to get their money's worth. In general, a proposal presents a plan to solve some problem that the people with the money have. Therefore, consider the following items for inclusion in this section: definition of the problem, immediate background of the problem, benefits that will come from the solution, and the feasibility of the solution.

Definition of the Problem. People and companies often suffer for years from a problem they have never defined and may not suspect they have. Exorbitantly high telephone bills, delays in invoicing and shipping, poor inventory control, poor customer relations, ineffective advertising, high maintenance costs—such "facts of life" may steadily grow worse through inattention and lack of recognition. An outsider, not having been blinded by day-to-day living with the problem, is usually in a far better position to detect its presence and to formulate a description of it.

Depending upon the scale of the proposal, you should spend from a paragraph to several pages in defining, locating, and describing the problem you propose to solve. By this means you may "shock" your intended client into sudden and full awareness that a problem really does exist. However, you should guard against overstatement and overdramatization, because these techniques can boomerang. While you are "selling" your readers, place yourself properly on the salesperson-scientist continuum (see pages 13–14). In proposals, you should probably be closer to the scientist end than the salesperson end.

Immediate Background of the Problem. Trash collection and disposal is a generally recognized problem today. Populous cities along the Atlantic Coast transport their refuse hundreds of miles where it is dumped in landfills in lightly populated areas. Citizens near these dumping grounds are increasingly raising their voices in protest. Who wants the natural countryside to be the garbage dump for the big cities?

How did we get into this sorry plight? We really do not know. It seems clear, however, that before we can devise solutions we must first discover what generated the problem in the first place. Discovering the origin and development of the problem may lead to means for its solution. Have Americans become an increasingly wasteful people? How much of the blame can be placed on marketing practices that emphasize bottles, cans, and cartons? We don't know, but we must find out.

A historical review can be quite enlightening and helpful in proving your grasp of the problem. However, your job is not to write history but to solve an existing problem. Therefore, keep the historical review brief and concentrate upon the recent past. Be selective. Make the main points stand out.

Need for Solution to the Problem.   In the 1920s and 1930s automobile exhausts were discharging contaminants into our atmosphere just as they are today—and possibly at a higher rate per vehicle mile. But it was not until the 1950s that a few clear-sighted individuals pointed with alarm to the extent of air pollution from this source alone. Respiratory irritation and pulmonary congestion, smog that has killed the frail and aged, soiling of buildings, and grime that settles everywhere are now costing us millions of dollars every year. The cost in health, comfort, and cleanliness cannot be stated in dollars.

Air pollution from automobile exhausts and other sources is now generally recognized as a problem by people in government at all levels, by automobile manufacturers, and by ordinary citizens. But the pioneers in the fight against air pollution have had to fight—and are still fighting—a lusty campaign to convince Americans that the need to do something about the problem is urgent and critical.

For the most part, people are not really aware of local problems they have grown up with and become accustomed to. In a proposal dealing with such a problem you may have to shock them into acknowledging that the problem is a crucial one that will adversely affect their health or well-being. Spell out the consequences of ignoring the problem, as the Surgeon General of the United States has done with cigarette smoking.

Benefits That Will Come from the Solution.   With respect to the trash disposal problem, you might stress that avoiding the long-distance shipping of trash would save hundreds of thousands of dollars a year for a city the size of Philadelphia. Freight cars and trucks would be released for more profitable work. One possible local solution might be to convert garbage into burnable pellets that could be used to heat public buildings or the downtown area. On the whole, such a solution might prove to be income producing rather than tax consuming.

Feasibility of the Solution.   Work plans and methods are fine and necessary for inclusion in proposals, but such plans are seldom taken at face value. Will they solve the problem? Arguments for their feasibility are to be found in the laws of physics and chemistry, the practices of engineering, mathematical models, and analyses. Such demonstrations tend to prove that the means to be used can be made to work; that is, the means are both practical and practicable in terms of present knowledge and attainments.

If you turn to your dictionary, you will find that the word *feasible* has two meanings. A feasible action or undertaking is one that

1. Can be accomplished in a purely practical sense (that is, not precluded by physical, economic, or similar concerns).
2. Is suitable or appropriate (that is, generally conceded to be desirable as well as practical).

The first meaning is more commonly applied than the second. However, because the second meaning overlaps the first—in the sense that a totally inappropriate project also ceases to be practical— our example will consider both.

Suppose you wish to determine whether it would be feasible to provide all students at a certain university access to word processors to be used for writing assignments. Furthermore, this equipment would be provided with programs that would check the students' spelling, punctuation, diction, and sentence style. First, the project would have to be assessed in practical terms. Could the university equip itself with enough word processors in view of purchase costs, maintenance costs, student population, room space, need for security, and so on?

Your research might establish that these practical obstacles are insurmountable, in which case there would be no reason to go any further. If the school can neither afford nor install the word processors, their desirability becomes irrelevant.

Suppose, however, that the practical problems appear to be manageable. At this point, you need to consider the more subtle issue of desirability. For instance, if the word processors and their accompanying programs were readily available to students, would they come to depend on them so much that they would forget the basic rules of mechanics and style? Would students forget how to spell? Many parents and teachers might raise this question to illustrate the seeming "unsuitability" of word processors. That response is a natural one, based on the tradition that tedious mental work is essential to learning anything. But suppose your research indicates that the word processor and its programs are excellent teaching tools that enhance rather than detract from the students' writing abilities— that the students actually learn the rudiments of spelling, punctuation, diction, and style faster when using the word processor than by older methods? Furthermore, the word processor could enable students who did have a sound grasp of these fundamentals to progress much more rapidly toward advanced concepts. You might then conclude that word processors are not only practical but appropriate. (For more detailed treatment of feasibility, see Chapter 17, "Feasibility Reports.")

Plan of Work

After you have established the problem and the need for its solution, you have to tell your readers how you are going to carry out the work. In your plan of work, you will give your readers information on scope, methods to be used, task breakdown, and your time and work schedule. You may also include statements concerning the likelihood of success and the products of the project.

Scope.   The scope statement sets boundaries and states what is to be done within these boundaries. You may state that you will use only off-the-shelf commercial equipment or that you will design, develop, and test equipment that has not been devised to date. You may sample public opinion or disregard the question entirely. You may adopt existing techniques and modify them or create entirely new techniques. You may limit study to estimates of technical feasibility, or you may agree to demonstrate its practicability. You may study all possible sites within your state, only one site, or a selected few. In other words, the scope statement establishes the depth, breadth, and means of your approach. Also, for everyone's protection, you may include in it negative observations covering what you will not do.

Methods to Be Used.   It is always heartwarming to have someone say that he or she intends to do something we very much want to have done. But immediately, realists that we are, we ask how. A plumber without tools can diagnose the leak in our shower faucet and sympathize with us about the drip. However, without knowledge gained from experience and without wrenches, screwdrivers, and washers, the plumber is powerless to stop the leak.

When you propose a problem for solution, you must demonstrate that you have the means required to solve it. A plan of attack, a method of operation, a systematic analysis of intended procedures will attest to your practical good sense and know-how. Frequently, in research projects you may be proposing methods that have been created for earlier research similar to your own. When such is the case, you may save space and time by referring to such methods and providing references to sources, usually scholarly journals, where complete details can be found. Be sure to refer, however, only to sources that your readers can easily obtain.

Task Breakdown.   All but the simplest projects require a breakdown into tasks. One task may consist of initial exploration and planning, another of a search of the literature, another of correspondence and interviews, and so on. (See Chapter 5, "Gathering and Checking Information.") In projects that extend over a long period of time, reporting progress will certainly be one of the tasks. (See Chapter 16, "Progress Reports.") To try to do everything at one time produces confusion and dispersion of effort. Furthermore, one or several tasks may have to be completed before others can be started. You cannot, for example, assemble equipment before you have collected parts, analyze data before you have acquired them, or report progress before you have made any. Breaking a project into its component tasks also gives some assurance that the total job will be done in an orderly manner and be completed on schedule.

Time and Work Schedule.   With few exceptions, work proposals and agreements specify the calendar period within which projects are to be completed. The period may extend for days, months, or years. Further, the schedule may stipulate that portions of the work are to be done in a stated order and are to be completed by a given date. For complicated projects, time and effort must be carefully allocated. It is common practice to prepare a time-based flow chart showing activities and their intended duration and to provide this chart as a graphic. Even limited projects, such as student research problems, require planning for efficient performance to guarantee that work will be completed by target or deadline dates.

Likelihood of Success.   At first thought, you may feel that you would be foolhardy to write a proposal concerning any problem that you could not guarantee to solve. But understand that every problem exists simply because it has not been solved. If the solution were self-evident, then the problem would very probably have been solved before your arrival on the scene. Furthermore, some problems—cancer prevention and cure, racial tensions, and nuclear war—must be attacked, and are being attacked, even though the leads are slim and the prospects for early and full solution are dim. The universality and severity of such problems warrant the expenditure of effort even though the solution may not be found in our lifetime. Polio would still be killing and maiming our children and young adults if researchers had not had the courage to gamble against the odds.

Given infinite time, money, and means, we can reasonably hope to solve all problems that now afflict humanity. But when you prepare a proposal, you must set bounds on time, money, and means. Often you simply guarantee a limited solution. Therefore, in part to protect yourself and your professional reputation against charges of misrepresentation and nonfulfillment, you should consider including an estimate of success in every proposal, perhaps in percentage form.

Products of the Project. Stating what you are to produce throughout a project and at its termination does commit you to definite performance and productiveness. Sometimes you may find it burdensome and costly to satisfy all points of your agreement. However, if you succeed in dodging this issue and do not specify what you will produce and supply, you may be placing yourself in a still worse situation. Your customer may "construe" your proposal as promising delivery of a working model or prototype equipment or as promising follow-up maintenance and consulting services free of extra cost. Therefore, a clear and specific statement of products to be delivered, although it commits you, also sets bounds on what you are to deliver. For the moral here, read the guarantee on the next roll of photographic film you buy. You will see that the manufacturer limits its obligation to replacing the original roll of film—and only if the original was defective in manufacture or processing. If you leave the lens cap on your camera, you, not the film manufacturer, will have to stand the loss.

Facilities and Equipment

For the simplest of projects you may need only a pad of writing paper and a ball-point pen. This problem you can solve by spending a few cents at the stationery counter. But many projects, apparently simple on first consideration, may shortly prove to require more than you bargained for. Facilities and equipment needed may include laboratory space, a typewriter, secretarial support, a desk calculator, a technical library, use of a car, surveying instruments, testing apparatus, materials for constructing models, photographic equipment— and so on. On a still larger scale, you might need a building, a truck, the use of a light aircraft, and computers.

Before you decide to prepare a proposal, you must realistically determine the facilities and equipment the proposed work will require. Some you may already have. For those items you do not already possess, all or some of the cost of acquiring them must be included in the cost and financing of the project. If you already have the items, some or all of their original cost should be paid off, amortized, during the life of the project proposed.

PERSONNEL QUALIFICATIONS

As with the person who is seeking a first job, someone's lack of experience in offering a proposal makes prospective employers pause, whereas someone's successful accomplishment in the past promises success in the future. Whenever possible, therefore, you should cite earlier successes with similar problems when you prepare the new proposal. The past and present problems need not be identical—but they should overlap and have major points in common. You should include dates, contract numbers, names and addresses, references to published research, and so forth, so that your prospective client can verify your statements.

Many persons in research, industry, and government enjoy national and international reputations for their knowledge and performance. Institutions seek to attract celebrated "names." If you qualify, or if people in your organization qualify, you would do well to include their names and biographical sketches in a proposal. Resumes much like those prepared for the job hunt may be suitable (see pages 305–309). Known names tend to be regarded as a known quantity. However, if you cannot guarantee the availability of such persons for the project, do not use wording that can be construed as a firm promise. Otherwise, your contract may be broken on the ground of misrepresentation.

To support your performance record, you would do well to include the names and addresses of individuals, companies, and agencies that are able and willing to testify in your behalf. Satisfied customers can write letters that spell the difference between a hit and a near miss with proposals. Be sure, however, that they hold you in favor and will respond promptly and generously to inquiries concerning you. Obtain their permission well in advance of using them as references.

BUDGET

Practices differ in handling money matters in connection with proposals. Sometimes the "cost accounting" is included in the proposal paper, usually labeled as "Budget." Sometimes it accompanies the proposal as an addendum or rider. And sometimes it is handled in a separate financial and legal contract. In any case, you must never commit yourself to working for others without a binding agreement covering dollar amounts, hours of labor, fees and profits, timing, and method of payment. Penalty clauses for nonperformance or late performance, limits on fees and profits as a percentage of the whole, and delayed payment practices should be scrutinized closely. If you are not adept in these matters, go to a qualified financial expert or attorney.

If you do set up a budget, classify your costs in some coherent way that clearly shows where the money is going by using headings such as these:

Materials
Labor
Test Equipment
Travel
Administrative Expense
Fee

In small proposals, the detailed itemization and cost breakdown of your major entries—such as the kinds and cost of test equipment and travel—will be presented right in the budget. In larger proposals, only the totals may be presented in the budget, with reference to detailed itemizations in an appendix.

APPENDIXES

Think carefully before you put anything into an appendix. There are decided advantages and disadvantages. An advantage may be that the ease of reading the proposal is increased if detailed analyses, budget itemizations, personnel resumes, and so forth are put into an appendix. The disadvantage may be that important material placed in an appendix may be overlooked. Also, material indiscriminately thrown into an appendix may add bulk to your proposal but not credibility. It's a judgment call.

In any event, appendixes are frequently used in proposals. Sometimes they include previously prepared exhibit literature that companies and experienced consultants keep on hand. They may include biographical sketches, descriptions of earlier projects, company organization charts, and statements concerning company policy, security clearances held, employment practices, and physical location and setting. (See also pages 227–228.)

URGE TO ACTION

As we comment at the start of this chapter, every proposal involves sales functions. A good product, of course, sells itself, but it does no harm to advertise its merits also. Therefore, at one or more places in a proposal, particularly at the end of the proposal or in the transmittal letter, you may wish to include several sentences whose purpose is to trigger acceptance of the proposed project. These statements should be given an objective basis, however. Horn blowing, banner waving, and self-praise may work against you rather than for you. If you have proposed the purchase of word processors, you may wish to state that adoption of the proposal this fall would make it possible to have the machines in operation by January 1. You may stress the need and the benefits. You may give evidence of your sincere intentions by requesting an interview or by stating your willingness to modify the proposal. Whatever techniques you use to precipitate acceptance of the proposal should reflect your appreciation of the reader's situation and should express your own tact and good taste. Perhaps you also have had the experience of an overzealous salesperson's killing a sale by talking long and loud after you were willing to buy.

STUDENT PROPOSALS

Students, while probably not ready for the 100-page proposals often found in industry and government, can use proposals in two ways. First, they can follow the general proposal format described in this chapter and by simulating a real proposal situation produce the letter or memo proposals frequently used in industrial and governmental settings. The student proposal in Figure 15-2 is such a simulation. That is, the student author, by starting with a real situation and projecting her imagination, did all the things she would have to do in working for hire as a researcher. Despite the fictional trappings, the proposal is realistic and professional.

Second, on a still smaller scale the student can submit a proposal for a report, as demonstrated in Figure 15-3. In this proposal, the student addresses not a client but the instructor to propose a report the student intends to write. The student should tailor the proposal for a report to fit the proposed paper and the instructor's needs but, in general, should include the following information:

- Subject, purpose, and scope for the proposed paper
- Task and time breakdown
- Resources available
- Qualifications for doing the report

20 September 1982

To: Mrs. Nancy Johnson, Chair
         Wyoming Lake Citizens Committee
         Marinette County Courthouse
         Wyoming Lake, WI 54126

From: Barbara J. Buschatz   *B. J. Buschatz*
         4062 Hoven Street
         St. Paul MN 55108

Subject: Proposal for a Feasibility Study of
         Rehabilitating Wyoming Lake

Project Summary

Earlier research shows that Wyoming Lake is clogged by
approximately 2-million cubic yards of sediment that
effectively destroy the lake's potential as a recreational
resource. This letter proposes that the Wyoming Lake
Citizens Committee commission a study to ascertain the
feasibility of dredging the lake and restoring its recreational use. The study would analyze the sediment deposits,
investigate needed dredging equipment and methods, and ascertain
feasibility of a dredging project by using the criteria of time,
cost, and community benefits and problems. Using the report,
the Committee would have a basis upon which to proceed in its
decisions concerning possible implementation of a dredging
project.

Project Description

In 1980, the Wyoming Lake Citizens Committee, concerned by the
increasing deterioration of Wyoming Lake, commissioned a study
of the sediment deposits in the lake. Since the completion of
that study, no further steps have been taken in the lake's
rehabilitation. Each year the sediment deposits grow thicker,
further impeding the use of the lake and increasing the ultimate
cost of their removal. I propose that your committee commission
me to carry out a study to evaluate the feasibility of removing
the sediment from Wyoming Lake. I will investigate and report
on the equipment and methods needed for removal of the deposits.
As a major outcome of the study, I will present an evaluation of
the feasibility of a dredging project based upon the criteria of
cost, time-to-completion, and community benefits and problems.
If the dredging project proves feasible and is completed, it will
restore Wyoming Lake as an important recreational resource to the
citizens of the area.

     Rationale and significance. Charles J. Anderson's report,
"Sediment Deposits in Wyoming Lake," commissioned by the Wyoming
Lake Citizens Committee in 1980, shows the extent of the lake's
problems. Mr. Anderson reports that there are over 1.2-million
cubic yards of sediment. As a result, in over 5 miles of shoreline there are few areas that can be used for swimming or wading.
Swimming is no longer allowed in the public park at the south end
of the lake.

FIGURE 15-2 Simulated Proposal

Wyoming Lake is typical of many lakes left in Wisconsin and Minnesota when the great glaciers of the ice age retreated some 11,000 years ago. Like many such lakes it is subject to sedimentation as silt washes into it from the surrounding area. Left alone, in time it will become a bog and ultimately a completely filled-in area. The longer the problem remains, the more costly a solution becomes. At some point, the cost will become prohibitive, and the lake will be lost as a resource for water sports. However, with timely intervention, the sediment can be removed and the lake kept clear. Rehabilitation projects restoring sediment clogged lakes have been carried out successfully by several nearby cities.

The experience of the neighboring cities indicates that a rehabilitation project is technologically feasible. Modern dredging equipment is available that will remove the sediment deposits of the size and type found in Wyoming Lake. More pertinent perhaps are the social and political questions of how long will the project take, how much will it cost, what problems will it create for the citizens of Wyoming Lake, and what benefits will the citizens receive in return for the cost of the project. Should the benefits outweigh the cost and problems created, the project would seem to be feasible. Again the experience of neighboring cities indicates the likely feasibility of the project on these social and political terms.

Plan of work. The proposed study breaks down into these three tasks:

1. Analyzing the location, depth, and content of the sediment deposits.
2. Investigating the equipment and methods needed for removal of the deposits.
3. Applying the criteria of time, cost, and community benefits and problems.

The Anderson report provides adequate information as to the location and depth of the sediment deposits. I'll refine his research by taking sediment samples from three locations in the lake. The samples will be sent to the soils laboratory of the University of Wisconsin-Madison for analysis. The results of the analysis will suggest possible uses for the removed sediments.

To ascertain equipment and methods, I'll correspond with dredging companies that work in the Midwest. I'll give them estimates of the amount of sediment to be removed and ask them to comment on the methods, equipment, time-to-completion, and cost. I'll also correspond with the Wisconsin Department of Natural Resources (DNR). They should be able to provide information concerning legal restrictions and procedures involved in such projects. I'll also query the DNR concerning possible state and federal assistance.

Research on all three tasks outlined will be carried out more or less simultaneously and should take eight weeks. Allowing an additional two weeks for me to organize and write my findings, I should be able to have a final report within ten weeks after you authorize work to begin.

As products of the project, I will furnish to you two monthly
progress reports and a final report. The final report will
be a feasibility study that reports and analyzes the data
gathered in the study. The report will make a recommendation
as to the feasibility of the project. Further, if the project
proves feasible, the report will recommend methods and equipment for carrying it out. The report will estimate the cost
and time-to-completion of the project. It will detail what
are likely to be the major problems to be encountered and
possible solutions to them. It will discuss the benefits
that may result from the project.

The report will not provide a detailed blueprint on how to
proceed with the project, but it will provide a basis for
the Wyoming Lake Citizens Committee to make such decisions
as are needed.

  Facilities and equipment. No expensive facilities or
equipment are needed to carry out this research. I will
provide a car and a boat as needed at nominal cost. Equipment for sampling the lake's sediment can be borrowed at no
cost from the Wyoming Lake Maintenance Department.

Personnel Qualifications

I am a senior student in agricultural engineering at the
University of Minnesota, where I carry a 3.3 grade point
average based upon a 4.0 scale. I am familiar through my
course work with dredging equipment and methods. In the
summer of 1980, I assisted Charles Anderson in his study
of the Wyoming Lake sediment deposits. In the summer of
1981, as an intern with the Wisconsin DNR, I assisted in
several eutrophication studies of Wisconsin lakes.
(Eutrophication is the process by which lakes become more
favorable to plant life than to animal life, leading to
their becoming clogged with plants.) I am familiar,
therefore, with the problems of Wisconsin lakes in general
and with Wyoming Lake in particular.

The people listed below will provide references for me if
requested:

    Professor Arnold Flikke
    Department of Agricultural Engineering
    University of Minnesota
    St. Paul, MN 55108

    Dr. Charles Murphy
    Department of Natural Resources
    Madison, WI 53702

Budget

In this project budget I indicate estimated expenses for several
items. Where estimates are indicated, the cost may be less than
estimated and will not be more. Costs are as follows:

Use of my boat and motor	$ 15
Use of my car @ 20¢ a mile (estimated)	$ 80
Cost of soil sample analysis	$ 10
Mailing and telephone expense (estimated)	$ 30
Fee	$350
TOTAL	$485

Payment is due within 30 days of the report's acceptance by the
Wyoming Lake Citizens Committee.

Date: 27 January 1983

TO: Professor Victoria Winkler
           Department of Rhetoric
           University of Minnesota

From: David M. Zellar *DMZ*

Subject: Proposal for a Feasibility Report

Professor Milton Weller, Head of the Department of Entomology, Fisheries and Wildlife, wishes to purchase a portable tape recorder for the Department. The recorder will be used on field trips, primarily to record bird vocalizations. I have volunteered to help the Department select an appropriate recorder.

Professor Weller and I have discussed the criteria to be used in comparing possible recorders. We agreed that for ease of operation the tape recorder should be a cassette recorder and not reel-to-reel. Other criteria to be used are the following:

    Frequency response: Bird vocalizations range from 100 to 18000 Hz. Therefore, the recorder should have a frequency response to match.

    Size and weight: Because the recorder is to be used on field trips, it should be small and light. Professor Weller and I agreed that 20 pounds was probably a reasonable limit for maximum weight.

    Battery operation: The recorder will be used in places where electricity is not available. Because it will be used on extended field trips of a week or more, even rechargeable batteries would not be practical. Therefore, the recorder must be capable of operating on dry cell batteries.

    Maintenance requirements: Ideally, the recorder should be available for purchase locally, and the availability of local maintenance is a must.

    Cost: Professor Weller, because of budgetary restrictions, has set a maximum limit of $500 for cost.

At least two tape recorders, the Marantz Supercope and the AIWA TPR-945 seem to fit the criteria. Undoubtedly, investigation will turn up others.

Figure 15-3 Proposal for a Report

Task and Time Breakdown

Professor Weller wishes to buy the recorder as soon as possible. Furthermore, to use the project as a feasibility report in your class, I must complete it in the next six weeks. Therefore, I propose the following task and time breakdown:

1. Visit stores where tape recorders are sold. Interview salespeople, gather literature on available recorders, and examine and record with potential selections (2 weeks).

2. Perform my analysis of available recorders, comparing them by our criteria (1 week).

3. Write and turn in to you a progress report that includes preliminary findings and an organizational plan for the final report, with a copy to Professor Weller (1 week).

4. Write a preliminary draft of the report and discuss it with you (1 week).

5. Write the final copy of the report and submit it to you no later than 10 March 1983.

Resources Available

For information about bird vocalizations, I'll use Birds of North America by Robbins, Bruun, Zim, and Singer. Professor Weller will continue to serve as a resource concerning criteria.

There are over one-hundred stereophonic and high fidelity equipment dealers in the Minneapolis-St. Paul area. I have dealt with at least ten of them on various occasions. All recorders of the caliber to be considered in this study have literature available that furnishes accurate information as to their size, weight, and capabilities.

For the last eight years I have owned and operated stereophonic and high fidelity equipment. I'm familiar with equipment terminology and can interpret sales literature with no difficulty.

I am a senior in Wildlife. Having been on two extended field trips--one a week long, the other two weeks--I understand the techniques and problems of capturing bird vocalizations and of carrying and caring for field equipment.

EXERCISES

1. Assume that you have been asked by the Community Planning Commission to identify what you believe to be your town's most urgent problem. Here are some likely candidates:

- Traffic and parking congestion
- Need for new business and industry
- A deteriorated business district
- Housing for low-income families
- Inadequate public transportation
- Need for a central downtown heating system

2. List the means you would use to gather information for a follow-up proposal you would submit to the Community Planning Commission. Consider questionnaires, interviews, library research, and the like. Be as specific as possible.

3. If you were hired to research the problem referred to in Exercises 1 and 2, how would you break down the whole project into tasks? How much calendar time do you estimate you would need to complete the investigation? What section titles do you foresee for the final report?

4. Based upon the information you have generated for Exercises 1–3, simulate a situation in which you can submit a proposal similar to the one in Figure 15-2. Applying your imagination to as realistic a situation as you can find, write a proposal to a client.

5. Write a proposal for a paper as demonstrated in Figure 15-3. Follow your instructor's directions as to what kind of paper it should be.

Chapter # 16

Progress Reports

DISTANCE, AS WELL as differences in main interests, often separates workers from those who make use of their work, whether that work is research, construction, design, or whatever. A progress report, submitted as a bound report, letter, or memorandum, helps to keep the client in touch with the work being done.

The main and obvious function of any progress report is to give the company, department, or individual an accounting of the work that has been done. It explains how you have spent your hours and the client's money and what you have accomplished as a result of the investment. Though this purpose is dominant, we must not lose sight of four other purposes that are discharged by a progress report:

- It enables the client to check on progress, direction of development, emphasis of the investigation, and general conduct of the work. Thus the client can alter the course of the work before too much time and money have been invested.
- It enables workers to estimate work done and work remaining with respect to the total time and effort available.
- It compels workers to evaluate their work and focus their attention.
- If the work includes a final report, as a research project or a feasibility study would, it provides a sample report that helps both the client and the workers to decide upon the tone, content, and plan of the final report.

Several different arrangements are used to report progress. A popular plan for a year-long project is shown in Figure 16-1. A progress letter is sent at the close of work each month, except at the end of a month closing the first, second, and third quarters of the year. At the end of each quarter except the last, a formal bound report of progress is sent to the customer. This quarterly report recapitulates the main contents of the preceding monthly letters and adds the work done during the month since the last letter. In short, the bound quarterly report is sent instead of a third letter for each of the first three quarters.

Toward the end of a project (about 11 months into a year's project), when affairs often have reached a critical stage, a preliminary copy of the proposed final report may be sent either in addition to or instead of the eighth monthly letter. Usually the prose has yet to be edited and polished; some or all of the illustrations may be roughly sketched or omitted entirely. The submission of this report enables the client to react and criticize and thereby get a final report more to his or her liking (see Figure 16-1).

Progress reports, like most pieces of a writing, have a beginning, a middle, and an end. The following plan for a progress report shows what these three parts may contain:

Beginning
Introduction
Project description
Middle
Summary of the work done in the preceding period(s)—included in all
    project reports after the first
Work done in period just closing
Work planned for next work period
Work planned for periods thereafter
End
Overall appraisal of work to date
Conclusions and recommendations concerning work

Recognize that the plan shown here is for a full-scale report. More modest progress reports may collapse some of the parts together. For

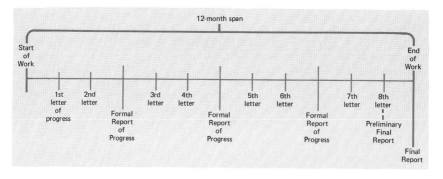

FIGURE 16-1 Two-Level Method of Reporting Progress

example, the project description may be part of the introduction or, if there are no changes in it, omitted altogether. "Work planned for next period" and "Work planned for periods thereafter" may be lumped together under a heading like "Work remaining." Conclusions and recommendations may be integrated into the overall appraisal. But, no matter how, all these elements should be presented.

THE BEGINNING

Progress report introductions are, in general, typical four-part introductions that make clear the subject, purpose, scope, and plan of the report (see pages 218–221). They should clearly relate the report to the work being done. Also, because progress reports are executive reports, give the readers some idea of your overall progress in the introduction.

The project description (also known by other names, such as work statement and contractual requirements) spells out what the workers are required to do and produce. It makes clear the purpose and scope of the project. If changes are made in the contractual agreement, the project description should reflect these changes.

THE MIDDLE

The middle section of every progress report must bring together two elements: time and the tasks accomplished or to be accomplished during that time. This fact suggests that the middle portion of the progress report can be organized around either time or tasks.

Time Plan

If the time plan is used for the overall plan of a progress report, the reader understands what time period is being discussed at any point in the report. Let us look at four possible headings for proof of this point:

- Work Previously Done
- Work Done in the Period Just Closing
- Work Scheduled for the Next Period
- Work Scheduled for Periods Thereafter

Obviously, this is the plan foreshadowed on pages 369–370 of this chapter. This often-used plan has a dynamic character. It gives the recital a cumulative effect. And it offers few problems in obtaining coherent flow from section to section.

We have commented that the time plan tends to be dynamic—that is, it gives the impression that something is being accomplished. Having said this, we should also remark that the time plan may lead to windy generalization and too much emphasis on procedure. However, with all its possible faults, the time plan is widely used and generally has proven effective.

Task Plan

The project description of the progress report often gives the task breakdown of the entire project. These tasks may be performed at different times, by separate individuals, working at different locations. If this, or something close to it, is the working arrangement, then it may be convenient to organize the progress report on the task plan.

When you were younger your mother very likely told you to "straighten up your room." Whether you or she used the term *task,* that is exactly what your mother was assigning: a specific action, chore, or undertaking. It is a task to prepare, administer, and tabulate a questionnaire. It is a task to install air conditioners or design a new wing for a hospital. It is a task to conduct library or empirical research. Once you have clearly defined and described your tasks, then the main headings of the body may be simply stated, as in the following example:

- Establishing Location, Depth, and Content of the Deposits
- Determining Equipment and Methods
- Applying Criteria to Proposed Project

The task plan is realistic and objective. The customer and project supervisor can readily and firmly estimate the amount of work done and remaining on each task. By the same token, this plan throws the glaring spotlight of unfavorable publicity on tasks that are not going well. Also, if the tasks are to be done in sequence, with little or no overlapping, then a given progress report may have solid prose to devote to only one of the tasks, with the other tasks being essentially blank for the time being. Therefore, the task plan seems most appropriate when several or all tasks are being performed concurrently.

Combination Plans

Clearly, these plans can be used in combination, the exact arrangement depending upon where you wish to put the emphasis. You could, for instance choose between these arrangements:

I. Work Done in Preceding Periods
   A. Establishing Location, Depth, and Content of the Deposits

B. Determining Equipment and Methods
C. Applying Criteria to Proposed Project
II. Work Done in Period Just Closing
A. Establishing Location, Depth, and Content of the Deposits -etc.-

---

I. Establishing Location, Depth, and Content of Deposits
A. Work Done in Preceding Periods
B. Work Done in Period Just Closing
C. Work Scheduled for Next Period
D. Work Scheduled for Periods Thereafter
II. Determining Equipment and Methods
A. Work Done in Preceding Periods -etc.-

We do not offer any one plan or combination of plans as being generally superior to others. We do urge that you collect the information you wish to include in the progress report and then select the plan that will best hold and present the information. And you must always be prepared to try a different plan if the first choice proves unwieldy.

THE ENDING

In the ending of the progress report you draw things together for your readers. You summarize and review your progress. You provide answers to the kinds of questions executives are likely to ask. Are you on time or ahead of time? Are you running into unexpected problems? How are you solving them? Is the scope of the work changing? If so, how and why? Is a consultation needed between you and the client? Are costs running higher than expected? How much higher and why? Would some new approach or procedure be more efficient and less costly? Is something unexpected and significant showing up in the research? Does it throw past findings into doubt? Are materials called for in construction specifications no longer available? What can be substituted for them?

And so on. Much of the material you cover in the overall appraisal will have already been presented in the middle section of the report, but here you highlight the significant facts. Then you draw conclusions from these facts and make recommendations (see pages 222–227).

Remember what you know about executive reading habits from pages 32–36. Because executives read selectively, they may skip the middle of the report and come directly to this part. Therefore, this section of the progress report should be able to stand on its own. Also, you may wish to consider moving the overall appraisal nearer the front of the report, perhaps placing it right after the introduction. In modest letter and memorandum reports, you might consider making it part of the introduction.

ADDITIONAL CONSIDERATIONS

The total impression a progress report conveys to the reader is, of course, important. But this total impression is the integrated product of more particular things, including those discussed in the following pages.

Physical Appearance

To put the reader-client in a receptive frame of mind, a progress report must be physically attractive. By this we do not mean expensiveness and glamour but rather neatness, appropriateness, and good design. Reports that exceed letter or memorandum length should have a protective cover both front and back. The title page should be tasteful and uncluttered. The print (or typing) should be clean and legible. We do not expect diamonds to come in a soiled paper bag, nor do readers expect logic, sound content, and honest performance to come in shoddy wrappings. Nevertheless, all the progress letters and reports arriving from a project should not exceed a very small percentage (perhaps 5 percent) of the total funding of the project. After all, we are paid to make progress and not to linger lovingly over the reporting of progress.

Style and Tone

Progress reports are a project's emissaries. If these emissaries seem tired, confused, and unhappy, what then is the customer to think of the workers "back home"? Progress reports therefore should read with vigor, firmness, and authority—one might risk optimism. Yet their forcefulness must lie in more than artful writing. Generalizations must be bolstered by recitations of detailed factual accomplishment. Snags, problems, and delays should be honestly discussed, but the accent should be on positive accomplishments. A neat balance between these two aspects of a project will prevent progress reports from reading like either a trail of disaster or an outpouring of giddy optimism. Excess in either direction will always have its day of reckoning.

Originality

The first progress report often leaves much to be desired. For one thing, the project just recently got under way, and whatever progress has been made cannot yet be crystallized. For another thing, the first progress report, lacking precedent, sometimes seems tentative and experimental. The next two or three progress reports usually represent a substantial improvement over the first, for they have accomplishments to report and they profit from earlier experience. However, after the third or fourth progress report of a series, the reports tend to hit a plateau or go downhill. A feeling of ennui and repetition may be disturbing to authors. This slacking off must be prevented at any cost. Bringing in new blood to the writing staff may help. A staff review of the project may help. But a clear recognition of the need to maintain reading interest, verve, and originality is a necessity. Keep out of the rut. Do not simply warm over last month's progress report like the proverbial Sunday roast. Take a fresh look at the whole problem of reporting progress.

Accomplishment and Foresight

Your client will find it heartening to learn all you have accomplished on his or her behalf. Past performance is probably the most reassuring promise of future performance. Yet investigators should not seem to be moving blindly into the future work periods, like an automobile driver about to run off the margin of his only road map. Therefore, progress reports, while stressing what has been done, should give adequate attention to plans for the future. For the next work period, the plans should be firm and detailed; for work periods thereafter, the plans may understandably be less specific. Showing your plans reveals you to be a professional and also makes it possible for your client to suggest modifications. Getting your client into the act usually works to everyone's benefit.

Exceeding Expectations

A progress report should give your client a pleasant but mild surprise. Notification that a new task has been started a few days before its scheduled beginning, three or four graphic aids, a technical appendix, some noticeable improvements in format are useful ways of cheering your client without undue labor or expense on your part. On the other hand, avoid sudden and excessive novelty, such as using multiple overlays and color printing that have no precedent in previous progress reports of the project. Again, do not increase the length of the previous progress reports by more than, say, 20 percent, for your client may suspect that you are trying to "cover up" and divert attention. In a research project, do not include a set of firm conclusions or ultimate recommendations in a progress report, for you will thereby steal the fire from your final report. At the worst, your client may end the project before its normal completion date.

SAMPLE PROGRESS REPORT

In Figure 16-2, we show you a sample progress report, based upon the task plan. It is linked both to the sample proposal on pages 363–365 and the excerpts from a student feasibility report used as exam-

15 November 1982

Mrs. Nancy Johnson, Chair
Wyoming Lake Citizens Committee
Marinette County Courthouse
Wyoming Lake, WI 54126

Subject: First Progress Report on Feasibility Study of
         Rehabilitating Wyoming Lake

Dear Mrs. Johnson:

On 15 October, after your acceptance of my proposal to study
the feasibility of rehabilitating Lake Wyoming by removing
its sediment deposits, I started work on all phases of the
study. As agreed upon, I am submitting my monthly progress
report.

This report details the work completed and the work remaining
on the three tasks of the study, as described next under
Project Description, and concludes with an overall appraisal
of the progress made.

Work on the study has progressed well. I also have the good
news that with prompt action we may be eligible for major
assistance in meeting the cost of the rehabilitation project.
See the report for further details on this.

Project Description

    Purpose. The purpose of the Lake Wyoming Study is to
evaluate the feasibility of removing sediment from Wyoming
Lake to rehabilitate the lake and restore it as a recreational
resource.

    Scope. The study breaks down into three major tasks:
        1. Analyzing the location, depth, and content of
           the sediment deposits
        2. Investigating the equipment and methods needed
           for removal of the deposits
        3. Applying the criteria of time, cost, and community
           benefits and problems

Location, Depth, and Content of the Deposits

    Work completed. Adequate data on location and depth of
the sediment deposits already exist in a report commissioned
by the Wyoming Lake Citizens Committee in 1980. The report--
Charles J. Anderson's "Sediment Deposits in Wyoming Lake--
provides detailed maps to the location of the sediment deposits

FIGURE 16-2   Sample Progress Report

Mrs. Nancy Johnson   -2-   15 November 1982

There are over 1.2 millon square yards of deposits. To
rehabilitate the lake may require the removal of up to
2 million cubic yards of sediment.

Work remaining. In order to determine the content of the
sediment deposits I will in the next week take sediment samples
from three locations in the lake. These samples will be sent
to the soils laboratory at the University of Wisconsin-Madison
for analysis. The potential use of the removed deposits will
depend to some extent on the results of this analysis.

Equipment and Methods

Work completed. I have written to the Madison Dredging
Company, Madison, Wisconsin; The Capital Dredging Company,
St. Paul, Minnesota; and Mississippi Dredges, Inc., St. Louis,
Missouri. I have given these companies estimates of the volume
of sediment to be removed and asked them to comment on the
method, equipment, time to completion, and approximate cost of
the project.

Telephone conversations with James H. Olson, Wisconsin
Department of Natural Resources (DNR) have provided useful
information concerning the legal restrictions and procedures
involved in the project. Mr. Olson will send me a bulletin
that should contain complete information regarding these
matters. He also said that if we act promptly the project
might qualify as a pilot project and be eligible for state
and federal assistance for up to half its cost.

Work remaining. The possibility exists that the City of
Wyoming Lake might be better off buying the necessary dredging
equipment and contracting the job on its own, rather than having a dredging firm do the job. I intend to explore this
possibility with several firms that sell equipment.

Time, Cost, and Community Benefits and Problems

Work completed. Some progress has been made in this area.
Mr. Olson of the DNR has provided a list of city managers in the
state whose cities have completed rehabilitation projects similar
to ours. The replies from the dredging company correspondence
should be helpful concerning time and costs. Also, Mr. Olson
told me that public communication and proper coordination with
the DNR often pose problems. He suggested that early appointment of a project coordinator would help with both these
problems.

Work remaining. This week I'll begin correspondence with
the city managers whose names were given to me by Mr. Olson.
I'll ask them questions concerning time, cost, problems, and
benefits. Using the replies received from them, I'll design
questionnaires to be administered to a random selection of
city residents to ascertain what they see as benefits, problems,
and problem solutions.

Mrs. Nancy Johnson   -3-   15 November 1982

Overall Appraisal

Good progress has been made. Data are available on location and depth of sediment deposits. I will shortly begin work to determine the content of the deposits. Expected letters from major dredging contractors and dredge manufacturers should provide information on equipment, method, time, and costs. Conversations with the DNR indicate that legal restrictions and procedures do not present insurmountable problems. Planned correspondence with city managers should provide a basis for judging benefits and problems.

I will call you next week concerning the possibility that prompt action on our part may make the city eligible for state and federal assistance for up to 50% of the cost of the project. With such a possibility, we may want to move our timetable forward. Work has progressed well enough that I would consider an earlier submission of my final report than originally planned.

Sincerely,

Barbara J. Buschatz
4062 Hoven Street
St. Paul, MN 55108

---

ples in Chapter 17, "Feasibility Reports." The sample progress report is a modest one that collapses some of the suggested parts of a full-scale progress report together. However, it contains all the essential elements of a progress report. It is addressed to the client for the feasibility study it discusses. Student progress reports can be addressed to a teacher or to a client—either real or fictional (see Exercise 6, page 379).

EXERCISES

1. In what ways would a diary of activities (log or journal) benefit the worker when the time came to compose a progress report? To what other uses might the worker or a superior put such a diary?

2. Referring to Figure 16-1, can you explain in your own words why the first progress report for a project is often relatively poor? Why the second and third progress reports tend to be very good? Why progress reports thereafter tend to decline in interest and readability?

3. What section or sections of the usual progress report hold the greatest interest for the project client?

4. In outlining a progress report, what considerations should guide you in selecting the time plan? The task plan?

5.   What elements of human psychology enter into the composition of progress reports?

6.   Write a progress report addressed to a client for some project you are currently working on, perhaps the project you developed for either Exercise 4 or 5 on page 368. Use a memorandum or letter format. (See Figure 16-2 and pages 311–316).

Chapter 17

Feasibility Reports

IN SIMPLEST TERMS, a feasibility report documents the results of a feasibility study. It defines the study in terms of its objectives and the criteria that determine whether something is feasible or not. Feasibility is determined by the answers to questions concerning technological possibility, economic practicality, social desirability, ecological soundness, and so forth. The feasibility report presents, interprets, and summarizes the data relevant to feasibility. It presents the conclusions of the study and recommends actions to be taken. Before we get into the parts of a feasibility report, let's examine first the feasibility study of which the report is the end product. Although exactly how you would conduct a feasibility study would depend upon your area and level of expert knowledge, we can, in general terms, describe the conduct of a study for you. (See also pages 356–357).

LOGIC OF THE FEASIBILITY STUDY

A feasibility study always involves a choice among options. The options may involve doing something or not doing it. For example, should the Department of Transportation require active restraint devices in all automobiles? Or given the decision that something should be done, the choice may lie among the options available to do it: Should the Department of Transportation require air bags activated by impact or seat belt restraints activated when a car door is shut?

At all levels of human activity, from the individual engrossed in personal and domestic problems to the highest level of policy making in government, we live in a society where such decision making goes on. Home owners may discuss whether to replace the worn-out furnace with a new conventional furnace or to switch to a heat pump. The town council may debate whether to install a downtown heating plant and sell heat to local businesses. A company may study the feasibility of manufacturing a new product. The state legislature may argue about where to locate a hazardous waste dump. The governments of the United States, Canada, and Mexico may study the feasibility of a $300 billion water system to bring water from Alaska and the Yukon to the rest of Canada and the United States and to Mexico.

Such questions—whether domestic and local or high level and global—are not easily answered. In an attempt to get the best possible answers, we study the problem, defining it and considering our options. We gather information that we hope will help us reach a wise decision. The home owners with the worn-out furnace may go to the library to read about heat pumps. They may call a friend who has one. They may write away for government documents and talk to salespeople. They expand their knowledge in the manner that we describe in Chapter 5, "Gathering and Checking Information."

Gradually, the home owners form criteria to judge the option of the conventional furnace against that of the heat pump. The criteria may involve cost, both initial and operating, longevity of the appliance, safety, comfort, tax credits, and so forth. When the criteria are formed and enough information is at hand, the home owners make an analysis. They run comparisons applying the criteria to the data. The heat pump costs more initially, but it costs less to operate. Will the lower operating cost offset the higher initial cost? The heat pump can also serve as an air conditioner. How much is that worth? And so forth. Gradually, they reach some conclusions; that is, they form opinions based upon their evidence and their reasoning that will guide them to their decision. Although it's unlikely that the home owners will write a feasibility report, they are engaging in what we can recognize as a feasibility study. At any level, a feasibility study involves these steps:

- Setting the purpose and scope of the study
- Gathering and checking information
- Analyzing data
- Reaching conclusions
- Arriving at a decision or recommendation

As in all creative mental exercises, there will be a good deal of back and forth movement in following these steps, but all the steps should be taken. Because the feasibility study is critical from start to finish, you have to be clearheaded at every step, and the first test comes when you formulate and state the purpose and scope of your study. Because this step is so critical, we discuss it at some length.

Before you do any research, define the precise purpose of all the work you will do. Usually, a single sentence is ideal for the purpose statement:

> The purpose of this investigation is to determine whether X Company of Old Town should establish a branch plant in New Town.

An announcement such as this may seem easy and self-evident, perhaps superfluous. But many investigators have floundered around and eventually bogged down simply because they did not clearly and consciously formulate the objective toward which they were striving.

If you do not know what you are trying to accomplish with a feasibility study, you have no sensor to tell you when you are on or off the right track. Get on the right track at the start and stay on it by constantly reminding yourself of the purpose.

Here are some additional examples of purpose statements:

- The purpose of this study is to dermine the feasibility of using particle board to sheathe the interior of houseboats.
- The purpose of this investigation is to select the best of several promising methods of obtaining the major supply of fresh water for the Pink Sea Horse Motel.
- Our primary objective is to decide which of several word processors would be the best choice for the Department of Mechanical Engineering to purchase.

Inexperienced investigators are sometimes inclined to resent and reject the notion that their work should have a statable and stated purpose. They are satisfied, rather, "to learn all they can about such-and-such a subject." This kind of open-ended purpose is not suitable for a feasibility study. Here an extended illustration will help to clarify our point.

One student investigator proposed to her instructor in technical writing that she be permitted to investigate the fossil spore content of a cranberry bog. "What bog do you have in mind?" her instructor asked. The student looked pained, but under pressure disclosed that she had no particular bog in mind. Her instructor immediately sent her to the Department of Geology to learn the name and location of a bona fide cranberry bog in the vicinity of the campus. When she returned with an answer, both student and instructor were pleased, for visible headway had been made.

"Now," the instructor asked, "why do you want to investigate the fossil spore population of this bog?" The student looked more pained than before. After some moments of cogitation, she volunteered this information: "Well, I want to find out what kinds of fossil spores are in this bog and how many there are."

"But what are you going to determine?" her instructor pressed.

"Why, just what I said," the student replied, "what kinds are there and how many."

"I'm afraid that won't do," said her instructor. "The effort would be interesting and informative, I'm sure, but you are really describing an activity rather than stating the purpose of that activity. The real question hinges on what you will do, or recommend that someone else do, as a result of what you learn."

It was a sultry August afternoon. Minds were working none too well, and tempers were brittle. But by 5:15 the student was at last equipped with an acceptable purpose, as follows:

> The purpose of my project is to decide whether Mrs. Rose's cranberry bog contains a fossil spore population large enough to justify the acquisition of exploration rights over the bog by the Department of Geology for field research by its graduate students.

Equipped with this concrete purpose, the student would not waste her time beating around cranberry bushes and plowing knee-deep in watery bogs. Rather, she was committed to making a decision, settling a question. In other words, she was now furnished with an outcome to be accomplished instead of a nebulous description of a messy activity.

Once the purpose of a feasibility study has been clearly and exactly decided, a fairly tight logical process has been set in motion. Certain other decisions follow with near inevitability. Ordinarily, the next major decision hangs on this question: Given my stated purpose, what must I do to accomplish it? The term *scope* is often applied to this concept. Whatever it is called, the decision consists of determining the actions to be taken, the range of data to be delved into, the bounds to be set to the problem, the criteria to be used, the emphasis to be made. To illustrate, let us use a previous example of purpose statement to find what it implies in terms of scope:

Purpose
- Should X Company of Old Town establish a branch plant in New Town?

Scope
- Does X Company now have, or can it develop, enough business in New Town to justify a branch there?
- Does New Town offer adequate physical facilities, utilities, and other ser-

vices for plant operation there—office space, transportation, communications, and so on?

- Can the required staff be obtained, whether by local hiring or moving personnel into the area, or both?
- Are local business practices and codes, tax structure, and so forth favorable for conducting business there?
- What impact, for better and worse, would opening a branch plant in New Town have upon overall company organization, operations, policy, financial condition?

Necessary as it is to generate and compile the chief items of the scope statement early in an investigation, we would be dishonest to pretend either that the process is easy and forthright or that first efforts will produce a scope statement which will hold unchanged throughout the investigation. Devising a scope statement requires insight, foresight, and hindsight—in other words, speculative thinking. But the ability to create serviceable scope statements improves with practice and with experience gained in the subject matter.

In any case, as an investigator you should not remain blindly committed to your initial statement but should reexamine it from time to time in the light of the information you gather. Look for holes, overlaps, superfluous items, and the like. Frequently, a person unacquainted with the study is in a far better position than you to spot shortcomings and illogicalities in the statement. Therefore, someone outside the study should be asked to review and react to the list of scope items you compile.

For the most part, feasibility studies are conducted by experts in the field of the study or, in many cases, teams of experts from several fields. For example, environmental impact studies are specialized forms of feasibility studies that decide whether some new project, perhaps a new highway, is environmentally sound or not. The study may bring together civil engineers, wildlife biologists, soil scientists, archeologists, and so forth.

ROUTINE OF THE FEASIBILITY REPORT

When the study is complete, the results, conclusions, and recommendations have to be reported to the ultimate users of the study. Generally speaking, the users are not experts in the field of the study. The users of an environmental impact study may be citizen groups and state legislators. In industry, the users of feasibility studies will be the executives charged with the decision-making responsibility. All of these diverse audiences are, in general, acting as executives, and reports for them should be written in the manner we describe as suitable for executives (see pages 32–36). As the writer of a feasibility report, you should write in plain language, avoiding technical jargon when possible. Give necessary definitions and background information. Use suitable graphics. Emphasize consequences and function over methodology and theory. Interpret your data and state clearly the conclusions and recommendations that your best professional judgment leads you to.

Feasibility reports do not have definite formats. They may include all or some of the following elements:

- Letter of transmittal or preface
- Title page
- Abstract
- Table of contents
- List of illustrations
- Glossary of terms
- Introduction
- Discussion
- Factual summary
- Conclusions
- Recommendations
- References
- Appendixes

How many of these elements you include will depend upon audience factors and the length and complexity of the report. For example, a long report aimed at a narrow audience of several people should have a letter of transmittal. A report for a more general audience would have a preface instead. A short feasibility report of only several pages may be cast as a memorandum and essentially consist of only an introduction, discussion, conclusions, and recommendations.

We discuss all the elements you may need for a feasibility report in Chapter 11, "Formal Elements of Reports" and on pages 311–316, "Letter and Memorandum Reports." Here we single out the key elements—introduction, discussion, factual summary, conclusions, and recommendations—for brief discussion and exemplification. In our discussion of these key elements, we'll draw upon a student report entitled "The Feasibility of Removing Sediment Deposits from Wyoming Lake."[1] We reproduce the table of contents from that report in Figure 17-1. A glance at it will help to orient you to the parts as we discuss them.

Introduction

Depending upon the length and complexity of the report, the introduction may range from a single paragraph to several pages. How-

TABLE OF CONTENTS

FIGURE 17-1 Table of Contents for Sample Report

ever, the introduction should be limited to essential preliminaries such as the following:

- Subject, purpose, and scope of the feasibility study
- Reasons for conducting the study

- Identification and characteristics of the person or company performing the study (if not given earlier in a preface or letter of transmittal)
- Definition and historical background of the problem studied (if this can be given briefly; if not, consider doing this in the discussion)
- Any limitations imposed upon the study
- Procedures and methods employed in the study (briefly!)
- Acknowledgement of indebtedness to others (if not given in an earlier preface or letter of transmittal)
- Preview of the report that follows

If all of these topics are separately treated, the introduction might run to considerable length. Some of these topics can be combined and handled very briefly. Others can be omitted altogether if they are not pertinent to the report that follows. Certainly, for a brief report of some ten pages or so, the introduction should seldom exceed one page.

Read over the following introduction to our student feasibility report:

> In recent years sediment deposits have built up along most of the shoreline area of Wyoming Lake making swimming and other recreational activities undesirable and even hazardous. Under contract to the Wyoming Lake Citizens Committee, I have studied the problem and evaluated the feasibility of removing these deposits in order to rehabilitate the lake and restore it as a recreational resource.
>
> The study (1) analyzed the location, depth, and content of the deposits; (2) investigated the equipment and methods needed for the removal of the deposits; and (3) using the criteria of time, cost, and benefits and problems to the community, evaluated the feasibility of the project.
>
> This report summarizes the results of the study, draws conclusions from the results, and recommends actions to be taken by the Citizens Committee. A detailed discussion of the results is presented in the annexes to this report.

Despite the brevity of this introduction, it accomplishes a good deal. Breaking it down, we see that it does the following:

- Defines the problem, relates historical background, and gives the reason for conducting the study
- Establishes the contractual relationship between the author and the Citizens Committee
- Gives the purpose and scope of the study, including the criteria used
- Relates the report to the study and previews the report

The prose in the introduction is businesslike and yet reads easily. In the first paragraph the author economically sets forth the rationale of her study. The second paragraph does double duty in that it depicts the author's logical procedure and also previews the organization used in the discussion. The last paragraph makes the transition from the actual study that was conducted to the presentation of the results of that study—that is, to the report the reader has at hand. The author has made a good beginning.

Discussion

The discussion of a feasibility report presents and discusses the results of the study—the information upon which the investigator bases the conclusions and recommendations. In logical progression, the discussion therefore precedes the summary, conclusions, and recommendations of the report. In traditional report formats, the logical order is followed. Also, in brief reports presented as memorandums, the logical order is frequently used on the grounds that the reader does not have to wait very long for the conclusions and recommendations. However, increasingly, in long reports the discussion is removed from this central position in the report. Feasibility reports are executive reports, and executives, for the most part, like the key data, conclusions, and recommendations early. Following this executive preference, many feasibility report writers these days present their reports in this order:

- Introduction
- Factual summary
- Conclusions
- Recommendations
- Discussion

When the discussion is removed from the center of the report and placed near the end of the report, it is often labeled "Annexes." The table of contents to our sample report presented in Figure 17-1 shows the discussion in this annex position, which is the format our student actually chose. Figure 17-2 shows a more traditional format with the discussion in a central position. In either central or annex position, the discussion must present enough evidence to justify the conclusions and recommendations reached.

There is no rigid organizational plan to be followed for the discussion itself. But three plans seem to be used more than others:

- Problem-solution-evaluation (see pages 354–357)
- Argument (see pages 141–144)
- Comparison (see pages 112–113)

No doubt, the heavy use of these three plans results from the nature of feasibility reports. In them you are either presenting and evaluating the solution to a problem, arguing for your conclusions and recommendations, or comparing two or more alternatives. Whichever plan you choose, remember where you are on our sales-

TABLE OF CONTENTS

FIGURE 17-2   Traditional Plan

person-scientist scale (see page 13). You have made your study as a scientist without preconceived bias, and you must report it in the same way. That is, you must present a balanced discussion, pointing out the risks in your recommendations as well as the benefits.

Another glance at our sample report's table of contents will show you that the author chose a problem-solution-evaluation approach:

A. LOCATION AND DEPTH OF SEDIMENT DEPOSITS
   [The definition of the problem, quite literally its extent and depth.]
B. LAKE REHABILITATION
   [The discussion of the solution to the problem in terms of equipment needed, method, legal restrictions and procedures, and the disposition of the removed deposits.]
C. EVALUATION OF PROJECT
   [The application of the criteria of time, cost, and benefits and problems to evaluate the solution. As part of the evaluation, two alternative ways of carrying out the solution are compared.]

We reproduce parts of our sample feasibility report here for your examination. The first segment is from Annex A in which the writer defines the problem. In brackets we refer to our reproductions of her figures and tables.

A. LOCATION AND DEPTH OF SEDIMENT DEPOSITS

The locations of the sediment deposits in Wyoming Lake are indicated by the shaded areas in Figure 1 [Figure 17-3], designated A, B, C, etc. The accompanying table indicates the square yardage in each area. The mapped areas indicate that a considerable portion of the lake has sediment deposits. In over five miles of shoreline there are few areas that can be used for swimming or wading. The only area that does not contain sediment deposits is near the shoreline along the park at the south end. However, swimming has been discontinued in this area also.

Depths of a selected number of deposits in Wyoming Lake are indicated in Figure 2 [Figure 17-4]. The readings

FIGURE 1. Sediment
Deposits in Wyoming
Lake

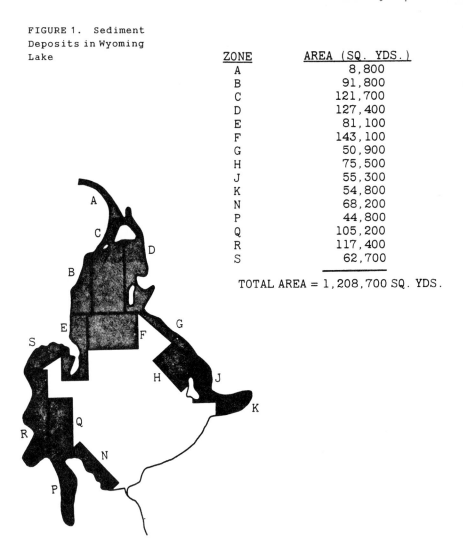

ZONE	AREA (SQ. YDS.)
A	8,800
B	91,800
C	121,700
D	127,400
E	81,100
F	143,100
G	50,900
H	75,500
J	55,300
K	54,800
N	68,200
P	44,800
Q	105,200
R	117,400
S	62,700

TOTAL AREA = 1,208,700 SQ. YDS.

FIGURE 17-3 Sample Report, Figure 1. [Figure based on data from the following unpublished report: Charles J. Anderson, *Sediment Deposits in Wyoming Lake* (A report for the Wyoming Lake Citizens Committee, Wyoming Lake, Wisconsin, 1980), p. 6–8.]

were taken at selected locations around the lake to give an

overall picture of the depth of the deposits beneath the

water. The depth readings show that deposits are generally

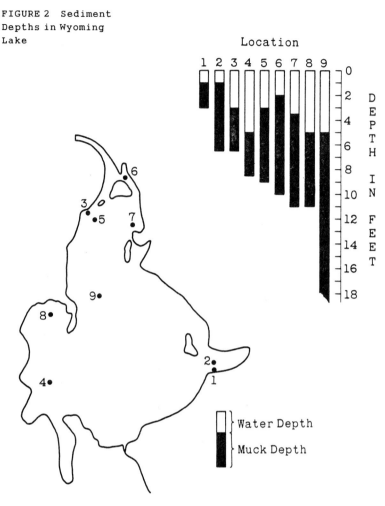

FIGURE 2 Sediment Depths in Wyoming Lake

Location

Water Depth
Muck Depth

FIGURE 17-4 Sample Report, Figure 2. [Figure based on data from the following unpublished report: Charles J. Andersen, *Sediment Deposits in Wyoming Lake* (A report for the Wyoming Lake Citizen's Committee, Wyoming Lake, Wisconsin, 1980), 8.]

6 to 11 feet deep below 3 to 5 feet of water. The deposits extend out over a large area, in some cases 300 to 400 yards from the shoreline. Removal of 6 feet of these deposits would provide water depth of 9 to 12 feet and eliminate most deposits down to the solid base.

> Table 1 [Figure 17-5] provides estimates of the volume
>
> of sediment removed and the volume of water obtained by
>
> removing 3 or 6 feet of the deposits. The elimination of the
>
> large quantity of sediment, coupled with the addition of
>
> 260 to 520 million gallons of water, would substantially
>
> improve the condition of Wyoming Lake.

In the discussion, enough detail is presented and analyzed so that the readers can comprehend the true extent of the problem. The analysis uses the logical techniques of generalizing from particulars and particularizing from generalizations—that is, induction and deduction (see pages 80–83). Recognize that every reader may not want this much detail. That is why our student writer has chosen to put these details in the subordinate position of an annex. But for those readers who want or need the details, she has presented them with care. Her use of figures and tables is particularly skillful.

In comparing two alternatives for dredging the lake, the writer presented a good many calculations showing how she arrived at her cost figures. We present one such section for you here. Notice the writer's use of lists and informal tables to make her data easily accessible to the reader.

TABLE 1. Sedimentation Volumes in Wyoming Lake

Depth of Sediment Removed	Volume of Sediment Removed	New Water Depth	Water Added to Lake Body
3 feet	1.2 million cubic yards	6 to 8 feet	262 million gallons
6 feet	2.4 million cubic yards	9 to 11 feet	524 million gallons

FIGURE 17-5 Sample Report, Table 1

<u>Purchase of dredging equipment.</u> Estimates of the cost of dredging equipment suitable for Wyoming Lake are based upon the following operational requirements:

Maximum distance pumped	6,000 feet
Maximum discharge elevation	82 feet
Total material removed	2,400,000 cubic yards
Maximum digging depth	12 feet

Based upon the above requirements, the proposed equipment consists of the following:

1	14″ Hydraulic dredge	$210,000
1	14″ Booster pump	$ 80,000
6,000 feet	16″ Discharge pipe	$ 50,000
	Pontoons, connectors, and joints	$ 50,000
Total Equipment Costs		$390,000

The booster pipe is required because of the long distance to the farthest deposit areas and the necessity of pumping to an elevation of 82 feet.

The 16″ discharge pipes are specified to reduce the frictional loss in the pipe, so that long distances and high elevations can be accommodated.

Factual Summary

With the factual summary we begin to sum up the feasibility study. In the discussion all the pertinent and concrete evidence has been presented. But what does it all add up to? The informed and dedicated readers might be able to draw their own conclusions, but the researcher-author occupies a position of special advantage: complete familiarity with the material and its ramifications. The author must draw upon that special position to report the outcome of the study clearly and succinctly.

In its full logical form, the outcome of a feasibility study is reported in three stages:

Factual Summary	*Conclusions*	*Recommendations*
Two or more facts given in meaningful context.	Interpretive generalizations, implications.	Action urged as a consequence of conclusions.

In this section we deal with the factual summary, in the next two with the conclusions and recommendations (see also pages 222–227).

Creating a good summary is one of the hardest challenges you face as a writer. In meeting that challenge, keep these principles in mind:

- You can summarize only that information appearing elsewhere in the report. A factual summary must not introduce new or additional formation.
- Make every statement a genuine assertion of fact. Unless this rule is observed, you muddy the waters and offer an unsure footing for conclusions to follow.

  No: Every American home should have a telephone.
  (Who says so? This statement sets up an arbitrary standard of the author's own making. Avoid preferential and opinionative wording such as *ought, should, better, best.)*
  Yes: In 1979, 98% of U.S. households had telephone service *(Statistical Abstract of the United States,* 1981, p. 562.)
- Integrate your facts for the reader. Gather facts together from various places in the report and put them together in a meaningful context. In your discussion you may in three separate statements offer these facts:
  Model A retails for $425.
  Model B retails for $385.
  Model C retails for $465.
  In your summary bring these three separate statements together in a statement like this one:

  Of the three models tested, model B at $385 costs the least.

  You may have to reword, rearrange, and otherwise process your data, but don't change their meaning.
- Make every statement pave the way for one or more conclusions to follow. The only valid reason for summarizing facts is to use them as a foundation for conclusions.

We reproduce for you here the summary from our student report. To demonstrate the ratio of summary data to discussion data, we have underscored the two summary statements that relate to the discussion sections we have already shown you on pages 390–394.

Throughout her summary, the writer has integrated her data and boiled down her data and calculations to those absolutely essential to support the conclusions that follow.

FACTUAL SUMMARY

Wyoming Lake was once a widely used recreational area. However, over the past decade the lake has stagnated, clogged by sediment, making the lake unfit for many recreational uses.

There are presently 2,400,000 cubic yards of sediment deposits averaging 6 to 11 feet in depth along most of the shoreline of the lake. In order to remove these deposits and rehabilitate the lake, certain steps have to be taken. First, the city of Wyoming Lake has to apply for a dredging permit from the Wisconsin Department of Natural Resources, Division of Resource Development and from the U.S. Army Corps of Engineers. After permission is granted, the lake can be dredged. With prompt application on the part of the city, the dredging project could be designated a pilot project by the Department of Natural Resources and receive a 50% subsidy.

Deposit areas where the removed sediment can be used as fill are available within pumping distance from the lake. These areas can actually benefit from this fill

because the lake bottom sediment tests out to be about the

same as most agricultural soils. Thus, the deposits would

improve subsoils and scalped areas and would be suitable

for filling and reclaiming nearby abandoned gravel pits.

A hydraulic dredge is used for dredging operations.

Two alternatives for accomplishing the dredging were

considered: the city's contracting with a dredging firm and

the city's buying a dredge and doing the job itself.

Informal bids for the project received from dredging

firms fell into a range from $720,000 to $1,200,000.

If the city bought a dredge and did the job itself, <u>the</u>

<u>cost of the dredge and associated equipment would be</u>

<u>$390,000</u>. The operating cost for the two years estimated to

be necessary for completing the project would be

approximately $304,000. The total of purchase price and

operating costs is $694,000, a cost lower than the lowest

bid. Also, the dredge would be available for future

projects or resale at approximately 60% of its $210,000

purchase price.

Other cities in the state that have dredged their

lakes have found that property values and tax revenues have

increased when their lakes once more provided a healthy

recreational environment in which to swim, fish, and water

ski.

Problems that other cities have encountered include

paying for the project, poor communication with the public

during the project, and gaining approval of the dredging

work by the Department of Natural Resources as the work

proceeds. Bond sales, local assessment, and state and

federal assistance are all available as means of paying for

the project. Appointment of an experienced project

coordinator is a way of solving the remaining problems.

Conclusions

As shown in the schematic on page 395 and as discussed on pages 223–224, conclusions act as the intermediate step or bridge between the factual summary statements and the recommendations. They are the inferences and implications you draw from your data through the use of induction and deduction. They are the answers you should be able to give if someone looks at your data and asks, not unkindly, "So what?"

Conclusions can be presented in normal essay style or in a series of short, separate statements. Perhaps for reasons of succinctness and to make the logic of the writer's reasoning more evident, the presentation of short statements is an often-used style. It is the style used by our student writer. However you present your conclusions, you should arrange them in a clear and premeditated order that reveals your thinking process to the reader. Do not insist that the order of events in the factual summary be exactly paralleled in the conclusions. Do not insist that each statement in the factual summary be represented by one and only one conclusion. The truth is that a single conclusion often embraces two or more factual summary statements—and sometimes not consecutive ones at that.

We can give you no foolproof guidelines to the number of conclusions you should have. Too few and your logic will not be adequately revealed. Too many and you run the risk of confusing your readers. However, we can point out that a prime way to make an error in writing conclusions is to assume that something that is obvious to you will be obvious to your readers. Remember that you have been living with your material for a long time and know it well. Not knowing the territory as well as you do, your readers may miss implications that you should point out. Do your readers a favor and wrap things up neatly for them.

If your readers can read your introduction, factual summary, conclusions, and recommendations and feel that you have justified your recommendations, you have probably presented your material well. The readers should not be left wondering why you chose alternative A over alternative B. Ideally, your last conclusion should serve as a stepping-off platform for the recommendations that follow.

We present the conclusions drawn by our student writer. The numbers in parentheses following each conclusion refer to pages in the annexes where complete support for the conclusions can be found—an excellent practice. As we frequently emphasize, help your readers to read your report selectively.

```
CONCLUSIONS

     1. Wyoming Lake will become completely useless as a

recreational resource if its sediment deposits are not

removed (pp. 1-5).

     2. Suitable dredging technology for removing the

sediment within two years and environmentally sound areas

for depositing the removed sediment are both available (pp.

5, 9-11).

     3. The total cost of the city's purchasing a dredge and

the associated equipment and running the dredging project

itself is lower than the lowest bid for contracting the

project with a dredging company (pp. 11-14). Therefore, the

city serving as its own contractor seems the sounder choice

of the two.
```

4. The appointment of an experienced project

coordinator should forestall potential problems of

communication with the public and poor coordination with

the Department of Natural Resources (p. 17).

5. Standard techniques for funding the project such as

bond sales and tax assessments are available. State and

federal assistance is also available (p. 17).

6. With prompt action, state and federal assistance

for up to 50% of the cost may be available (p. 11).

7. The city should benefit from a healthy recreational

environment, and property values should increase resulting

in increased tax revenues (p. 15).

8. The project, given the right decisions and prompt

action, is feasible. The first step is application for

approval of the project by the Wisconsin Department of

Natural Resources and the Army Corps of Engineers.

Recommendations

In contrast to conclusions, a recommendation is an action statement. That is, it recommends that the report users take some proposed action or refrain from taking it. *Most important, the first (or only) recommendation discharges the purpose set forth early in the report.*

If the purpose is to determine whether X Company should establish a branch in Memphis, Tennessee, and the investigator-reporter has determined the plan is feasible, the recommendation (first or only) should read something like the following:

X Company should establish a branch plant in Memphis, Tennessee.

Any further recommendations usually implement the first. If the first recommendation favors establishment of a branch plant, succeeding recommendations may detail the necessary steps to carry out the plan. If the first recommendation opposes the establishment of a branch, succeeding recommendations may propose alternative plans or urge that the whole question be reinvestigated several years hence.

Recommendations are opinions, professional judgments. No logic, no array of evidence—nothing under the sun—can make them anything else. Let them therefore be sound and well considered.

The recommendations offered by the investigator-reporter may or may not be accepted by the customer or report user. However, for the feasibility study report the author is obliged to arrive at one or more recommendations. If the research has been properly conducted, then these recommendations should carry great weight.

Our student-writer's recommendation section consists of one primary recommendation followed by five implementing recommendations.

```
RECOMMENDATIONS

     City of Wyoming Lake should take the necessary steps

to rehabilitate Wyoming Lake as follows:

     • Apply promptly to the Department of Natural

Resources and the Army Corps of Engineers for permission

to begin dredging.

     • Take necessary steps to qualify the rehabilitation

project as a pilot project eligible for state and federal

assistance at the 50% level.

     • Appoint a project coordinator.

     • Appoint a committee to investigate and report on the

best method of funding the project.

     • When funding is assured and all permissions secured,
```

```
purchase a dredge and the associated equipment and begin

operations under the leadership of the project

coordinator.
```

When as the writer you have stated your recommendations, you have completed your objectives as an investigator. Your task is over. For an example of a complete feasibility report, see Appendix A, pages 489–504.

EXERCISES

1. The feasibility study is a necessary prelude to any important action. In your local newspaper identify any news items that call for feasibility studies prior to action. Is your community or school considering any actions that should be preceded by feasibility studies?

2. Prepare a slip like the one shown here, but substitute your own subject matter following the four colons:

    General field: Meterology
    Specific topic: Short-term weather forecasting
    Purpose: To determine the feasibility of making 24-hour forecasts for specific areas to expedite road crews for highway storm treatment
    Customer: Richard Ferguson, Road Maintenance Engineer, state Department of Transportation

3. Using the slip you prepared for Exercise 2, determine the scope of your treatment (see pages 383–384). Rough out the areas of information you will need and the methods you will use to gather the information (see Chapter 5, "Gathering and Checking Information."

4. Using the Proposal for a Report shown in Figure 15-3 on pages 366–367 as a model, submit a proposal for the feasibility study and report described in Exercises 2 and 3.

5. Prepare an organizational plan for your report. Begin with an updated purpose, scope, and audience statement. Decide on an appropriate format for the report. Will it be a memorandum or a full report? If a full report, which elements of formal reports will you include? How many and what kinds of graphics do you anticipate using in your report? Justify your choices (see pages 161–165).

6. Write the report planned in Exercise 5.

Chapter **18**

Empirical Research Reports

CAN AMERICAN CATERPILLARS be crossbred with oriental silkworms? Is baby talk good for children's language development? Can wine be made from milk? What is a feasible way to store heat gathered by solar collectors? Can optical fibers be used as acoustic sensors?

To get answers to questions of this kind, you can do two things:

- You can find, usually through library research, the answers that previous researchers have obtained.
- You can obtain firsthand answers for yourself by the direct empirical methods of experimentation and observation.

Of course, you can—and probably should—use these two methods in combination. However, in studies and reports of the kind treated in this chapter, the emphasis falls upon empirical methods.

An illustration may help to clarify our point. Suppose that we have a chunk of glass, crude and irregular, dumped out of the ladle and unmolded. We desire to find the impact strength of the chunk of glass, that is, how many pounds of force will be required to shatter it. We may approach our solution in two ways:

- We may read up on the chemical makeup of the chunk of glass. We may measure its geometric properties. We may pass white light through it to obtain a reading of its internal structure. By turning to suitable handbooks we may then estimate the minimum impact force required to shatter the chunk.

- We can whack the chunk with a hammer, hitting harder and harder until it shatters. A pressure gauge or similar accessory will tell us how hard we had to hit to get the result we wanted. This is the pragmatic test, pure and simple.

The empirical research study places the emphasis on the second approach.

Empirical research is very common in our daily lives. Take the bachelor keeping house for himself as a prime example. Suppose he has just had a new picture window installed, 5 feet high and 12 feet wide.

Since buying a home six years ago, he has become a faithful reader of the "Household Hints" column in the *Morning Mirror*. He tries to recall what the contributors to the column have had to say about cleaning large windows. What did his parents do? He recalls a mishmash of advice: Start at the bottom. Start at the top. Use a circular cleaning motion. Use broad horizontal strokes. Never clean with the sun on the glass. Never clean the inside when it's dark outside; you'll be sorry in the morning. Squeaky glass is clean glass. Scouring powders will dull the surface. For guaranteed results, call your Handi-Dandi Window Cleaning Service. Helpful or not, this body of "theory" runs through his head as the time comes to clean the new picture window. What cleaning agents should he use?

A commercial window spray?
Detergent and water?
A powder such as Belle de Jour?
Ammonia and vinegar diluted with water?

And what should he use to wipe it? His cabinet under the sink holds an extensive array of likely tools: natural sponges, synthetic sponges, rolls of cotton gauze, paper towels, old Turkish towels, torn bed sheets, and chamois skins.

Though he may never state his objectives in so many words, he senses them keenly. He wants to keep the cleaning costs down. He wants a window that is clear and sparkling clean. Washing windows is not his favorite pastime, so he wants the glass to stay clean as long as possible. And he wants to get the job done quickly and efficiently, avoiding spillage, fuss, and excitation of any allergies he may have.

Being a conscientious home owner, he now tries out the various cleaning agents, wiping materials, and application techniques. It's a long haul. Every cleaning session is followed by hard stares at the window and by stock taking and debate. He concludes that every method is less than perfect but that some are better than others.

During coffee breaks at work, he may report to his colleagues on his efforts. He may explain the problem he faced, summarize what the *Morning Mirror* said about window cleaning, describe his cleaning apparatus and methods, detail which methods left streaks and which didn't, and, finally, tell how his experiments helped him decide which method was best. In his informal conversation, he has followed, quite naturally, what has come to be the standard format for reports of empirical research:

Introduction and Literature Review
Materials and Methods
Results
Discussion

When put into final form, the empirical research report may also contain the usual elements found in reports such as tables of content, abstracts, and references. Where the research is reported—usually either in a journal article or in a student thesis—will determine the exact format. Because we cover the additional parts in Chapter 11, "Formal Elements of Reports," we deal here only with the sections in the list above.

To illustrate the attributes that these sections should have, we draw upon several well-written research reports. Although the reports used are written by experts for their fellow experts (see 36–40), we have selected passages that you should be able to read regardless of your specialization. On several occasions, however, we do define terms that you as a nonspecialist might not know. We place these definitions in brackets to distinguish them from the authors' work.

INTRODUCTION AND LITERATURE REVIEW

When research reports are presented in journal articles, most often the introduction and literature review are integrated. The major function of the introduction is to describe the subject, purpose, and scope of the research. The literature review, as the name implies, reviews the scientific literature pertinent to the research being reported. In a journal article, the main function of the literature review is to define the problem that has been investigated. It usually leads into the statement of purpose. Because space in journals is expensive, the integrated literature review is held to material absolutely necessary to the investigation. Generally, relevance is measured by whether the literature cited helps to define the problem or to explain the choice of methodology or materials.

When a research report is presented as a student thesis rather than an article, the literature review is often quite extensive and presented separately from the introduction. When such is the case, its purpose is not so much to introduce the research as to demonstrate the writer's mastery of a certain subject matter. In writing a thesis, always check with your advisor to determine the type of literature review required and the subject matter coverage desired.

Read over now the introduction and literature review of the article, "Sex Determining Temperatures in Turtles: A Geographic Comparison."

> The sex of many turtles is determined by the incubation temperature of the egg (Bull, 1980). Laboratory studies of six genera of turtles, for example, show that a developmental temperature of 25 C produces all males, 31 C or higher produces all females, and survival is sufficiently high in some studies to conclude that this is not due to differential death of the sexes (Pieau, 1975; Yntema, 1976; Bull and Vogt, 1979; Yntema and Mrosovsky, 1980). The laboratory studies have been corroborated by field studies, suggesting the nest temperature is the sex determining agent in these species (Bull and Vogt, 1979, and unpubl.).
>
> Temperature-dependent sex determination is unique because a variety of factors interact to determine the sex ratio: (1) maternal behavior in choosing a nest site, (2) the zygote's [the fertilized egg] response to temperature in becoming male or female, and (3) environmental effects on the temperature of the nesting area. As a result of the environmental component to sex determination, temporal and spatial variation of the environment may cause the sex ratio to vary. On the one hand, this sex ratio variation may jeopardize the evolutionary stability of temperature-dependent sex determination (Bull, 1980, 1981; Bulmer and Bull, 1982; see also Charnov and Bull, 1977). Here, however, we consider how the sex ratio evolves in response to the environment, rather than evolution of the sex determining mechanism per se. Long term environmental changes may lead to temporary sex ratio biases, but from Fisher (1930), selection acting on maternal behavior and/or zygotic sex determination would be expected to restore the equilibrium sex ratio. A straightforward prediction of this hypothesis is that, in the warmer parts of a species range, (i) mothers should choose relatively cooler nests, and/or (ii) embryos should develop as male/female at higher temperatures than in the cooler localities. We have investigated the latter possibility by comparing sex determining temperatures among turtles from different geographic areas.[1]

As is typical of a scientific research article, this article is documented by parenthetical references (see pages 239–241). Notice how the authors use the literature cited to define the nature and scope of the problem. They state the precise objective of the research clearly in the form of a hypothesis: "in the warmer parts of a species' range ... embryos should develop as male/female at higher temperatures than in the cooler localities." We know now what question the researchers have asked of nature. We will expect before the article is finished to have an answer to the question: to know whether the hypothesis has been proved or disproved.

The objectives need not always be presented as a hypothesis. Frequently, they are expressed as questions, as in this passage from another report:

> We then seek to answer the following questions raised by the model: (1) Is the spatial arrangement of nectar rewards within the inflorescence [pattern of flowers on a stalk or in a cluster] exploited by foraging bees in the manner predicted by our model? (2) Does the plant's pattern of nectar rewards elicit bee behavior which promotes pollen transfer?[2]

Sometimes, the writer simply states what was studied, as in this statement from "Sexual Selection in a Brendit Weevil":

> I report here the results of a study of the influence of body size on mate preference and on success in intrasexual competition in a natural aggregation of the neotropical brentid weevil, *Brentus anchorago* L. (Coleoptera, Curculionoidea, Brentidae).[3]

No matter how you present your purpose or objectives, be sure to present them with absolute clarity. No doubt should exist in the reader's mind concerning what you were up to in your research.

The weevil article also demonstrates how the literature review can be used to explain the choice of materials used in the research:

> Brentids make good candidates for studies of sexual selection and individual variation because most species of the family exhibit pronounced sexual dimorphism [differences between the sexes in characteristics such as size and color] (Muizon, 1960; Haedo Rossi, 1961; Damoiseau, 1967, 1971). The males generally possess greater body length, a stouter rostrum [beak or snout], and more powerful mandibles [jaws], one of which may be grossly enlarged (Darwin, 1871). Within each sex there is impressive phenotypic variation in body size, especially in males, which fight one another with snout and mandible for access to females (Wallace, 1869; Meads, 1976). The most size-variable brentid may be *B. anchorago:* after examining a large series of this species, Sharp (1895) commented that "the variation in length is enormous, and perhaps not equalled in the case of any other species of Coleoptera, small males being only 10–11 mm long, while large examples of the same sex attain 52 mm." Such variation in size is common within a single aggregation, and is important in male mating success, in female choice, and in patterns of mating in the aggregation as a whole.[4]

Choosing proper verb tense is frequently a problem in writing a literature review. It will help to keep these principles in mind. When referring to the actual work that researchers have done in the past, use the past tense. When referring to the knowledge their research produced, if the knowledge is still considered to be true, use the present tense. Thus the literature review usually mixes together past and present tense as shown in this passage in which we have italicized the verb forms:

> The best evidence for this theory *was provided* by Paulson (1974). He *pinned* live females of a number of species of coenagrionid damselflies to sedge stems and *presented* them to males of the various species. He then *scored* whether the male *approached* and *attempted* to clasp the female, and whether he *was* successful in clasping her. . . . Thus, for most of the species, reproductive isolation *is* normally *maintained* by the inability of males to clasp non-conspecific females with their anal appendages, presumably because their differently shaped and sized appendages *are* not compatible with the thoracic structure of the non-conspecific females[5].

MATERIALS AND METHODS

The major criterion by which you can measure the success of a materials and methods (M & M) section is simply stated: Using this section, an experienced researcher in the discipline should be able to duplicate the research. For a second criterion, an experienced researcher should be able to evaluate the research by using this section. If these criteria are not met, the M & M section fails.

M & M sections follow a fairly definite pattern. As pointed out in the excellent *Council of Biology Editors Style Manual*, "The usual sequence for experimental studies is design of the experiment, subjects (plant, animal, human), materials, procedures, and methods for observation and interpretation."[6] Not every M & M section will contain all these parts, but every one should contain all of the parts needed to meet the criteria of duplication and evaluation.

Design of the Experiment

Give your readers an overview of your research design before you plunge them into the details. The overview need not, usually should not, be elaborate. Our turtle researchers give it in one sentence and, indeed, include in the same sentence information about the materials used:

> The effect of temperature on sex determination was studied in turtles of the subfamily Emydinae, genera *Graptemys* (map turtles), *Pseudemys* (sliders), and *Chrysemys* (painted turtles), from populations in the northern U.S. (Wisconsin) and southern U.S. (Alabama, Mississippi, and Tennessee) (Table 1).[7]

Materials

We combine the *CBE Style Manual*'s "subjects and materials" into one part: materials can be human, animal, vegetable, or mineral.

They are whatever you used by way of subjects, material, or equipment to do your research. In a report for the social sciences, instruments such as questionnaires would be described in this section. Remember that your descriptions of your materials have to be accurate enough that your readers could obtain similar materials. In the case of animals and plants this usually means using their scientific as well as their common names. If you have a good deal of necessary information about your subjects or materials, use a table to display some of it. All these attributes are exhibited in the passage quoted above from the turtle research. (See our Figure 18-1 for the Table 1 referred to in the passage.)

Sometimes the descriptions of materials can be precise indeed, as in this passage:

> The organism used in this study was the wild type Chicago strain of the confused flour beetle, *Tribolium confusum* Duval, maintained by Prof. Thomas Park at the University of Chicago for over four decades on a single-sifted flour medium.[8]

Equipment used throughout the experiment can be described at this point, as in this passage:

> A bank of five parallel G.E. G8T5 germicidal lamps were used to generate ultraviolet light predominantly at 254 nm. The uncovered dishes to be irradiated were placed on a rotating platform 82 cm from the light source. A 10 cm diameter aperture midway between the light source and the rotating platform was used to collimate the incident light and to reduce shielding by the sides of the culture dishes.[9]

Equipment used only at specific times in the experiment will more likely be described at the appropriate place in the procedure, as in this sentence: "Thirty young adults drawn from a common stock were placed in a rearing jar."[10]

Procedure

In the procedure part, you tell your readers step by step how you did your experiment. The description should be as complete as necessary, but remember you are writing for an expert audience. When you are working with a procedure or equipment common in the discipline, you do not need to describe it in detail. For instance, in the sentence just quoted the writers do not define or describe a rearing jar. However, if you anticipate that your readers might have some question about why you conducted some step as you did, take time to explain.

You can save a great many words by referring to procedures

TABLE 1. Sex ratio as a function of incubation temperature in turtles. (Data presented as percentage male, with the sample size in parentheses.)

Species	Loc[1]	Temperature (degrees C)						
		28.0	28.3	29.0	29.3	29.5	30.0	30.6
Chrysemys picta	TN	2 (41)	9 (11)	0 (12)	—	0 (5)	0 (16)	0 (14)
Chrysemys picta	WI	98 (94)	—	63 (38)	—	0 (7)	0 (56)	0 (22)
Graptemys geographica	WI	100 (26)	—	33 (6)	—	—	0 (28)	—
G. ouachitensis	WI	100 (93)	—	83 (64)	—	—	1 (89)	—
G. pseudogeographica	TN	100 (7)	96 (25)	0 (5)	28 (47)	16 (25)	0 (5)	0 (22)
G. pseudogeographica	WI	100 (70)	100 (14)	92 (24)	58 (57)	33 (15)	11 (82)	0 (17)
G. pulchra	MS	100 (17)	—	0 (4)	—	—	0 (14)	—
Pseudemys scripta	AL	100 (21)	—	37 (16)	—	—	0 (17)	—
Pseudemys scripta	TN	—	92 (36)	—	—	30 (40)	—	5 (42)

[1]Locality—AL: Alabama, Perry Co., Lakeland Farms; MS: Mississippi, Greene Co., Chickasawhay River; TN: Tennessee, Obion Co., Reelfoot Lake; WI: Wisconsin, Vernon Co., Mississippi River.

FIGURE 18-1 Table 1 from "Sex Determining Temperatures in Turtles: A Geographic Comparison"

described elsewhere rather than repeating the information found in the original source. This is an excellent practice so long as you don't refer your readers to journals unobtainable either by reason of their geographic location or by their obscurity. All of these techniques are evident in these excerpts from our turtle report:

> Eggs were removed from gravid [full of ripe eggs] females or from fresh nests and placed under controlled incubation temperatures within two weeks of removal, usually within one week.
>
> . . . .
>
> In nearly all cases eggs from each clutch were divided between three different incubation temperatures in the range 28–31 C and incubated at constant temperature throughout development; preliminary work on *Graptemys* had indicated that this range included male determining as well as female determining temperatures. The temperatures 28.3, 29.5, and 30.6 C were studied at Carnegie; 28.0, 29.0, 29.3, and 30.0 C were studied at Wisconsin, with separate clutches. Since incubators were precise only to within ±0.2 C degrees of the intended temperature, there was little to be gained from further subdividing this temperature range into narrower intervals.
>
> Hatchling sex was diagnosed by inspection of gonads under a dissecting microscope (Bull and Vogt, 1979); specimens have been deposited in the Carnegie museum. All species studied here are known to develop as male at 25 C and female at 30–31 C with sufficiently high survival that the possibility of differential mortality of the sexes can be ruled out as the cause of the sex ratio biases (Bull and Vogt, 1979, and unpubl.).[11]

Methods for Observation and Interpretation

Finally, tell your readers how your observed your materials during the experiment and how you interpreted your results. Because these methods of observation and interpretation are often standardized, this part can frequently be quite short, as is this passage from our weevil report:

> The behavioral sampling techniques I used are described in Altmann (1974). Sequences of male and female behavior that occurred around drilling females were dictated into a tape recorder, using Focal-Animal sampling. Focal animals were selected sequentially along the log to reduce repeat samplings to a minimum. Periodically I scanned the entire log and recorded the lengths of all weevils involved in copulations, fights, and guardings by an All Occurence of Same Behaviors sampling technique. All major types of behavior were videotaped, and the verbal descriptions of motor patterns were prepared from these tapes.[12]

Notice that in the last passage the author uses the first-person *I*. However, also notice that in most of the passages we have quoted from M & M sections the authors have seldom used the first person and have used far more passive voice than active voice sentences. In most cases it is either obvious that the researchers performed the steps described, or it is unimportant who performed them. Under such circumstances, the passive voice is as good a choice or perhaps even a better choice than the active voice. But don't fear using the active voice and first person when they seem appropriate to you. Most modern style books encourage such practices, and an occasional *I* or *we* reminds your readers that real people are at work.

Also, remember that the passive voice used carelessly creates a great many dangling modifiers: "After drawing the blood, the calf was returned to the pen." Here the case of the blood-letting calf can be cleared up with a judicious use of the active voice: "After drawing the blood, I returned the calf to the pen." (See page 131 and 180–181).

RESULTS

Because your results section answers the questions you have asked, it is the most important section of your report. Nonetheless, it is often the shortest section of an empirical report. It often takes a great deal of work to gain only a few bits of knowledge.

Begin your results section with an overview of what you have learned. The first sentence or two in the results section should be like a lead in a newspaper story where the main points of the story are quickly given, to be followed in later paragraphs by the details. In a results section, the following details are presented in tables and graphs if at all possible. You do not then need to restate such details. But you may want to refer to key data, both to emphasize their significance and to help your readers comprehend your tables and graphs. The following excerpts from our turtle report demonstrate these attributes:

> The hatchling sex ratios obtained are shown in Table 1 [see our Figure 18-1]. A striking result in all populations is that there is an abrupt change from virtually all males to all females over a narrow range of temperatures. In every sample except Tennessee *C. picta*, all or nearly all turtles are male at 28 C, female at 30 C and above. (Tennessee *C. picta* are nearly all females at 28 C and above, all males at 25 C, so there is an abrupt change in this population as well.) This abrupt change in sex ratio is referred to as a threshold (Bull, 1980), and the temperature producing a sex ratio (proportion male) of ½ is defined here as the *threshold temperature.*
>
> . . . .
>
> The samples at 29 C offer the most extensive set of interspecific comparisons for a temperature within the range of intermediate sex ratios. Among these there is no indication that threshold temperatures are higher in southern species, and some comparisons even suggest that

threshold temperatures can at most be only marginally higher in the southern species.

. . . .

Even if threshold temperatures are higher in most of these southern populations (having escaped detection because of the small samples or possible experimental errors), in several cases the maximum difference is of the order of 0.4 C degrees.[13]

At this point, probably even a nonbiologist could tell whether the turtle researchers' hypothesis is proved or disproved. If not, the discussion should tell us.

DISCUSSION

The discussion interprets the results. It answers questions such as these:

- Do the results really answer the questions raised?
- Are there any doubts about the results? Why? Was the methodology flawed? How could it be improved?
- Were the research objectives met?
- Was the hypothesis proved or disproved?
- How do the results compare with results from previous research? Are there areas of disagreement? Can disagreements be explained?
- What are the implications for future work?

Though the discussion section may cover a lot of ground, it should not ramble until the editor cuts it down. Therefore, keep it tightly organized around the answers to the questions that need to be asked. Our turtle researchers did exactly that, as these excerpts from their discussion show:

According to our hypothesis, the southern populations should have higher threshold temperatures, perhaps by as much as 2 C, to compensate for the presumed warmer nests.

The data overwhelmingly indicate that (1) threshold temperatures do not increase to nearly the same magnitude as climatic temperatures between the Wisconsin and southern populations. In many cases threshold temperatures are at most only slightly higher (0.3–0.4 C) in the southern population, if indeed higher. We therefore must reject the hypothesis that differences in threshold temperature between populations are of the same magnitude and direction as differences in climatic temperature. Furthermore, (2) in the two northern intraspecific comparisons, the threshold temperatures differ in the opposite direction as climatic temperatures, and not even one North-South interspecific comparison provides significant evidence of a higher threshold in the southern population. We therefore reject even the less stringent form of our hypothesis—that, regardless of the magnitude of difference, climatic and threshold temperatures merely covary in the same direction. Of course data on additional populations may show these results to be atypical of turtles, but even so, these results require explanations in themselves.

If threshold temperatures do not increase with ambient temperature, there might be greater female-based sex ratios in the South than in the North. Preliminary evidence is to the contrary, however. In a study of three U.S. turtle communities, McCoy and Vogt (unpubl.) observed an adult sex ratio of nearly ½ in *Pseudemys* and female-based sex ratios in several species of *Graptemys*. However, Vogt (1980) observed even more heavily female-based sex ratios in two species of *Graptemys* at the Wisconsin site, and the sex ratio biases were documented using a variety of collecting techniques (Vogt, 1980a). Thus although southern localities are warmer than northern ones and threshold temperatures do not compensate, sex ratios are not obviously more female-based in the South.

We consider three possible explanations for a lack of a positive association between environmental temperature and threshold temperature in these turtles. (1) The characteristics we diagnose as male and female in hatchlings may be irrelevant to sex in adults, if the sex phenotype changes later in life. Although numerous observations support a hypothesis of constant sex phenotype (Yntema, 1976, 1981; Bull and Vogt, 1979; Vogt, 1980), unequivocal data are lacking, and this should be entertained as a possibility. (2) The threshold temperatures in this study may not accurately reflect sex determination in nests. Nest temperatures fluctuate on a daily basis, and the fluctuations possibly affect sex determination in a way not evident from threshold temperatures.

Perhaps the most likely explanation for the lack of correspondence between climatic temperature and threshold temperature is that (3) nest temperatures do not correspond to climatic temperatures, either because differences in nesting behavior compensate (overcompensate) for the climatic differences or because climatic data are misleading when extrapolated to subsurface soil temperatures. If maternal behavior compensates, this means that southern turtles choose nest sites which are relatively cool, whereas northern turtles choose nest sites which are relatively warm in their local environment. This has not been studied directly. However, southern turtles begin nesting a few weeks earlier than northern ones (Vogt, unpubl.), and this will partly compensate for the climatic differences, provided the southern turtles also cease nesting earlier or at the same time as the northern ones. Comparison of June temperatures for the southern localities with July temperatures for the northern ones removes only about half the difference in climatic temperatures, and this is likely an overestimate of the magnitude of compensation from early laying. Other means of maternal behavior compensating for climatic differences are possible as well, such as changes in nest position relative to shading vegetation. And it is further possible that the female has only a limited range of nest habitats available, so that nest temperature depends more upon the habitat than on ambient temperature. The resolution of why these threshold temperatures do not (strongly) accord with climatic temperatures clearly requires supplementation with data of nest temperatures in the different populations.[14]

A FINAL WORD

In this chapter we have given you general advice about reporting empirical research. If you are to become a professional in any field that requires such reporting, doing it well will be of vital importance to you. Therefore, we strongly urge you to examine representative journals and student theses in your discipline. Observe closely their format and style. Most journals have a section labeled something like "Information for Contributors." This section will inform you about manuscript preparation and style. Often it will refer you to the style manual, such as the *Council of Biology Editors Style Manual*, that governs the journal. Likely your library will have a copy of the manual you need. Check it out and read it carefully. It will supplement what you have learned here.

EXERCISES

1. Empirical research is primarily concerned with fact finding and interpretation. Do you see any similarity between empirical research and feasibility studies (Chapter 17)? In what major respects are they different?

2. Referring to Chapter 5 of this book, "Gathering and Checking Information," determine how the methods one uses to gather information are affected by the nature and purpose of the investigation. How do the techniques discussed in Chapter 5 relate to empirical research?

3. Choose a research problem in your discipline, perhaps with the help of an instructor in that field. Research the literature in the problem sufficiently so that you can formulate an empirical study to deal with some aspect of the problem. Then write an introduction and literature review and a materials and methods section for the study. Submit your work to both your writing teacher and the teacher in the discipline.

4. Choose an empirical research report published in a journal in your discipline. With the help of this chapter and the style manual that governs the journal, if one is available to you, analyze the report's format, style, organization, and content. How closely does it follow the principles of this chapter and the style manual. Present your analysis as a written report for your writing instructor.

Chapter **19**

Oral Reports and Group Conferences

YOU WILL MAKE oral reports for many reasons. You will have to report committee work, laboratory experiments, and research projects. You will give reports to learned societies. You will instruct, if not in a teacher-student relationship, perhaps in a supervisor-subordinate relationship. You may have to persuade a group that a new process your section has devised is better than the present process. You may have to brief your boss about what your department does to justify its existence. You will participate in group conferences.

In this chapter we divide the speech process into two parts: preparation and delivery. Under *preparation* we tell you how to organize and practice oral reports. Under *delivery* we concentrate on the physical aspects of oral reports—voice and body movement—and upon interacting with your audience. Following our discussion of the speech process, we include material about the use of visual aids in speaking, and about group conferences.

ORAL REPORT PREPARATION

In many ways preparing an oral report is much like writing a paper. Researching an oral report and a paper are identical processes. You will use the rhetorical forms we explain in Chapters 6 and 7—exposition, narration, description, and argument—as much in speaking as you do in writing.

You will analyze your audience as we described in Chapter 3. In addition, find out as much as you can about the physical layout. Inquire about the size of the room you will speak in, and the size of the audience. If you have to speak in a large area to a large group, will a public address system be available to you? Find out if you will have a lectern for your notes. If you plan to use visual aids, inquire about the equipment. Does the sponsoring group have projectors to show 35 mm slides or vugraphs? More than one speaker has arrived at a hall and found all of his vital visual aids worthless because projection equipment was not available. Find out if there will be someone to introduce you. If not, you may have to work your credentials as a speaker into your talk. Consider the time of day and day of the week. An audience listening to you at 3:30 on Friday afternoon will not be nearly as attentive as an audience earlier in the day or earlier in the week. Feel free to ask the sponsoring group any of these questions. The more you know beforehand, the better prepared and therefore the more comfortable you will be.

Choosing Delivery Techniques

A vital part of your preparation is choosing your delivery technique. There are four basic techniques, but you really need to think about only two of them. The four are (1) impromptu, (2) speaking from memory, (3) extemporaneous, and (4) reading from a manuscript. Impromptu speaking involves speaking "off the cuff." Such a method is too risky for a technical report where accuracy is so vital.

In speaking from memory, you write out a speech, commit it to memory, and then deliver it. This gives you a carefully planned speech, but we cannot recommend it as a good technique. The drawbacks are (1) your plan becomes inflexible; (2) you may have a memory lapse in one place that will unsettle you for the whole speech; (3) you think of words rather than thoughts, which makes you more artificial and less vital; and (4) your voice and body actions become stylized and lack the vital spark of spontaneity.

We consider the best methods to be extemporaneous speech and the speech read from a manuscript, and we will discuss these in more detail.

The Extemporaneous Speech. Unlike the impromptu speech, with which it is sometimes confused, the extemporaneous speech is carefully planned and practiced. In preparing for an extemporaneous speech, you go through the planning and organizing steps described in Chapter 8. But you stop when you complete the outline stage. You do not write out the speech. Therefore, you do not commit yourself to any definite phraseology. In your outline, however, include any vital facts and figures that you must present accurately.

You will want no lapses of memory to make you inaccurate in presenting a technical report.

Before you give the speech, practice it, working from your outline. Give it several times, before a live audience preferably, perhaps a roommate or a friend. As you practice, fit words to your outlined thoughts. Make no attempt to memorize the words you choose at any practice session, but keep practicing until your delivery is smooth. When you can go through the speech without faltering, you are ready to present it. When you practice a speech, pay particular attention to timing. Depending upon your style and the occasion, plan upon a delivery rate of 120–180 words per minute. Nothing, *but nothing*, will annoy program planners or an audience more than to have a speaker scheduled for 30 minutes go for 40 minutes or an hour. The long-winded speaker probably cheats some other speaker out of his or her allotted time. Speakers who go beyond their scheduled time can depend upon not being invited back.

We recommend that you type your outline. Use capitals, spacing, and underlining generously to break out the important divisions. But don't do the entire outline in capitals. That would make it hard to read. As a final refinement, place your outline in a looseleaf ring binder. By so doing you can be sure that it will not become scattered or disorganized.

There are several real advantages of the extemporaneous speech over the speech read from manuscript. With the extemporaneous speech you will find it easier to maintain eye contact with your audience. You need only glance occasionally at your outline to keep yourself on course. For the rest of the time you can concentrate on looking at your audience.

You have greater flexibility with an extemporaneous speech. You are committed to blocks of thought but not words. If by looking at your audience you see that they have not understood some portion of your talk, you are free to rephrase the thought in a new way for better understanding. If you are really well prepared in your subject, you can bring in further examples to clarify your point. Also, if you see you are running overtime, you can condense a block by leaving out some of your less vital examples or facts.

Finally, because you are not committed to words, you retain conversational spontaneity. You are not faltering or groping for words, but neither are you running by your audience like a well-oiled machine.

The Manuscript Speech. Most speech experts recommend the extemporaneous speech above reading from a manuscript. We agree in general. However, speaking in a technical situation often requires the manuscript speech. Papers delivered to scientific societies are almost always written and then read to the group. Often the society will later publish your paper. Often technical reports contain complex technical information or extensive statistical material. Such reports do not conform well to the extemporaneous speech form and you should plan to read them from a manuscript.

Planning and writing a speech is little different from writing a paper. Follow the advice in Chapters 2 through 9. However, in writing your speech try to achieve a conversational tone. Certainly in speaking you will want to use the first person and active voice. Remember that speaking is more personal than writing. Include phrases like "it seems to me," "I'm reminded of," "Just the other evening, I," and so forth. Such phrases are common in conversation and give your talk extemporaneous overtones. Certainly, prefer short sentences to long ones.

Type the final draft of your speech. Just as you did for the extemporaneous speech outline, be generous with capitals, spacing, and underlining. Plan on about three typed pages per five minutes of speech. Put your pages in order and place them in a looseleaf binder.

When you carry your written speech to the lectern with you, you are in no danger of forgetting anything. Nevertheless, you must practice it, again preferably aloud to a live audience. As you practice, remember that because you are tied to the lectern, your movements are restricted. You will need to depend even more than usual on facial expression, gestures, and voice variation to maintain audience interest. Do not let yourself fall into a sing-song monotone as you read the set phrases of your written speech.

Practice until you know your speech well enough to look up from it for long periods of good eye contact. Plan an occasional departure from your manuscript to speak extemporaneously. This will aid you to regain the direct contact with the audience that you so often lose while reading.

Organizing Report Content

For the most part you will organize your speech as you do your written work. The guidelines we lay down in Chapter 8 for organizing written work apply here as well. However, the speech situation does call for some differences in organization and even content, and we will concentrate on these differences. We will discuss the organization in terms of introduction, body, and conclusion.

Introduction. A speech introduction should accomplish three jobs: (1) create a friendly atmosphere for you to speak in, (2) interest the audience in your subject, and (3) announce the subject, purpose, scope, and plan of development of your talk.

Be alert before you speak. If you can, mingle and talk with members of the group to whom you are going to speak. Listen politely to their conversation. You may pick up some tidbit that will help you to a favorable start. Look for bits of local color or another means to establish a common ground between you and the audience. When you begin to speak mention some member of the audience or perhaps a previous speaker. If you can do it sincerely, compliment the audience. If you have been introduced, remember to acknowledge and thank the speaker. Unless it is a very formal occasion, begin rather informally. If there is a chairperson and a somewhat formal atmosphere, we recommend no heavier beginning than, "Mr. Chairman, ladies and gentlemen."

Gain attention for your subject by mentioning some particularly interesting fact or bit of illustrative material. Anecdotes are good if they truly tie in with the subject. But take care with humor. Avoid jokes that really don't tie in with the subject or the occasion. Forget about risqué stories.

Some speakers try to startle their audience, but there are pitfalls to this method. Shortly after World War II, one of the authors of this text began a speech by pretending to be Adolf Hitler for a few minutes. (The purpose of the speech was to expose the wickedness of the big lie in propaganda.) He raged and ranted and denounced Wall Street, Franklin Roosevelt, and the Jews. The audience froze into resentment, and he never got them back. When he stopped his play acting and explained why he had imitated Hitler, most of the audience never even heard him—they were already too angry.

Be careful also about what you draw attention to. Do not draw attention to shortcomings in yourself, your speech, or the physical surroundings. Do not begin speeches with apologies.

Announce your subject, purpose, scope, and plan of development in a speech just as you do in writing. (See pages 218–221.) If anything, giving your plan of development is more important in a speech than in an essay. Listeners cannot go back in a speech to check on your organization the way that a reader can in an essay. So, the more organizational guideposts you give an audience, the better. No one has ever disputed this old truism about speech organization: (1) Tell the audience what you are going to tell them. (2) Tell them. (3) Tell them what you just told them. In instructional situations, some speakers provide their audiences with a printed outline of their talk.

Body.   When you organize the body of a speech, you must remember one thing: a listener's attention span is very limited. Analyze honestly your own attention span—be aware of your own tendency to let your mind wander. You listen to the speaker for a moment, and then perhaps you think of lunch, of some problem, or an approaching date. Then you return to the speaker. When you become the speaker, remember that people do not hang on your every word.

What can you do about the problem of the listener's limited attention span? In part, you solve it by your delivery techniques. We will discuss these in the next section of this chapter. It also helps to plan your speech around intelligent and interesting repetition.

Begin by cutting the ground you intend to cover in your speech to the minimum. Build a five-minute speech around one point, a 15-minute speech around two. Even an hour-long talk probably should not cover more than three or four points.

Beginning speakers are always dubious about this advice. They think, "I've got to be up there for 15 minutes. How can I keep talking if I only have two points to cover? I'll never make it." Because of this fear they load their speeches with five or six major points. As a result, they lull their audience into a state of somnolence with a string of generalizations.

In speaking, even more than in writing, your main content should be masses of concrete information—examples, illustrations, little narratives, analogies, and so forth—supporting just a few generalizations. As you give your supporting information, repeat your generalization from time to time. Vary its statement, but cover the same ground. The listener who was out to lunch the first time you said it may hear it the second time or the third. You use much the same technique in writing, but you intensify it even more in speaking.

We have been using the same technique here in this chapter. We began this section on the speech body by warning you that a listener's attention span is short. We reminded you that your listening span is short: same topic but a new variation. We asked you what you can do about a listener's limited span: same topic with only a slight shift. In the next paragraph we told you not to make more than two points in a 15-minute speech. We nailed this point down in the next paragraph by having a dubious speaker say, "I've got to be up there for 15 minutes. How can I keep talking if I have only two points to cover?" In the paragraph just preceding this one we told you to repeat intelligently so that the reader "who was out to lunch the first time you said it may hear it the second time or the third." Here we were slightly changing an earlier statement that "You listen to the speaker for a moment, and then perhaps you think of lunch. . . ." In other words, we are aware that the reader's attention sometimes wanders. When you are paying attention we want to catch you. Try the same technique in speaking, where the listener's attention span is even more limited than the reader's.

Creating suspense as you talk is another way to generate interest in your audience. Try organizing a speech around the inductive method. That is, give your facts first and gradually build up to the generalization that they support. (The deductive method states the generalization first and then supports it.) If you do this skillfully, using good material, your audience hangs on wondering what your point will be. If you do not do it skillfully or use dull material, your audience will tune you out and tune into their own private worlds.

Another interest-getting technique is to relate the subject matter to some vital interest of the audience. If you are talking about water pollution, for example, remind the audience that the dirtier their rivers get, the more tax dollars it will eventually take to clean them up.

Visual aids often increase audience interest. Remember to keep your graphics big and simple. No one is going to see typewritten captions from more than three or four feet away. Stick to big pie and bar graphs. If you have tables, print them in letters from two to three inches high. If you are speaking to a large group, put your graphic materials on transparencies and project them on a screen. Prepare your transparencies with care. Don't just photocopy typed or printed pages or graphics from books. No one behind the first row will see them. To work, letters and numbers on transparencies should be at least twice normal size. Large-type typewriters are available. If you need an assistant to help you project visual aids, get one or bring one with you.

Do not display a visual aid until you want the audience to see it. While the aid is up call your listeners' attention to everything you want them to see. Take the aid away as soon as you are through with it. If using a projector, turn it off whenever it is not in use. Be sure to key every visual aid into your speaker's script. Otherwise you may slide right by it. (See also the section, "Visual Aids," pages 429–441.)

Conclusion.   In your speech conclusion, as in your essay conclusions, you have your choice of several closes. You can close with a summary, or a list of recommendations including a call for some sort of action, or what amounts to "Good-bye, it's been good to talk to you." We stress again here what we said in Chapter 11 and add two points. In speaking, never suggest that you are drawing to a close unless you really mean it. When you suggest that you are closing, your listeners perk up and perhaps give a happy sigh. If you then proceed to drag on, they will hate you.

Second, remember that audience interest is usually highest at the beginning and close of a speech. Therefore, you will be wise to provide a summary of your key points at the end of any speech. Give your listeners something to carry home with them.

REPORT DELIVERY

After you have prepared your speech you must give it. For many people giving a speech is a pretty terrifying business. Before speaking they grow tense, have hot flashes and cold chills, and experience the familiar butterflies in the stomach. Some people tremble before and even during a speech. Try to remember that these are normal reactions, for both beginning and experienced speakers. Most people can overcome them, however, and it is even possible to turn this nervous energy to your advantage.

If your stage fright is extreme, or if you are the one person in a hundred who stutters, or if you have some other speech impediment, seek clinical help. The ability to communicate ideas through speech is one of humanity's greatest gifts. Do not let yourself be cheated. Winston Churchill had a speech impediment as a child. Some of the finest speakers we have ever had in class were stutterers who admitted their problem and worked at it with professional guidance. Remember, whether your problems are large or small, the audience is on your side. They want you to succeed.

The Physical Aspects of Speaking

What are the physical characteristics of good speakers? They stand firmly but comfortably. They move and gesture naturally and emphatically but avoid fidgety, jerky movements and foot shuffling. They look directly into the eyes of people in the audience, not merely in their general direction. They project enthusiasm into their voices. They do not mumble or speak flatly. We will examine these characteristics in detail—first movement and then voice.

Movement. A century ago a speaker's movements were far more florid and exaggerated than they are today. Today we prefer a more natural mode of speaking, closer to conversation than oratory. To some extent, electronic devices such as amplifying systems, radio, and television have brought about this change. However, you do not want to appear like a stick of wood. Even when speaking to a small group or on television (or, oddly enough, on the radio) you will want to move and gesture. If you are speaking in a large auditorium, you will want to broaden your movements and gestures. From the back row of a 2,500-seat auditorium you look about three inches tall.

Movement during a speech is important for several reasons. First, it puts that nervous energy we spoke of to work. The inhibited speaker stands rigid and trembles. The relaxed speaker takes that same energy and puts it into purposeful movement.

Second, movement attracts attention. It is a good idea to emphasize an idea with a pointing finger or a clenched fist; and a speaker who comes out from behind the lectern occasionally and walks across the stage or toward the audience awakens audience interest. The speaker who passively utters ideas deadens the audience.

Third, movement makes you feel more forceful and confident. It keeps you, as well as your audience, awake. This is why good speakers while speaking over the radio will gesture just as emphatically as though the audience could see them.

What sorts of movements are appropriate? To begin with, movement should closely relate to your content. Jerky or shuffling motions that occur haphazardly distract an audience. But a pointing finger combined with an emphatic statement reinforces a point for an audience. A sideward step at a moment of transition draws attention to the shift in thought. Take a step backward and you indicate a conclusion. Step forward and you indicate the beginning of a new point. Use also the normal descriptive gestures that all of us use in conversation: gestures to indicate length, height, speed, roundness, and so forth.

For most people, gesturing is fairly normal. They make appropriate movements without too much thought. Some beginning speakers, however, are body inhibited. If you are in this category, you may have to cultivate movement. In your practice sessions and in your classroom speeches, risk artificiality by making gestures that seem too broad to you. Oddly enough, often at the very point where your gestures seem artificial and forced to you, they will seem the most natural to your audience.

Allow natural gestures to replace nervous mannerisms. Some speakers develop startling mannerisms and remain completely oblivious of them until some brave but kind soul points them out. Some that we have observed include putting glasses off and on; knocking a heavy ring over and over on the lectern; fiddling with a pen, pointer, chalk, cigar, microphone cord, ear, mustache, nose, you name it; shifting from foot to foot in time to some strange inner rhythm; and pointing with the elbows while the hands remain in the pants pockets. Mannerisms may also be vocal. Such things as little coughs or repeating comments such as "OK" or "You know" to indicate transitions may become mannerisms.

Listeners are distracted by such habits. Often they will concentrate on the mannerisms to the exclusion of everything else. They may know that a speaker put her glasses on and off 22 times but not have the faintest notion of what she said. If someone points out such mannerisms in your speaking habits, do not feel hurt. Instead, work to remove the mannerisms.

Movement includes facial movement. Do not be a deadpan. Your basic expression should be a relaxed, friendly look. But do not hesitate to smile, laugh, frown, or scrowl when such expressions are called for. A scowl at a moment of disapproval makes the disapproval that much more emphatic. Whatever you do, do not freeze into one expression, whether it be the stern look of the man of iron or the vapid smile of a smoker in a magazine ad.

Voice.   Your voice should sound relaxed, free of tension and fear. In a man, people consider a deep voice to be a sign of strength and authority. Most people prefer a woman's voice to be low rather than shrill. If you do not have these attributes, you can develop them to some extent. Here we must refer you to some of the good speech books in our bibliography (Appendix C) where you will find various speech exercises described. If, despite hard work, your voice remains unsatisfactory in comparison with the conventional stereotypes, do not despair. Many speakers have had somewhat unpleasant voices and through force of character or intellect directed their audiences to their ideas and not their voices. Eleanor Roosevelt, for example, had a somewhat shrill, distracting voice all of her life, but her warmth and brilliance shone through, and she captivated most members of her audience.

Many beginning speakers speak too fast, probably because they are anxious to be done and sit down. A normal rate of speech falls between 120 and 180 words a minute. This is actually fairly slow. Generally, you will want a fairly slow delivery rate. When you are speaking slowly, your voice will be deeper and more impressive. Also, listeners have trouble following complex ideas delivered at breakneck speed. Slow up and give your audience time to absorb your ideas.

Of course, you should not speak at a constant rate, slow or fast. Vary your rate. If you normally speak somewhat rapidly, slowing up will emphasize ideas. If you are speaking slowly, suddenly speeding up will suggest excitement and enthusiasm. As you speak, change the volume and pitch of your voice. Any change in volume, whether from low to loud or the reverse, will draw your listeners' attention and thus emphasize a point. The same is true of a change in pitch. If your voice remains a flat monotone and your words come at a constant rate, you deprive yourself of a major tool of emphasis.

Many people worry about their accent. Normally, our advice is *don't*. If you speak the dialect of the educated people of the region where you were raised, you have little to worry about. Some New Englanders, for example, put *r*'s where they are not found in other regional dialects and omit them where they are commonly found.

Part of America's richness lies in its diversity, and President Kennedy reminded all of America that an educated man could say "Cuber" and "idear." Accents vary in most countries from one region to another, but certainly not enough to hinder communication.

If, however, your accent is slovenly—"Ya wanna cuppa coffee?"—or uneducated, do something about it. Work with your teacher or seek other professional help. Listen to educated speakers and imitate them. The musical *My Fair Lady* has probably made the need for correct speech, and some of the methods used to attain it, known to most Americans.

Whatever your accent, there is no excuse for mispronouncing words. Before you speak, look up any words you know you must use and about which you are uncertain of the pronunciation. Speakers on technical subjects have this problem perhaps more than other speakers. Many technical terms are jawbreakers. Find their correct pronunciation and practice them until you can say them easily.

Audience Interaction

One thing speakers must learn early in their careers is that they cannot count on the audience's hanging on every word. Some years ago an intelligent, educated audience was asked to record its introspections while listening to a speaker. The speaker was an excellent one. Despite his excellence and the high level of the audience, the introspections revealed that the audience was paying something less than full attention. Here are some of the recorded introspections:

> God, I'd hate to be speaking to this group. . . . I like Ben—he has the courage to pick up after the comments. . . . Did the experiment backfire a bit? Ben seems unsettled by the introspective report. . . . I see Ben as one of us because he is under the same judgment. . . . He folds his hands as if he was about to pray. . . . What's he got in his pocket he keeps wriggling around. . . . I get the feeling Ben is playing a role. . . . It is interesting to hear the words that are emphasized. . . . This is a hard spot for a speaker. He really must believe in this research. . . .

> Ben used the word "para-social." I don't know what that means. Maybe I should have copied the diagram on the board. . . . Do not get the points clearly . . . cannot interrupt . . . feel mad . . . More words. . . . I'm sick of pedagogical and sociological terms. . . . Slightly annoyed by pipe smoke in my face. . . . An umbrella dropped. . . . I hear a funny rumbling noise. . . . I wish I had a drink. . . . Wish I could quit yawning. . . . Don't know whether I can put up with these hard seats for another week and a half or not. . . . My head itches. . . . My feet are cold. I could put on my shoes, but they are so heavy. . . . My feet itch. . . . I have a piece of coconut in my teeth. . . . My eyes are tired. If I close them, the speaker will think I'm asleep. . . . I feel no urge to smoke. I must be interested. . . .

Backside hurts ... I'm lost because I'm introspecting.... The conflict between introspection and listening is killing me. Wish I didn't take a set so easily.... If he really wants me to introspect, he must realize himself he is wasting his time lecturing.... This is better than the two hour wrestling match this afternoon.... This is the worst planned, worst directed, worst informed meeting I have ever attended.... I feel confirmation, so far, in my feelings that lectures are only 5% or less effective.... I hadn't thought much about coming to this meeting, but now that I am here it is going to be O.K. ... Don't know why I am here.... I wish I had gone to the circus.... Wish I could have this time for work I should be doing.... Why doesn't he shut up and let us react.... The end of the speech. Now he is making sense.... It's more than 30 seconds now. He should stop. Wish he'd stop. Way over time. Shut Up.... He's over. What will happen now? ... [1]

As some of the comments reveal, perhaps being asked to record vagrant thoughts as they appeared made some members of the audience less attentive than they normally would have been. But most of us know that we have very similar thoughts and lapses of attention while we attend classes and speeches.

Reasons for audience inattention are many. Some are under the speaker's control; some are not. The speaker cannot do much about such physical problems as hard seats, crowded conditions, bad air, and physical inactivity. The speaker can do something about psychological problems such as the listeners' passivity and their sense of anonymity, their feeling of not participating in the speech.

Even before they begin to speak, good speakers have taken audience problems into account. They have analyzed the audience's education and experience level. They have planned to keep their points few and to repeat major points through carefully planned variations. They plan interesting examples. While speaking they attempt to interest the audience through movement and by varying the speech rate, pitch, and volume.

But good speakers go beyond these steps and analyze their audience and its reactions as they go along. In an extemporaneous speech and even to some extent in a written speech, you can make adjustments based on this audience analysis.

To analyze your audience, you must have good eye contact. You must be looking at Ben, Bob, and Irma. You must not merely be looking in the general direction of the massed audience. Look for such things as smiles, scowls, fidgets, puzzled looks, bored expressions, interested expressions, sleepy eyes, heads nodded in agreement, heads nodded in sleep, heads shaken in disagreement. You will not be 100 percent correct in interpreting these signs. Many students have learned to smile and nod in all the proper places without ever hearing the instructor. But, generally, such physical actions are excellent clues as to how well you are getting through to your audience.

If your audience seems happy and interested, you can proceed with your speech as prepared. If, however, you see signs of boredom, discontent, or a lack of understanding, you must make some adjustments. Exactly what you do depends to some extent on whether you are in a formal or informal speaking situation. We will look at the formal situation first.

In the formal situation you are somewhat limited. If your audience seems bored, you can quickly change your manner of speaking. Any change will, at least momentarily, attract attention. You can move or gesture more. With the audience's attention gained, you can supply some interesting anecdotes or other illustrative material to support better your abstractions and generalizations. If your audience seems puzzled, you know you must supply further definitions and explanations and probably more concrete examples. If your audience seems hostile, you must find some way to soften your argument while at the same time preserving its integrity. Perhaps you can find some mutual ground upon which you and the audience can agree and move on from there.

Obviously, such flexibility during the speech requires some experience. Also it requires that the speaker have a full knowledge of the subject. If every bit of material the speaker knows about the subject is in the speech already, the speaker has little flexibility. But do not be afraid to adjust a speech in midstream. Even the inexperienced speaker can do it to some extent.

Many of the speaker's problems are caused by the speech situation's being a one-way street. The listeners sit passively. Their normal desires to react, to talk back to the speaker, are frustrated. The problem suggests the solution, particularly when you are in a more informal speech situation, such as a classroom or a small meeting.

In the more informal situation you can stop when a listener seems puzzled. Politely ask him where you have confused him and attempt to clarify the situation. If a listener seems uninterested, give him an opportunity to react. Perhaps you can treat him as a puzzled listener. Or, you can ask him what you can do to interest him more. Do not be unpleasant. Put the blame for the lack of interest on yourself, even if you feel it does not belong there. Sometimes you may be displeased or shocked at the immediate feedback you receive, but do not avoid it on these grounds. And do not react unpleasantly to it. You will move more slowly when you make speaking a two-way street, but the final result will probably be better. Immediate feedback reveals areas of misunderstanding or even mistrust of what is being said.

In large meetings where such informality is difficult, you can build in some audience reaction through the use of informal subgroups. Before you talk, divide your audience into small subgroups, commonly called buzz groups. Use seating proximity as the basis for your division if you have no better one. Explain that after your talk the groups will have a period of time in which to discuss your speech. They will be expected to come up with questions or comments. People do not like to seem unprepared, even in informal groups. As a result, they will be more likely to pay attention to your speech in order to participate well in their buzz groups.

Whether you have buzz groups or not, often you will be expected to handle questions following a speech. If you have a chairperson, he or she will field the questions and repeat them, and then you will answer them. If you have no chairperson you will perform this chore for yourself. Be sure everyone understands the question. Be sure you understand the question. If you do not, ask the questioner to repeat it and perhaps to rephrase it.

Keep you answers brief, but answer the questions fully and honestly. When you do not have the answer, say so. Do not be afraid of conflict with the audience. But keep it on an objective basis: talk about the conflict situation, not personalities. If someone reveals through his question that he is becoming personally hostile, handle him courteously. Answer his question as quickly and objectively as you can and move on to another questioner. Sometimes the bulk of your audience will grow restless while a few questioners hang on. When this occurs release your audience and, if you have time, invite the questioners up to the platform to continue the discussion. Above all, during a question period be courteous. Resist any temptation to have fun at a questioner's expense.

VISUAL AIDS

Today, good speakers increasingly use visual aids during their talks. You will use a visual aid for three reasons: (1) *to support* and *expand* the content of your message; (2) *to focus the audience's attention* on a critical aspect of your presentation; and (3) *to clarify* your meaning.[2]

To Augment Your Message.   The first purpose of any visual support material, then, is *to augment your message*—to enlarge on the main ideas and give substance and credibility to what you are saying. Obviously, the material must be relevant to the idea being supported. Too often a speaker gives in to the urge to show a visually attractive or technically interesting piece of information that has little or no bearing on the subject.

Suppose, for the purposes of our analysis, that you were asked to meet with government people to present a case for your company's participation in a major federal contract. Your visual support would probably include information about the company's past performances with projects similar to the one being considered. You would show charts reflecting the ingenious methods used by the company's development people to keep costs down; performance statistics to indicate your high quality standards; and your best conception-to-production times to show the audience how adept you are at meeting target dates.

In such a presentation, before an audience of tough-minded officials, you wouldn't want to spend much of your time showing them aerial views of the company's modern facilities or mug shots of smiling employees, antiseptic production lines, and the company's expensive air fleet. Such material would hardly support and expand your arguments that the company is used to working and producing on a Spartan budget.

To Focus Attention.   Your second reason for using visual aids is *focus of attention*. A good visual can arrest the wandering thoughts of your audience and bring their attention right down to a specific detail of the message. It forces their mental participation in the subject.

When you are dealing with very complex material, as you often will be, you can use a simple illustration to show your audience a single, critical concept within your subject.

For Clarity.   Finally, the overall purpose of any support—verbal or visual—is to help *clarify* the message. Visual material clarifies by adding information, by expanding on the ideas provided by the speaker verbally. Obviously, to accomplish this the visual material must be designed to provide *further* meaning to the message.

Criteria for Visual Aids

But what about the visual aids themselves? What makes one better than another for a specific kind of presentation? Before we consider individual visual aids, let's look at the qualities that make a visual aid effective for the technical speaker.

Visibility.   First, a visual aid must be *visible*. If that seems so obvious that it hardly need be mentioned at all, it may be because you haven't experienced the frustration of being shown something the speaker feels is important—and not being able to read it, or even make out detail. To be effective, your visual support material should be clearly visible from the most distant seat in the house. If you have any doubts, sit in that seat and look. Remember this when designing visual material: *Anything worth showing the audience is worth making large enough for the audience to see.*

Clarity. The second criterion for a good visual is *clarity*. The audience decides this. If they're able to determine immediately what they are seeing, the visual is clear enough. Otherwise, it probably calls for further simplification and condensation. Such obvious mistakes as pictorial material out of focus, or close-ups of a complex device that will confuse the audience, are easy to understand. But what about the chart that shows a relationship between two factors on *x*- and *y*-axes when the axes are not clearly designated or when pertinent information is unclear or missing? *Visual material should be immediately clear to the audience, understandable at a glance, without specific help from the speaker.*

Simplicity. The third criterion for good visual support is *simplicity*. No matter how complex the subject, the visual itself should include no more information than absolutely necessary to support the speaker's message. If it is not carrying the burden of the message, it need not carry every detail. Limit yourself to *one* idea per visual— mixing ideas will totally confuse an audience, causing them to turn you off midsentence.

When using words and phrases on a visual, limit the material to key words that act as visual cues for you and the audience. If a visual communications expert wished to present the criteria for a good visual he might *think* something like this:

A good visual must manifest visibility.

A good visual must manifest clarity.

A good visual must manifest simplicity.

A good visual must be easy for the speaker to control.

What would he show the audience? If he knows his field as well as he should, he'll offer the visual shown in Figure 19-1.

The same information is there. The visual is being used appropriately to provide emphasis while the speaker supplies the ideas and the extra words. The very simplicity of the visual has impact and is likely to be remembered by the audience.

Control. The fourth quality a visual aid needs is *control*, speaker control. You should be able to add information or delete it, to move

CRITERIA

Visibility

Clarity

Simplicity

Control

FIGURE 19-1 Criteria for Visual Aids

forward or backward to review, and, finally, *to take it away* from the audience to bring their attention back to you.

Some very good visual aids can meet the other criteria and prove almost worthless to a speaker because they cannot be easily controlled. The speaker, who must maintain a flow of information and some kind of rapport with his audience while he is doing it, can't afford to let his visual material interfere with his task. Remember, visual material is meant to *support* you as a speaker, not to replace you.

Visual Design—What and How?

So far we have discussed the *why* of visual support material. The remaining two questions of concern to you are What do I use? How do I use it? Let's consider them in that order, applying the criteria already established as we go. Types of visual material can be roughly classified in six categories: (1) graphs, (2) tables, (3) representational art, (4) words and phrases, (5) cartoons, and (6) hardware. Graphs, tables, and representational art are discussed in detail in Chapter 12. You'll want to apply the suggestions made there to the visual materials you use when speaking. In the next few pages, we'll discuss using words and phrases, cartoons, and hardware.

Words and Phrases. There will always be circumstances in which you will want to emphasize key words or phrases by visual support. This type of visual can be effectively used in making the audience aware of major divisions or subdivisions of a topic, for instance.

There is danger, however, in the overuse of words—too many with too much detail. Some speakers tend to use visuals as a "shared" set of notes for their presentation, a self-limiting practice.

Audiences who are involved in reading a long, detailed piece of information won't recall what the speaker is saying.

With technical presentations, there is still another problem with the use of words. Too often, because they may be parts of a specialized vocabulary, they do more to confuse the audience than increase their understanding. Such terms should be reserved for audiences whose technical comprehension is equal to the task of translating them into meaningful thoughts.

Cartooning.   Cartooning is no more than illustrating people, processes, and concepts with exaggerated, imaginative figures—showing them in whatever roles are necessary to your purpose. (See Figure 19-2.) Not only does it heighten audience interest, but cartooning can be as specific as you want it to be in terms of action or position.

Some situations in which you might choose to use cartooned visual material are these:

1. When dealing with subjects that are sensitive for the audience.
2. When showing people-oriented action in a stationary medium—any visual aid outside the realm of motion pictures.

The resourceful speaker will use cartoons to help give additional meaning to other forms of visual support. The use of cartoons as elements in a block diagram tends to increase viewer interest.

Cartooning, like any other technique, can be overdone. There are circumstances in which the gravity of the situation would suggest that you consider only the most formal kinds of visual support material. On other occasions, cartooning may distract the audience or call

FIGURE 19-2   Cartoon Used to Illustrate a Computer Process

too much attention to itself. Your purpose is not to entertain but to communicate.

Hardware.   After all this analysis of visual support material, you may wonder if it wouldn't be somewhat easier to show the real thing instead.

Certainly there will be times when the best visual support you can have is the actual object. Notably, the introduction of a new piece of equipment will be more effective if it is physically present to give the audience an idea of its size and bulk. If it is capable of some unique and important function, it should definitely be seen by the audience. (The greatest difficulty with the use of actual hardware is control. The device that is small enough for you to carry conveniently may be too small to be seen from the audience.)

Even when the physical presence of a piece of equipment is possible, it is important to back it with supplementary visual materials. Chances are the audience will not be able to determine what is happening inside the machine, even if they understand explicitly the principle involved. With this in mind, you will want to add information with appropriate diagrams, graphs, and scale drawings.

In this discussion of visual support, the points of visibility, clarity, simplicity, and control have been stressed over and over. The reason for this is their importance to the selection and use of visual support by the technical communicator. In the end, it is you who can best decide which visual support form is required by your message and your audience.

You are also faced with the choice of visual tools for presenting your visual material. The next section will deal with popular visual tools, their advantages, disadvantages, and adaptability to the materials we've already discussed.

The Tools of Visual Presentation

The major visual tools used today are these:

1. Chalkboards   4. Movies
2. Charts   5. Overhead projection
3. Slides

In the next few pages we'll discuss them individually, with an eye on the advantages each one offers the technical speaker.

The Chalkboard.   Anyone who has attended school in the past half century is familiar with the chalkboard. It has been a standard source of visual support for much longer than that, and often the only means of presenting visual information available to the classroom teacher.

As a visual aid, it leaves something to be desired. In the first place, preparing information on a chalkboard, especially technical information in which every sliding scrawl can have significance, takes time. And after the material is in place, it cannot be removed and replaced quickly.

Second, the task of writing on a surface that faces the audience requires that you turn your back toward them while you write. And people don't respond well to backs. They want you to face them while you're talking to them.

Add to these problems the difficulties of moving a heavy, semipermanent chalkboard around, and you begin to wonder why anyone bothers.

Low cost and simplicity are the reasons. The initial cost of a chalkboard is higher than you may think; but the cost of erasers and chalk is minimal. In spite of its drawbacks, a chalkboard is also easy to use. It may take time, but there's nothing very complicated about writing a piece of information on a chalkboard. This simplicity, of course, gives it a certain flexibility, makes it essentially a spontaneous visual aid on which speakers can create their visual material as they go.

There are specific techniques for using a chalkboard that make it a more effective visual tool and help overcome its disadvantages. Let's consider them one at a time.

1. Plan ahead. Unless there is a clear reason for creating the material as you go, prepare your visual material before the presentation. Then cover it. Later, you can expose the information for the audience at the appropriate point in your speech.

Interrupting your presentation and the flow of ideas to write on the board can ruin the impact of your material. For this reason, it's a good idea to prepare the board before you begin your speech. Then, because you normally won't want the audience staring at the information you've written until the right moment during your speech, cover it with a cloth or sheet of paper. In certain instances, the mystique of the hidden information will actually work to your advantage, whetting the audience's appetite and sharpening their curiosity.

2. Be neat and keep the information simple and to the point. If your material is complex, find another way of presenting it.
3. Prime the audience. Before showing your information, tell them what they are going to see and why they are going to see it.

This last point is especially important when you are creating your visual support as you go. Priming your listeners will allow you to maintain the flow of information and, at the same time, prepare them mentally to understand and accept your information.

Charts. Charts take a couple of forms. The first is the individual *hardboard* chart, rigid enough to stand by itself and large enough to be seen by the audience from wherever they might be seated in the room. It is always prepared before the presentation, sometimes at considerable cost.

The second chart form is the *flip* chart, a giant-sized note pad that may be prepared before or during the presentation. When you have completed your discussion of one visual, you simply flip the sheet containing it over the top of the pad as you would the pages of a tablet.

The two types of charts have a common advantage. Unlike the chalkboard, they allow for reshowing a piece of information when necessary—an important aid to speaker control.

The following techniques will help you use charts more effectively during your presentation. They're really rules of usage, to be followed each time you choose this visual form for support.

1. Keep it simple. Avoid complex, detailed illustrations on charts. A three-by-five-foot chart is seldom large enough for detailed visibility.
2. Ask for help. Whenever possible, have an assistant on one side of your charts to remove each one in its turn. This avoids creating a break in your rapport with the audience while you wrestle with a large cardboard chart or a flimsy flip sheet.
3. Predraw your visuals with a very light-colored crayon or chalk. During the presentation, you can simply draw over the original lines in darker crayon or ink. This allows you to create an accurate illustration a step at a time for clarity.
4. Prime the audience. Tell them what they are going to see and why before you show each visual, for the same reason you would do it with the chalkboard.

Slides. The 35 mm slide, with its realistic color and photographic accuracy, has always been a popular visual tool for certain types of technical presentations. Where true reproduction is essential, no better tool is available.

Modern projectors have two notable advantages over their predecessors. First, the introduction of slide magazines has made it possible for you to organize your presentation and keep it intact. Second, remote controls allow you to operate the projector—even to reverse the order of your material—from the front of the room.

In order to use slide projectors effectively, however, you must turn off the lights in the room. And any time you keep your audience in the dark, you risk damaging the direct speaker-listener relationship on which communication hinges. In a sense, it takes the control of the presentation out of your hands. Long sequences of slides tend to develop a will and a pace of their own. They tire an audience and invite mental absenteeism.

There are ways to handle the built-in problems of a slide presentation, simple techniques that can greatly increase audience attention and the effectiveness of your presentation.

1. When using slides in a darkened room, light yourself.

A disembodied voice in the dark is little better than a tape recorder; it destroys rapport and allows the audience to exit into their own thoughts. To minimize this effect, arrange your equipment so you may stay in the front of the room and use a podium light or some other soft, nonglaring light to make yourself *visible* to the audience.

2. Break the presentation into short segments of no more than five or six slides.
3. Always tell the audience what they're going to see, and what they should look for.

Everything considered, slides are an effective means of presenting visual material. But like any visual tool, they require control and preparation on your part. The important thing to remember is that they are there only to support your message—not to replace you.

Movies. Whenever motion and sound are important to the presentation, movies are the only visual form available to the speaker that can accomplish the effect. Like slides, they also provide an exactness of detail and color that can be critical to certain subjects. There is really no other way an engineer could illustrate the tremendous impact aircraft tires receive during landings, for instance. The audience would understand the subject only if they were able to view, through the eye of the camera, the distortion of the rubber when the plane touches down.

But movies *are* the presentation. They cannot be considered visual support material in the sense of the term developed in this book. They simply replace the speaker as the source of information, at least for the duration of the film. If the film is long, say the major part of the presentation, the speaker is reduced to an announcer with little more to do than introduce and summarize the content of the film.

This makes the movie the most difficult visual form to control. Yet it can be controlled and, if it is to perform the support functions we've outlined, it must be. Some effective techniques are given here.

1. Prepare the audience. Explain the significance of what you are about to show them.
2. If a film is to be used, it should make up only a small part of the total presentation.
3. Whenever possible, have the film cut into short three or four minute sections—preferably separated by a few seconds of blank film. During the breaks introduced by these periodic insertions, you can reestablish rapport with the audience by summarizing what they have seen and refocusing attention on the important points in the next segment.

Overhead Projection. Throughout this discussion of visual support, we've intentionally stressed the importance of maintaining a good speaker-audience relationship. It's an essential in the communication process. And it's fragile. Any time you turn your back to the audience, or darken the room, or halt the flow of ideas for some other reason, this relationship is damaged.

The overhead projector alone effectively eliminates all of these rapport-dissolving problems. The image it projects is bright enough and clear enough to be used in a normally lighted room without noticeable loss of visibility. And just as important from your point of view, it allows you to remain in the front of the room *facing* your audience throughout your presentation. The projector itself is a simple tool, and like all simple tools it may be used without calling attention to itself.

Visual material for the overhead projector is prepared on transparent sheets the size of typing paper. In recent years, the methods for preparing these transparencies have become so simplified and made so inexpensive that the overhead has become a universally accepted visual tool in both the classroom and industry.

Perhaps the most important advantage of the overhead projector is the total speaker control it affords. With it, you may add information or delete it in a variety of ways, move forward or backward to review at will, and *turn it off* without altering the communicative situation in any way. It is the last capability that makes the overhead projector unique among visual tools. By flipping a switch, you can literally "remove" the visual material from the audience's consideration, bringing their attention back to you and what you are saying. Because the projector is used in a fully lighted room, this on-and-off process seldom distracts the audience or has any effect on the speaker-audience relationship.

There are three ways to add information to a visual while the audience looks on—an important consideration when you want your listeners to receive information in an orderly fashion. In order of their discussion, they are (1) overlays, (2) revelation, and (3) writing on the visual itself.

The *overlay* technique (Figure 19-3) combines the best features of preparing your visuals in advance and creating them at the moment of their need. It is the simple process of beginning with a single positive transparency and adding information with additional transparencies by "overlaying" them, placing each one over the first so they may be viewed by the audience as a single, composite illustration. Ordinarily, no more than two additional transparencies should be used this way, but it's possible to include as many as four or five.

The technical person, who must usually present more complex concepts a step at a time to ensure communication, can immediately see the applications of such a technique.

The technique of *revelation* is simpler. (See Figure 19-4.) It is the process of masking off the parts of the visual you don't want the audience to see. A plain sheet of paper will work. By laying it over the information you want to conceal for the moment, you can block out selected pieces of the visual. Then when you're ready to discuss this hidden information, you simply remove the paper. The advantage is clear enough. If you don't want the audience to read the bottom line on the page while you're discussing the top line, this is the way to control their attention.

FIGURE 19-3 Using Overlays with an Overhead Projector

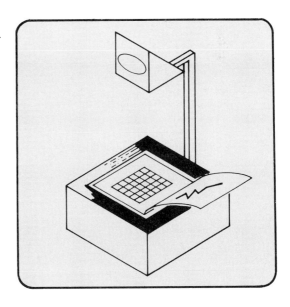

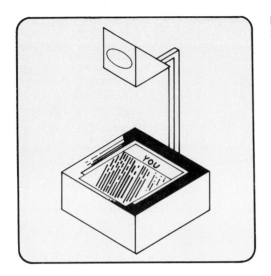

FIGURE 19-4 The Technique for Revelation

Writing on Transparencies. Writing information on an overhead-projection transparency (Figure 19-5) is nearly as easy as writing on a sheet of paper at your desk. Several felt-tipped pens available for this purpose may be used to create visual material in front of the audience. Often you can achieve your purposes by simply underlining or circling important parts of your visual—a means of focusing audience attention on the important aspects of your message.

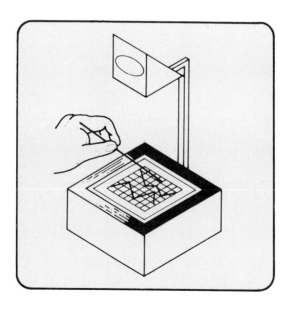

FIGURE 19-5 Writing on Transparencies

A final way of directing audience attention with overhead-projection transparencies is simply to use your pencil as a pointer. The "profile" shadow of the pencil will be seen on the screen, directing the audience's attention to the proper place.

Summary

When considering a visual aid, there are certain things you must keep in mind. First, does it satisfy the fundamental purposes of a visual aid? Will it *augment* the message? Will it provide a *focus of attention* for the audience? And will it *clarify* the subject it is supporting?

If it does these things effectively, and it meets the four criteria of a good visual aid—visibility, clarity, simplicity, and control—you have a legitimate piece of visual support material.

GROUP CONFERENCES

Most industries and government agencies today use the group conference for training and for problem solving. Therefore, the group conference is another speech situation you may find yourself in.

Group conferences do have their limitations. They are time consuming, and sometimes they spread responsibility too thin. They sometimes fail at truly being a conference when one member of the group has direct control over some of the other members. And if you do not have skilled conferees, you end up pooling ignorance rather than knowledge.

However, most organizations feel that the advantages outweigh the limitations. And the advantages are considerable. Conferences share judgments. Often several people together will have an insight that they would miss individually. Conference judgments are more temperate than individual judgments. Perhaps the biggest advantage of the conference is that people are more eager to carry out and support actions that they have planned themselves. Leaders can make decisions much more quickly alone than in a conference. But they will quickly lose the saved time if they have to spend hours or even days selling their decisions to their subordinates.

In this section we will give you some of the *do's* and *don'ts* of conference behavior and describe the useful roles that conferees can play. We will conclude by describing in some detail the methods of the problem-solving conference, probably the most common type of conference in industry.

Conference Behavior

A good group conference is a pleasure to observe. A bad conference distresses conferees and observers alike. In a bad group conference the climate is defensive. Conferees feel insecure, constantly fearing a personal attack and preparing to defend themselves. The leader of the bad conference cannot talk without pontificating; advice is given as though from on high. The group punishes those members who deviate from the majority will. As a result the ideas offered are tired and trite. Creative ideas are rejected. People compete for status and control, and they consider the rejection of their ideas as a personal insult. They attack those who reject their contributions. Everyone goes on the defensive, and energy that should be focused on the group's task flows needlessly in bitter debate. The group accomplishes little. As a rule, the leader ends up dictating the solutions, perhaps what was wanted all along.

In a good group conference the climate is permissive and supportive. Members truly listen to one another. People assert their own ideas, but they do not censure the opinions of others. The general attitude is "We have a problem. Let's put our heads together and solve it." Members reward each other with compliments for good ideas and do not reject ideas because they are new and strange. When members do reject an idea they do it gently with no hint of a personal attack on its originator. People feel free to operate in such a climate. They come forward with more and better ideas. They drop the defensive postures that waste so much energy and put the energy instead into the group's task.

How do the members of a group arrive at such a supportive climate? To simplify things we will present a list of *do's* and *don'ts*. Our rules cannot, of course, guarantee a good conference, but if they are followed they can help contribute to a successful outcome.

Do's

1. Do be considerate of others. Stimulate people to act rather than pressuring them.
2. Do be loyal to the conference leader without saying yes to everything. Do assert yourself when you have an idea or when you disagree.
3. Do support the other members of the group with compliments and friendliness.
4. Do be aware that other people have feelings. Remember that conferees with hurt feelings will drag their feet or actively disrupt a conference.
5. Do have empathy for the other conferees. See their point of view. Do not assume you know what they are saying or are going to say. Really listen and *hear* what they are saying.
6. Do conclude contributions you make to the group by inviting criticism of them. Detach yourself from your ideas and see them objectively as you hope others do. Be ready to attack your own ideas.

7. Do understand that communication often breaks down. Do not be shocked when you are misunderstood or when you misunderstand others.
8. Do feel free to disagree with the ideas of other members, but never attack people personally for the ideas they hold.
9. Do remember that most ideas that are not obvious seem strange at first. Yet they may be the best ideas.
10. Do remember that trying to monopolize or dominate a conference is the mark of the insecure "little person." The confident person feels secure and is willing to listen to the ideas of others. Being secure, confident people do not fear to adopt the ideas of others in preference to their own, giving full credit when they do so.

Don'ts

1. Don't continually play the expert. You will annoy other conferees with constant advice and criticism based upon your expertise.
2. Don't pressure people to accept your views.
3. Don't make people pay for past mistakes with continuing punishment. Instead try to change the situation to prevent future mistakes.
4. Don't let personal arguments foul a meeting. Stop arguments before they reach the personal stage by rephrasing them in an objective way.

Perhaps the rule "Do unto others as you would have them do unto you" best summarizes all these *do's* and *don'ts*. When you speak you want to be listened to. Listen to others.

Group Roles

You can play many roles in a group conference. Sometimes you bring new ideas before the group and urge their acceptance. Perhaps at other times you serve as information giver and at still others as harmonizer, resolving differences and smoothing ruffled egos. In this section we will discuss the useful roles that you as a conference leader or member can play. We purposely do not distinguish between leader and member roles. In a well-run conference, an outsider would have difficulty knowing who the leader is. We divide the roles into two groups: (1) *Task roles:* roles designed to move the group toward the accomplishment of its task; and (2) *Group maintenance roles:* roles designed to maintain the group in a harmonious, working condition.

We pause here to point out that it does not pay to analyze too closely *why* any particular group member plays some particular role. Few of us have perfectly pure motives for any role we play. Members introducing new ideas may have many motives. They may be seeking power: people often think of the idea giver as the leader of a group. They may be acting from a sincere desire to help the group solve its problem. They may also be building up their own egos or seeking promotions. At any one time you will be aware of only a few of a person's motives—including your own. So long as a person's role or roles are useful ones, do not turn the spotlight of analysis too brightly. Few of us can stand the glare. Now on to the roles.

Task Roles. When you play a task role you help the group accomplish its set task. Some people play one or two of these roles almost exclusively, but most people slide easily in and out of most of them.

1. *Initiators.* The initiators are the idea givers, the starters. They get the group moving toward its task, perhaps by proposing or defining the task, or by suggesting a solution to a problem or a way of arriving at the solution.
2. *Information seekers.* The information seekers see where needed facts are thin or missing. They solicit the group for facts relevant to the task at hand.
3. *Information givers.* The information givers provide data and evidence relevant to the task. They may do so on their own or in response to the information seekers.
4. *Opinion seekers.* The opinion seekers canvass group members for their beliefs and opinions concerning the problem. They might encourage the group to state the value judgments that form the basis for the criteria of the problem solution.
5. *Opinion givers.* The opinion givers volunteer their beliefs, judgments, and opinions to the group or respond readily to the opinion seekers. They help set the criteria for the problem solution.
6. *Clarifiers.* The clarifiers act when they see the group is confused about a conferee's contribution. They attempt to clear away the confusion by restating the contribution or by supplying additional relevant information, opinion, or interpretation. They may restate and thus clarify a problem or furnish a needed definition.
7. *Elaborators.* The elaborators further develop the contributions of others. They give examples, analogies, and additional information. They might carry a proposed solution to a problem into the future and speculate about how it would work.
8. *Summarizers.* The summarizers draw together the ideas, opinions, and facts of the group into a coherent whole. They may state the criteria that the group has set or the solution to the problem agreed upon. Often, after a summary, they may call for the group to move on to the next phase of the task.

Group Maintenance Roles. When you play a group maintenance role, you help to build and maintain the supportive group climate. Some people are so task oriented that they ignore the feelings of others as they push for a solution. Without the proper climate in a group, the members will often fall short of completing their task.

1. *Encouragers.* The encouragers respond warmly to the contributions of others. They express appreciation for ideas and reward conferees by

complimenting them. They go out of their way to encourage and reward the reticent members of the group when they do contribute.

2. *Feeling expressers.* The feeling expressers sound out the group for its feelings. They sense when some members of the group are unhappy and get their feelings out into the open. They may do so by stating the unhappiness as their own and thus encourage the others to come into the discussion.

3. *Harmonizers.* The harmonizers step between warring members of the group. They smooth ruffled egos and attempt to lift conflicts from the personality level and objectify them. With a neutral digression, they may lead the group away from the conflict long enough for tempers to cool, allowing people to see the conflict objectively without further help from them.

4. *Compromisers.* The compromisers voluntarily withdraw their ideas or solutions in order to maintain group harmony. They freely admit error. With such actions they build a climate within which conferees do not think their status is riding on their every contribution.

5. *Gatekeepers.* The gatekeepers are alert for blocked-out members of the group. They subly swing the discussion away from the forceful members to the quiet ones and give them a chance to contribute.

The Problem-Solving Conference

The type of conference most commonly held in industry is the problem-solving conference. Certainly, it is also the conference type most likely to create tension and argument among the conferees. For both these reasons we will discuss the problem-solving conference in some detail.

The problem-solving process involves five steps:

1. Defining the problem
2. Creating solutions
3. Testing solutions
4. Choosing a solution
5. Planning for action

We will consider each of these steps in detail.

Defining the Problem. At the beginning of the conference, the group must define the problem as objectively as it can. Move slowly at this point. Particularly avoid the temptation to assume that everyone understands the problem, and that the group can rush on to the solution stages. Often, some members of the group may be part of the problem or even a cause of the problem. Avoid statements that punish them. Do not drive them from the group. If not driven to the defensive, they may be in a better position to define and solve the problem than anyone else.

As you attack the problem, do not oversimplify. The problem may consist of many parts. For example, in a common industrial situation, some shop in a large plant may not be meeting its production quota. The low output is only the surface problem. The real one lies underneath and may involve such things as outmoded equipment, poor lighting, faulty training, inadequate rest periods, or many other factors. You cannot attempt solutions until you know what the real problem or problems are. Often, finding the true obstacles will make the solution step quite easy. Therefore, you must never hurry the problem-defining step.

Take your time and gather as much useful information as you can about each of the subproblems. In the production-quota problem, find out the educational level of the workers. Who trains them? Are they trained in classes or on the job? What are their incentives to learn their jobs thoroughly, and so forth? Somewhere in the mass of data gathered may lie the clue needed to suggest the solution.

In the problem-definition stage you will also set the criteria and limitations within which you must solve the problem. You may have some control over the criteria and limitations, or they may be forced on you from above. In either case the group must thoroughly understand them before it can suggest solutions. As a criterion for the production problem we have been using, you might have the production quota itself. The shop must turn out x number of units per hour or else the rest of the plant slows down. As a limitation you may have the stipulation that at this time no additional workers may be hired.

When the group has thoroughly defined the problem and has explored the criteria and limitations, it may move on to the next stage.

Creating Solutions. There are several good reasons for keeping the step of creating solutions separate from the steps of testing solutions and choosing a solution. First of all the separation slows the process down. The slower pace keeps the group from grasping at the first plausible-sounding solution and rushing into action with it. The best solution may not come until many suggestions have been put forth. The best solution may be a combination of several ideas.

Second, separating the steps reduces the threat of ridicule for the unusual solution, the solution that at first glance may seem too far out even to be considered. Remember that if the problem has existed long enough to reach the conference stage most of the obvious solutions have most likely already been tried. The answer may lie in the odd or the unusual. The nonconformists of the group should be encouraged, not discouraged.

The solutions offered should stem from the data gathered during the problem-defining period. In our production-quota problem, discussion may have revealed that the instruction manuals used in the shop were written in complicated language with little regard for the educational level of the audience. Suggested solution: Rewrite the manuals. This, of course, is a fairly obvious solution given such a well-defined problem. But a group with a success-oriented attitude doesn't let the idea drop with its first statement. The members pick it up and embellish it. Why not instructional movies instead of manuals? Put the manuals in comic book form. Use 35 mm slides. Use closed-circuit TV in the training. How about programmed-learning texts? Once started, the ideas will flow. Some will be practical; some will be impractical. But don't stop the flow at this point by testing them.

As a practical matter someone should keep track of the ideas. The group should appoint a recorder. If a chalkboard is available, record on it. Everyone can see the solutions in this way. The ideas are out in the open; this greatly aids the process of one conferee's building on the ideas of another.

By taking solution testing out of this step, you avoid the fear of punishment. But do not overlook the possibility of reward. When someone advances an idea that sounds good to you, say so. A warm, supportive atmosphere generates ideas; a cold, formal one freezes them out.

When the group has obviously exhausted the solutions it has to offer, it is time to move on to the next stage.

Testing Solutions.   In the problem-definition stage, the group sets up its criteria and limitations. In the testing stage, the group applies the criteria and limitations to the proposed solutions. Very often this step immediately knocks out a number of the proposed solutions. Probably the group will find it tougher to test the ones that remain. At this point several blocks to the testing process arise.

Although at this stage the group is supposedly only testing solutions, some members are already looking forward to choosing the solution. Personal interests begin to get in the way of the problem-solving process. People seeking status will find favorable arguments for solutions that increase their status and find fault with solutions that decrease their status. Some conferees feel bound to defend their own solutions. People will attack solutions that threaten them or their friends with change. People may even favor a solution that punishes an enemy.

Most of these attitudes result in bitter debate. Oddly enough, a block to solution testing can also arise from a lack of conflict. If the conferees are determined to maintain an atmosphere of sweetness and light, they will not attack ideas sufficiently to test them. Without proper testing of the solution, the group will flounder at the next stage of choosing a solution or the last stage of planning the action.

The group must argue about the solutions. But it must argue on objective and not on personal grounds. Insofar as humanly possible, personalities and personal interests must not influence the testing and the later decision. Such objectivity is difficult, if not impossible, in a group of human beings. Therefore, those members of the group who do not have personal axes to grind must be particularly alert for the arguments that camouflage personal interests. When such arguments arise, the impartial member must steer them as quickly as possible to more objective ground.

Consider whether a solution will create new problems. Suppose that in our production-quota problem, the solution definitely seems to lie in improving the worker's training and instruction. One of the proposed solutions is that instructions be put in comic book form. On the surface this sounds like a good solution. The combination of pictures and simple prose will enable the workers to understand the involved processes. The cost of the books does not seem prohibitive. "But," someone says, "the average worker in this section of the plant has a high school diploma. Will we insult the workers with a comic book approach?" The group kicks this idea around. In objective arguments like this, solutions gain or lose ground. If most of the group agree that comic books will insult the workers, the comic book approach will fade away. Unless someone in the group is attached to the idea for personal reasons, there will be little argument. The idea has been tested and found wanting. In this subtle way, the stage of testing solutions moves easily into the next stage, that of choosing a solution.

Choosing a Solution. As solutions are tested, most of them will gradually lose favor and drop from the discussion. Perhaps only one will remain and the group has therefore made the decision. But often several solutions will survive the testing period. How does the group decide which is best?

Essentially, there are two yardsticks to use in the final measurement before a choice.

1. *The objective test:* Is the solution objectively best? Does it best fit the facts of the situation? Which solution will do the most for the least cost? Which solution will gain the most support from the workers? Which solution will disrupt current practices the least? Such questions decide the objective worth of a solution. When the members of the conference do not have to carry out the solution personally, the objective approach is the best.
2. *The subjective approach:* Often some of the members of the conference

must carry out the decision. When this is the case, the other members must pay special attention to their preferred solution. The group should choose the solution the implementers favor unless it is obviously much poorer than the best objective solution. When the implementers favor a solution, they will work harder and more enthusiastically at it. This harder work and enthusiasm may very well make their solution the best.

Whether it uses the objective or subjective method, the group should take its time. The group should be patient with those who hold out against the majority. Sometimes, given enough time, the minority will prove to be right. In any event, don't rush the decision by voting. The final decision should be one that, if at all possible, the whole group approves. The term for this general approval is *consensus*. Consensus will be difficult to reach if the group takes an early vote. Once conferees commit themselves to a solution by a vote, they find it more difficult to change their mind later.

Finally, however, the group must reach its decision. (Remember that indecision is itself a form of decision.) After the group reaches its decision it must plan its action.

Planning the Action. What action the group plans depends entirely on the responsibilities of the group. If the group does not have the responsibility for carrying out the decision, probably the only action needed is a report to higher authority. In this case, you will find the rhetorical forms for argument (pages 141–144) and comparison (pages 112–113) useful to you.

If the group has the responsibility for carrying out the action, then it must begin detailed planning. Primarily, the planning will involve assigning responsibilities to different members of the group. We cannot go much beyond this step in advising you how to plan. The content and extent of the planning depend on the situation and the responsibilities of the group.

We will close with one final hint. Do not be surprised if you run into blocks while planning the action. Failure to test the solution thoroughly or failure to obtain a true consensus while choosing a solution may slow you up at this stage. When such blocks occur, back up to the stage where you failed and work through it again.

EXERCISES

1. Deliver a speech in one of the following speech situations:
   a. You are an instructor at your college. Prepare a short extemporaneous lecture on a technical subject. Your audience is a class of about 20 upperclassmen.
   b. You are the head of a team of engineers and technicians who have developed a new mechanism or process. Your job is to persuade a

group of senior managers from your own firm to accept the process or mechanism for company use. Assume these managers to have a lay person's knowledge about your subject. Speak extemporaneously.

c. You are a known expert on your subject. You have been invited to speak about your subject at the annual meeting of a well-known scientific association. You are expected to write out and read your speech. You are to inform the audience, which is made up of knowledgeable research scientists and college professors from a diversity of scientific disciplines, about your subject or to persuade them to accept a decision you have reached.

2. Change one of your written reports into an oral report. Deliver the report extemporaneously. Prepare several visual aids to support major points.

3. Focus upon some problem at your school and have a problem-solving conference concerning it. You may choose some technical problem such as parking, inadequate cafeteria space, poor lighting in classrooms or dormitories, or inadequate study facilities in the library. Or you may choose a more abstract issue such as required class attendance, student representation in school decisions, lower tuition rates, over- (or under-) emphasis on athletics, and so forth. You might even have a conference on the subject of "What Problem Shall We Consider?"

Often it is a valuable experience—particularly at early conferences—to assign roles. Before the conference begins, write the names of the various roles explained on pages 444–445 on slips of paper. Fold the slips and have every conferee choose one or more of them. The conferees will not tell anyone else what roles they have picked, but they will play the roles as defined.

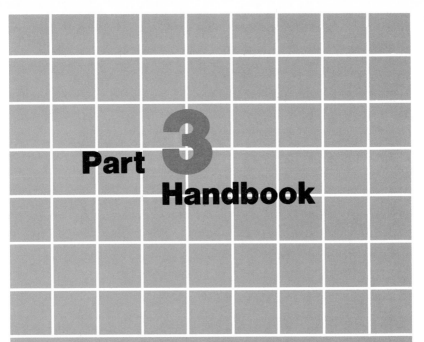

Part 3

Handbook

Any language is a growing, flexible instrument with rules that are constantly changing by virtue of the way it is used by its live, independent speakers and writers. Only the rules of a dead language are unalterably fixed.

Nevertheless, at any point in the development of a language certain conventions are in force. Certain constructions are considered errors, and the person who uses them is considered illiterate or uneducated. It is with these conventions and constructions that this Handbook deals.

Because we use continuous numbering, we do not divide the Handbook into chapters. But we do divide it into two sections. The first deals with eleven common errors. The second deals with the conventions of punctuation, italicization, capitalization, numbers, and abbreviations.

Each error and convention dealt with has a number and an abbreviated reference tag. The tags are reproduced on the back endpapers along with references to Chapter 9 on style and some of the more important proofreading symbols. If you are in a college writing course, your instructor may use some combination of these tags and symbols to indicate revisions needed in your reports.

Common Errors

1. FRAGMENTARY SENTENCE (FRAG)

Rewrite or repunctuate the sentence to make it a complete sentence or to join it to a complete sentence.

Most fragmentary sentences are either verbal phrases or subordinate clauses that the writer mistakes for a complete sentence.

A verbal phrase has in the predicate position a participle, gerund, or infinitive, none of which functions as a complete verb:

Norton, *depicting* the electromagnetic heart. *(participle)*

The *timing* of this announcement about Triptycene. *(gerund)*

Braun, in order *to understand* tumor cell growth. *(infinitive)*

When your fragment is a verbal phrase, either change the verb to a complete verb or repunctuate the sentence so that the phrase is joined to the complete sentence of which it is actually a part.

Fragment
Norton, depicting the electromagnetic heart. She made a mock-up of it.

Rewritten
Norton depicted the electromagnetic heart. She made a mock-up of it.

Norton, depicting the electromagnetic heart, made a mock-up of it.

Subordinate clauses are distinguishable from phrases in that they have complete subjects and complete verbs (rather than verbals) and are introduced by relative pronouns *(who, which, that)* or by subordinating conjunctions *(because, although, since, after, while).*

The presence of the relative pronoun or the subordinating conjunction is a signal that the clause is not independent but is part of a more complex sentence unit. Any independent clause can become a subordinate clause with the addition of a relative pronoun or subordinating conjunction.

Independent Clause
Early transistors were only the size of a pea.

Subordinate Clause
Although early transistors were only the size of a pea.

Repunctuate subordinate clauses so that they are joined to the complex sentence of which they are a part.

Fragment
Although early transistors were only the size of a pea. They were enormous compared to modern transistors.

Rewritten
Although early transistors were only the size of a pea, they were enormous compared to modern transistors.

Various kinds of elliptical sentences minus a subject and/or a verb do exist in English, for example, "No!" "Oh?" "Good shot." "Ouch!" "Well, now." These constructions may occasionally be used for stylistic reasons, particularly the representation of conversation, but they are seldom needed in technical writing. If you do use such constructions, use them sparingly. Remember that major deviations from normal sentence patterns will probably jar your readers and break their concentration upon your thought, the last thing that any writer wants.

2. COMMA SPLICE OR FUSED SENTENCE (CS OR FS)

Rewrite or repunctuate the sentence to make it a more effective sentence and to remove the comma splice or fused sentence error.

Punctuate two independent clauses placed together with a period, semicolon, or a comma and a coordinating conjunction (*and, but, for, nor,* or *yet*). Infrequently the colon or dash is used also. (In modern usage, there are some exceptions to these rules. See Rule 16-1.) The following three examples are punctuated correctly, the first with a period, the second with a semicolon, the third with a comma and a coordinating conjunction:

Check the hydraulic pressure. If it reads below normal, do not turn on the aileron boost.

We will describe the new technology in greater detail; however, first we will say a few words about the principal devices found in electronic circuits.

Ground contact with wood is particularly likely to cause decay, but wood buried far below the ground line will not decay because of a lack of sufficient oxygen.

If the example sentences had only commas between the independent clauses, the error would be a *comma splice.* If the examples lacked even commas, the error would be a *fused sentence.*

Writers most frequently commit comma splices when they mistake conjunctive adverbs for coordinating conjunctions. The most common conjunctive adverbs are *also, anyhow, besides, consequently, furthermore, hence, however, moreover, nevertheless, therefore,* and *too.*

When a conjunctive adverb is used to join two independent clauses, the mark of punctuation most often used is a semicolon (a period is used infrequently), as in this correctly punctuated sentence:

Ice fish are nearly invisible; however, they do have a few dark spots on their bodies.

Often the sentence will be more effective if it is rewritten completely, making one of the independent clauses a subordinate clause or a phrase.

Comma Splice
The students at the university are young Californians, they are between the ages of 18 and 24.

Rewritten
The students at the university are young Californians between the ages of 18 and 24.

3. VERB-SUBJECT AGREEMENT (V/AG)

Make the verb and subject agree in number and person.

Most of the time verb-subject agreement presents no difficulty to the writer. For example, in the sentence, "He *speaks* for us all," only a child or a foreigner learning English might say, "He *speak* for us all." However, various constructions exist in English that do present agreement problems even for the adult, educated, native speaker of English. These troublesome constructions are examined here.

The following words take singular verbs: *each, everyone, either, neither, anybody, somebody.*

Writers rarely have trouble with a sentence such as "No one is going to the game." Problems arise when, as is often the case, a prepositional phrase with a plural object is interposed between the simple subject and the verb as in this sentence: *"Each* of the freshmen is going to the game." In this sentence the temptation is to let the object of the preposition, *freshmen,* govern the verb and write *are.*

Divided usage: The pronouns *any* and *none* are now commonly treated as plural subjects, particularly when they are followed by a prepositional phrase with a plural object. Such usages as *"None* of the freshmen are going to the basketball game" or "Are *any* of the freshmen going to the basketball game?" are now respectable and even preferable.

When a compound subject is joined by *or* or *nor* the verb agrees with the closer noun or pronoun:

Either the designer or the builders are in error.

Either the builders or the designer is in error.

In informal and general usage one might now commonly hear, or see, the second sentence above as "Either the builders or the designer are in error." In writing you should hold to the more formal usage of the example.

Parenthetical expressions introduced by such words as *accompanied by, with, together with,* and *as well as* do not govern the verb:

Mr. Roberts, *as well as his two assistants,* is working on the experiment.

Two or more subjects joined by *and* take a plural verb. Inverted word order does not affect this rule:

Close to the Academy are Cathedral Rock and the Rampart Range.

Collective nouns such as *team, group, class, committee,* and many others take either plural or singular verbs, depending upon the meaning of the sentence. The writer must be sure that any subsequent pronouns agree with the subject and verb:

The team *is* going to receive *its* championship trophy tonight.

The team *are* going to receive *their* football letters tonight.

Note well: When the team was considered singular in the first example the subsequent pronoun was *its.* In the second example the pronoun was *their.*

4. PRONOUN-ANTECEDENT AGREEMENT (P/AG)

Make the pronoun agree with its antecedent.

Pronoun-antecedent is closely related to Rule 3. For example, the problem area concerning the use of collective nouns explained in Rule 3 is closely related to the proper use of pronouns. When a collective noun is considered singular it takes a singular pronoun as well as a singular verb. Also, such antecedents as *each, everyone, either, neither, anybody, somebody, everybody,* and *no one* take singular pronouns as well as singular verbs:

Each of the students had *his* assignment ready.

However, our growing sensitivity about using male pronouns exclusively when the reference may be to both men and women makes the choice of a suitable pronoun increasingly difficult. Many people might object to the use of *his* as the pronoun in the above example. Do not choose to solve the problem by introducing a grammatical error as in this example of incorrect usage:

Each of the students had *their* assignment ready.

The use of male and female pronouns together is grammatically correct, if a bit awkward at times:

Each of the students had *his* or *her* assignment ready.

Perhaps the best solution, one that is usually applicable, is to use a plural antecedent that allows the use of a neutral plural pronoun, as in this example:

All of the students had *their* assignments ready.

The same problem presents itself when we use such nouns as *student* or *human being* in their generic sense, that is, when we use them to stand for all students or all human beings. If used in the singular, such nouns must be followed by singular pronouns:

> The *student* seeking a loan must have *his* or *her* application in by 3 September.

Again the best solution is to use a plural antecedent:

> *Students* seeking loans must have *their* applications in by 3 September.

5. IMPROPER VERB FORM (VB)

Use the proper verb form.

Improper verb form includes a wide variety of linguistic errors ranging from such nonstandard usages as "He seen the show" for "He saw the show" to such esoteric errors as "He was hung by the neck until dead" for "He was hanged by the neck until dead." Normally a few minutes spent with any collegiate dictionary will show you the correct verb form. College level dictionaries list the principal parts of the verb after the verb entry.

6. IMPROPER PRONOUN FORM (PRON)

Use the proper pronoun form.

Almost every adult can remember being constantly corrected by parents and elementary school teachers in regard to pronoun form. The common sequence is for the child to say,"Me and Johnny are going swimming," and for the teacher or parent to say patiently, "No, dear, 'Johnny and *I* are going swimming.'" As a result of this conditioning, all objective forms are automatically under suspicion in many adult minds, and the most common pronoun error is for the speaker or writer to use a subjective case pronoun such as *I*, *he*, or *she* when an objective case pronoun such as *me*, *him* or *her* is called for.

Whenever a pronoun is the object of a verb or the object of a preposition, it must be in the objective case:

> It occurred to my colleagues and *me* to check the velocity data on the earthquake waves.

> Just between you and *me*, the man who says, "He grabbed he and I and shook us," is ludicrous.

However, use a subjective case pronoun in the predicate nominative position. This rule slightly complicates the use of pronouns after the verb. Normally the pronoun position after the verb is thought of as objective pronoun territory, but when the verb is a linking verb (chiefly the verb *to be*) the pronoun is called a predicate noun rather than an object and is in the subjective case.

It is *she*.

It was *he* who discovered the mutated fruit fly.

7. SPELLING ERROR (SP)

Correct your spelling error.

No easy road to spelling perfection exists. The condition of English spelling is chaotic and likely to remain so. George Bernard Shaw once illustrated this chaos by spelling *fish* as *ghoti*. To do so he took the *gh* from *rough*, the *o* from *women*, and the *ti* from *condition*. You may obtain some help from the spelling section in a collegiate dictionary where the common rules of spelling are explained. You can also buy, rather inexpensively, books that explain the various spelling rules and provide exercises to fix the rules in your mind. Certainly, you should memorize the spelling of words you commonly use and learn to look up in your dictionary difficult words that you use only rarely.

8. DANGLING MODIFIER (DM)

Give the modifier something to modify.

Many curious sentences result from the failure to provide the modifier something to modify:

Having finished the job, the tarpaulins were removed.

In this example it seems as though the tarpaulins have finished the job. As is so often the case, a passive voice construction has caused the problem (see pages 180–181). If we recast the sentence in the active voice we remove the problem:

Having finished the job, the workers removed the tarpaulins.

9. MISPLACED MODIFIER (MM)

Place the modifier as close as possible to the element modified.

As in the case of dangling modifiers (Rule 8), curious sentences result from the modifier's not being placed next to the element modified:

> An engine may crack when cold water is poured in *unless it is running.*

Probably, with a little effort, no one will misread this example; but, undeniably, it says that the engine will crack unless the water is running. Move the modifier and the sentence is clear:

> *Unless it is running,* an engine may crack when cold water is poured in.

It should be apparent from the preceding examples that a modifier may be in the wrong position to convey one meaning but in the perfect position to convey a different meaning. In the next example, the placement of *for three years* is either right or wrong. It is in the right position to modify *to work* but in the wrong position to modify *have been trying.*

> I have been trying to place him under contract to work here for three years. *(three-year contract)*

As the examples should suggest, correct placement of modifiers sometimes amounts to more than mere nicety of expression. It can mean the difference between stating falsehood and truth, between saying what you mean and saying something else.

A *which* clause that must look over several intervening words before finding its antecedent is sometimes called a *squinting which clause:*

> We found a pile of picture frames in the corner, which had not been dusted for years.

This construction is objectionable only when the intervening words include a substantive (that is, a noun or noun phrase) that could logically be the antecedent of *which.* Usually it is a simple matter for the writer to revise and find a way out of the difficulty:

> In the corner we found a pile of picture frames, which had not been dusted for years.

Of course, the sentence was all right to begin with if a dusty corner is being referred to.

10. PARALLELISM (PARAL)

Make the elements in a series grammatically parallel.

When you link elements in a series, they must all be in the same grammatical form. Link an adjective with an adjective, a noun with a noun, a clause with a clause, and so forth. Look at the italicized portion of the following sentence:

A good test would *use small amounts of plant material, require little time, simple to run, and accurate.*

The series begins with the verbs *use* and *require* and then abruptly switches to the adjectives *simple* and *accurate.* All four elements must be based on the same part of speech. In this case, it's simple to change the last two elements:

A good test would use small amounts of plant material, require little time, *be simple to run, and be accurate.*

Always be careful when you are listing to keep all the elements of the list parallel. In the following example, the third item in the list is not parallel to the first two:

The process has three stages: (1) the specimen is dried, (2) all potential pollutants are removed, and (3) atomization.

The error is easily corrected:

The process has three stages: (1) the specimen is dried, (2) all potential pollutants are removed, and (3) the specimen is atomized.

When you start a series, keep track of what you are doing and finish the series the same way you started it. Nonparallel sentences are at best awkward and off-key. At worst, they can lead to serious misunderstandings.

11. DICTION (D)

Choose a word or words that are more accurate, effective, or appropriate to the situation.

Many different kinds of linguistic sins are covered by the general term *diction.* All involve a faulty choice of words. Poor diction can involve a choice of words that are too heavy or pretentious: *utilize* for *use*, *finalize* for *finish*, *at this point in time* for *now*, and so forth. Tired old clichés are poor diction: *with respect to, with your permission, with reference to*, and many others. We talk about such language in Chapter 9, particularly in the section on pomposity (pages 184–185).

Sometimes the words chosen are simply too vague to be accurate: *inclement weather* for *rain, too hot* for *600°C*. See the section on specific words (pages 182–184) in Chapter 9 for more on this subject.

Poor diction can mean an overly casual use of language when some degree of formality is expected. One of the many synonyms for *intoxicated*, such as *bombed, stoned,* or *smashed*, might be appropriate in private conversation but totally wrong in a police or laboratory report.

Poor diction can reflect a lack of sensitivity to language—to the way one group of words relates to another group. Someone who writes that "The airlines are beginning a crash program to solve their financial difficulties" is not paying attention to relationships. The person who writes that "The Steelworkers' Union representatives are getting down to brass tacks in the strike negotiations" has a tin ear, to say the least.

Make your language work for you, and make it appropriate to the situation.

Conventions

12. PUNCTUATION (PN)

The chief marks of punctuation total 36. If from that list we exclude
the marks that are technical, foreign, and diacritical, we have 13.

Apostrophe	Hyphen
Brackets	Parentheses
Colon	Period
Comma	Question mark
Dash	Quotation marks
Ellipses	Semicolon
Exclamation point	

These few marks satisfy all the constantly recurring demands for
punctuating page after page of normal prose.

13. APOSTROPHE (APOS)

The apostrophe has three chief uses: (1) to form the possessive, (2)
to stand for missing letters or numbers, and (3) to form the plural of
certain expressions.

13.1 Possessives

Add an apostrophe and an *s* to form the possessive of most singular
nouns, including proper nouns, even when they already end in an *s*

or another silibant, such as *x*:

man's

spectator's

jazz's

Marx's

Charles's

Exceptions to this rule occur when adding an apostrophe plus an *s* would result in an *s* or *z* sound that is difficult to pronounce. In such cases, usually just the apostrophe is added:

Xerxes'

Moses'

conscience'

appearance'

To understand this exception, pronounce *Marx*'s and then a word like *Moses*'s or *conscience*'s. (Note also that though a proper noun such as *Marx* may be italicized its possessive *s* remains in roman type.)

To form plurals into the possessive case, add an apostrophe plus *s* to words that do not end in an *s* or other silibant and an apostrophe only to those that do:

men's

data's

spectators'

agents'

witnesses'

To show joint possession, add the apostrophe and *s* to the last member of a compound or group; to show separate possession, add an apostrophe and *s* to each member:

Gregg and Klymer's experiment astounded the class.

Gregg's and Klymer's experiments were very similar.

Of the several classes of pronouns, only the indefinite pronouns use an apostrophe to form the possessive.

Possessive of Indefinite Pronouns	Possessive of Other Pronouns
anyone's	my (mine)
everyone's	your (yours)
everybody's	his, her (hers)
nobody's	our (ours)
no one's	their (theirs)
other's	whose
neither's	

13.2 Missing Letters or Numbers

Use an apostrophe to stand for the missing letters in contractions and to stand for the missing letter or number in any word or set of numbers where for one reason or another a letter or number is omitted:

can't, don't, o'clock, it's (it is), etc.

We were movin' downriver, listenin' to the birds singin'.

The class of '49 was Colgate's best class in years.

13.3 Plural Forms

An apostrophe is often used to form the plural of letters and numbers, but this style is gradually dying, particularly with numbers.

6's and 7's (but also 6s and 7s)

A's and b's

14. BRACKETS (BRACKETS)

Brackets are chiefly used when a word or comment is inserted into a quotation:

"The result of this [disregard by the propulsion engineer] has been the neglect of the theoretical and mathematical mastery of the engine inlet problem."

"An ideal outlet require [sic] a frictionless flow."

"Last year [1982] saw a partial solution to the problem."

Sic, by the way, is Latin for *thus*. Inserted in a quotation, it means that a mistake is the original writer's, not yours. Use it with discretion.

15. COLON (COLON)

The colon is chiefly used to introduce quotations, lists, or supporting statements and between clauses when the second clause is an example or amplification of the first. Additionally, it is used in certain conventional ways with numbers and in correspondence and bibliographical entries.

15.1 Introductory

Place a colon before a quotation, a list, or supporting statements and examples that are formally introduced:

> Mr. Smith says the following of wave generation:
> The wind waves that are generated in the ocean and which later become swells as they pass out of the generating area are products of storms. The low pressure regions that occur during the polar winters of the Arctic and Antarctic produce many of these wave-generating storms.

The various forms of engine which might be used would operate within the following ranges of Mach number;

M-0 to M-1.5	Turbojet with or without precooling
M-1.5 to M-7	Reheated turbojet, possibly with precooling
M-7 to M-10+	Ramjet with supersonic combustion

Engineers are developing three new engines: turbojets, reheated turbojets, and ramjets.

Do not place a colon between a verb and its objects or a linking verb and the predicate nouns.

Objects
The engineers designed turbojets, reheated turbojets, and ramjets.

Predicate Nouns
The three engines the engineers are developing are turbojets, reheated turbojets, and ramjets.

Do not place a colon between a preposition and its objects:

The plane landed at Detroit, Chicago, and Rochester.

15.2 Between Clauses

If the second clause consists of an example or an amplification of the first clause, then the colon may replace the comma, semicolon, or period:

> The docking phase involves the actual "soft" contact: the securing of lines, latches, and air locks.
>
> The difference between these two guidance systems is illustrated in Figure 2: The paths of the two vehicles are shown to the left and the motion of the ferry as viewed from the target station is shown to the right.

You may follow a colon with a capital or a small letter. Generally, a complete sentence beginning after a colon is given a capital.

15.3 Styling Conventions

Place a colon after a formal salutation in a letter, between hour and minute figures, between the elements of a double title, and between chapter and verse of the Bible:

> Dear Mrs. Jones:
>
> 3:45
>
> at 7:15 P.M.
>
> *Mary Russell Mitford: Her Life and Writing*
>
> I Samuel 7:14–18

16. COMMA (C)

The most used—and misused—mark of punctuation is the comma. Writers use commas to separate words, phrases, and clauses. Generally, commas correspond to the pauses we use in our speech to separate ideas and to avoid ambiguity. As writing style has grown plainer, using shorter sentences and simpler constructions, the use of commas has lessened. In many places where your parents would have used a comma you will not. Nevertheless, you will use the comma often: about two out of every three marks of punctuation you use will be commas. Sometimes your use of the comma will be essential for clarity; at other times you will be honoring grammatical conventions. (See also Rule 2.)

16.1 Main Clauses

Place a comma before a coordinating conjunction *(and, but, or, nor, for, yet)* that joins two main (independent) clauses:

> During the first few weeks we felt a great deal of confusion, but as time passed we gradually fell into a routine.

> We could not be sure that the plumbing would escape frost damage, nor were we at all confident that the house could withstand the winds of almost hurricane force.

If the clauses are short, have little or no internal punctuation, or are closely related in meaning, then you may omit the comma before the coordinating conjunction:

> The wave becomes steeper but it does not tumble yet.

In much published writing there is a growing tendency to place two very short and closely related independent clauses (called contact clauses) side by side with only a comma between:

> The wind starts to blow, the waves begin to develop.

Sentences consisting of *three* or more equal main clauses should be punctuated uniformly:

> We explained how urgent the problem was, we outlined preliminary plans, and we arranged a time for discussion.

In general, identical marks are used to separate equal main clauses. If the equal clauses are short and uncomplicated, commas usually suffice. If the equal clauses are long, internally punctuated, or if their separateness is to be emphasized, semicolons are either preferable or necessary.

16.2 Clarification

Place a comma after an introductory word, phrase, or clause that might be overread or that abnormally delays the main clause:

> As soon as you have finished polishing, the car should be moved into the garage. *(Comma to prevent overreading)*

> Soon after, the winds began to moderate somewhat, and we were permitted to return to our rooms. *(First comma to prevent overreading)*

If the polar ice caps should someday mount in thickness and weight to the point that their combined weight exceeded the equatorial bulge, the earth might suddenly flop 90 degrees. *(Introductory clause abnormally long)*

After a short introductory element (word, phrase, or clause) where there is no possibility for ambiguity, the use of the comma is optional. Generally let the emphasis you desire guide you. A short introductory element set off by a comma will be more emphatic than one that is not.

16.3 Nonrestrictive Modifiers

Enclose or set off from the rest of the sentence every nonrestrictive modifier, whether a word, a phrase, or a clause. How can you tell a nonrestrictive modifier from a restrictive one? Look at these two examples:

Restrictive
A runway *that is not oriented with the prevailing wind* endangers the aircraft using it.

Nonrestrictive
The safety of any aircraft, *whether heavy or light,* is put in jeopardy when it is forced to take off or land in a crosswind.

The restrictive modifier is necessary to the meaning of the sentence. Not just any runway but "a runway that is not oriented with the prevailing wind" endangers aircraft. The writer has *restricted* the many kinds of runways he could talk about to one particular kind. In the nonrestrictive example, the modifier merely adds descriptive details. The writer does not restrict *aircraft* with the modifier, but rather makes the meaning a little clearer.

Restrictive modifiers cannot be left out of the sentence if it is to have the meaning the writer intends; nonrestrictive modifiers can be left out.

16.4 Nonrestrictive Appositives

Set off or enclose every nonrestrictive appositive. As used here the term *appositive* means any element (word, phrase, or clause) that parallels and repeats the thought of a preceding element. According to this view, a verb may be coupled appositively with another verb, an adjective with another adjective, and so on. An appositive is usually more specific or more vivid than the element that it is an appositive to; an appositive makes explicit and precise something that has not been clearly enough implied.

Some appositives are restrictive and, therefore, are not set off or enclosed.

Nonrestrictive
A crosswind, *a wind perpendicular to the runway,* causes the pilot to make potentially dangerous corrections just before landing.

Restrictive
In some ways, Eisenhower *the president* had to behave differently from Eisenhower *the general.*

In the nonrestrictive example, the appositive merely adds clarifying detail. The sentence makes sense without it. In the restrictive example, the appositives are essential to the meaning. Without them we would have, "In some ways, Eisenhower had to behave differently from Eisenhower."

16.5 Series
Use commas to separate members of a coordinate series of words, phrases, or clauses if *all* the elements are not joined by coordinating conjunctions:

Instructions on the label state clearly how to prepare the surface, how to apply the contents, and how to clean and polish the mended article.

To mold these lead figures you will need a hot flame, a two-pound block of lead, the molds themselves, a file or a rasp, and an awl.

Under the microscope the sensitive, filigreelike mold appeared luminous and transparent and faintly green.

16.6 Other Conventional Usages

Date
On November 26, 1950, all the forces of nature seemed to combine to wreak havoc upon the Middle Atlantic States.

Note: When you write the month and the year without the day, it is common practice now to omit the comma between them—as in June 1984.

Geographical Expression
During World War II Middletown, Pennsylvania, was the site of a huge military airport and supply depot.

Title After Proper Name
A card in yesterday's mail informed us that Carleton Williams, M.D., Pediatrician, would soon open new offices on Beaver Street, State College.

Noun of Direct Address
Clifton, do you suppose that we can find our way back to the cabin before nightfall?

Informal Salutation
Dear Jane,

17. DASH (DASH)

In technical writing, you will use the dash almost exclusively to set off parenthetical statements. You may, of course, use commas or parentheses for the same function, but the dash is the most emphatic separator of the three. You may also use the dash to indicate a sharp transition, but some editors may object to this use. On the typewriter you make the dash with two hyphens. You do not space between the words and the hyphens or between the hyphens themselves.

Typewriter

```
    The first phase in rendezvous--sighting and rec-
ognizing the target--is so vital that we will treat it
at some length.
```

Typeset
The target must emit or reflect light the pilot can see—but how bright must this light be?

18. ELLIPSES (ELL)

Use three spaced periods (called ellipses or ellipsis points) to indicate words omitted within a quoted sentence, four spaced periods if the omission occurs at the end of the sentence:

"As depth decreases, the circular orbits become elliptical and the orbital velocity . . . increases as the wave height increases."

"As the ground swells move across the ocean they are subject to headwinds or crosswinds. . . ."

You need not show ellipses if the context of the quotation makes it clear that the quotation is not complete:

Wright said the accident had to be considered a "freak of nature."

19. EXCLAMATION POINT (EXC)

Place an exclamation point at the end of a startling or exclamatory sentence.

On 9 January 1952, the S. S. *Pennsylvania* sank with no survivors in the North Pacific in a storm that produced 50-knot winds for 33 hours and waves up to 48 feet high!

Obviously, with the emphasis in technical writing on objectivity, you will seldom use the exclamation point.

20. HYPHEN (HYPHEN)

Hyphens are used to form various compound words and in breaking up a word that must be carried over to the next line.

20.1 Compound Numbers

See 28.1.3 and 28.3.

20.2 Common Compound Words

Observe dictionary usage in using or omitting the hyphen in compound words.

governor-elect	acid resistant
ex-treasurer	Croesus-like
Russo-Japanese	drill-like
pro-American	self-interest

But:

glasslike	wrist watch
neophyte	sweet corn
newspaper	weather map
newsstand	sun lamp
housewife	prize fight

20.3 Compound Words as Modifiers

Use the hyphen between words joined together to modify other words:

a half-spent bullet

an eight-cylinder engine

their too-little-and-too-late methods

Be particularly careful to hyphenate when omitting the hyphen may cause ambiguity:

two-hundred-gallon drums

two hundred-gallon drums

a pink-skinned pig

Sometimes you have to carry a modifier over to a later word, creating what is called a *suspended hyphen:*

GM cars come with a choice of four-, six-, or eight-cylinder engines.

20.4 Word Division

Use a hyphen to break a word that must be carried over to the next line.

Words compounded of two roots or a root and an affix are divided at the point of union:

self-/important	wind/jammer
cross-/pollination	desir/able
bladder/wort	anti/dote
summer/time	manage/able

Note: The first two words in this list are always spelled with a hyphen; the remaining words use the hyphen only when the word is syllabicated at the end of a line.

In general, noncompound words of more than one syllable are divided between any two syllables but only between syllables:

soph/o/more	con/clu/sion
sat/is/fac/tion	sym/pa/thy

A syllable of one letter is never set down alone; a syllable of only two letters is seldom allowed to stand alone unless it is a prefix or a suffix, and then only if it is pronounced as spelled:

hello/	elec/tro/type
method/	de/mand
pilot/	ac/cept
many/	walked/
saga/	start/ed

If a consonant is doubled because a suffix is added, include the second consonant with the suffix:

spin/ner	slip/ping
stir/ring	slot/ted

But:

stopped/	pass/ing
lapped/	stall/ing

Because -*ped* is not pronounced as a syllable it should not be carried over. In words like *passing* and *stalling* both consonants belong to the root and, therefore, only the suffix -*ing* is carried over.

21. PARENTHESES (PAREN)

Parentheses are used to enclose supplementary details inserted into a sentence. Commas and dashes may also be used in this role, but with some restrictions. You may enclose a complete sentence or several complete sentences within parentheses. But such enclosure would confuse the reader if only commas or dashes were used for the enclosure:

The violence of these storms can scarcely be exaggerated. (Typhoons and hurricanes generate winds over 75 miles an hour and waves 50 feet high.) The study . . . .

21.1 Lists

Parentheses are also used to enclose numbers or letters used in listing:

This general analysis consists of sections on (1) wave generation, (2) wave propagation, (3) wave action near a shoreline, and (4) wave energy.

21.2 In Sentences

Within a sentence place no mark of punctuation before the opening parenthesis. Place any marks needed in the sentence after the closing parenthesis:

If a runway is regularly exposed to crosswinds of over 10 knots (11.6 mph), then the runway is considered unsafe.

Do not use any punctuation around parentheses when they come between sentences. Give the statement *inside* the parentheses any punctuation it needs.

22. PERIOD (PER)

Periods have several conventional uses.

22.1 End Stop

Place a period at the end of any sentence that is not a question or an exclamation:

Find maximum average daily temperature and maximum pressure altitude.

22.2 After Abbreviations

Place a period after abbreviations:

M.D.   etc.

Ph.D.   Jr.

22.3 Decimal Point

Use the period with decimal fractions and as a decimal point between dollars and cents:

.4

.05%

$5.60

$450.23

23. QUESTION MARK (QUES)

Place a question mark at the end of every sentence that asks a direct question:

What is the purpose of this report?

A request that you politely phrase as a question may be followed by either a period or a question mark:

Will you be sure to return the experimental results as soon as possible.

Will you be sure to return the experimental results as soon as possible?

When you have a question mark with quotation marks, you need no other mark of punctuation:

"Where am I?" he asked.

24. QUOTATION MARKS (QUOT)

In technical writing you will use quotation marks to set off short quotations and certain titles. You will have little other need for them.

24.1 Short Quotations

Use quotation marks to enclose quotations that are short enough to work into your own text (normally, less than three lines):

According to Dr. Smith, "Magnetic tape is now so reliable, it is accepted as courtroom evidence."

When quotations are longer than three lines, set them off by single spacing and indenting them. See 15.1 for an example of this style. Do not use quotation marks when quotations are set off and indented.

24.2 Titles

Place quotation marks around titles of articles from journals and periodicals:

> Beller's article "Air Breathing Boosters Gain Favor" appeared in *Missiles and Rockets.*

24.3 Within Quotes

When you must use quotation marks within other quotation marks use single marks (the apostrophe on your typewriter):

> "Do you have the same trouble with the distinction between 'venal' and 'venial' that I do?" asked the copy editor.

24.4 With Other Punctuation Marks

The following are the American conventions for using punctuation with quotation marks:

Commas and periods: Always place commas and periods inside the quotation marks. There are no exceptions to this rule:

> G. D. Brewer wrote "Manned Hypersonic Vehicles."

Semicolons and colons: Always place semicolons and colons outside the quotation marks. There are no exceptions to this rule:

> As Dr. Damron points out, "The inlet designer has to satisfy both the aircraft designer and the propulsion engineer"; the aircraft designer need satisfy only himself.

Question marks, exclamation points, and dashes: Place question marks, exclamation points, and dashes inside the quotation marks when they apply to the quote only *or to the quote and the entire sentence at the same time.* Place them outside the quotation marks when they apply to the entire sentence only.

Inside
When are we going to find the answer to the question, "What causes clear air turbulence?"

Outside
Did you read Kelly's "Scientists Consider Aerospace Transporter"?

25. SEMICOLON (SEMI)

The semicolon lies halfway between the comma and the period in force. Its use is quite restricted. (See also Rule 2.)

25.1 Main Clauses

Place a semicolon between two closely connected main clauses that are not joined by a coordinating conjunction *(and, but, or, nor, for, or yet):*

> The expanding gases formed during burning drive the turbine; the gases are then exhausted through the nozzle.

If the clauses are long, have internal punctuation, or if separate emphasis is desired, then the comma before the coordinating conjunction may be increased to a semicolon:

> The front lawn has been planted with a Chinese beauty tree, a Bechtel flowering crab, a mountain ash, and assorted small shrubbery, including barberry and cameo roses; but so far nothing has been done to the rear beyond clearing and rough grading.

25.2 Series

When a series contains commas as internal punctuation within the parts, use semicolons between the parts:

> Included in the experiment were Peter Moody, a freshman; Jesse Gatlin, a sophomore; Burrel Gambel, a junior; and Ralph Leone, a senior.

26 ITALICIZATION (ITAL)

Italic print is a distinctive typeface, like this sample: *Scientific American.* When you type or write, you represent italics by underlining, like this:

> Scientific American

26.1 Foreign Words

Italicize foreign words that have not yet become a part of the English language:

> We suspected him always of holding some *arrière pensée.*

> Corey's *joie de vivre* was contagious.

Also italicize Latin scientific terms:

Cichorium endivia (endive)

Percopsis omiscomaycus (trout-perch)

But do not italicize Latin abbreviations or foreign words that have become a part of the language (including such scholarly abbreviations as "ibid." and "op. cit."):

etc.	bourgeois
vs.	status quo

Your college dictionary will normally indicate which foreign words are still italicized and which are not.

26.2 Words, Letters, and Numbers Used as Such

The words *entrance* and *admission* are not perfectly interchangeable.

Don't forget the *k* in *picnicking*.

His 9s and 7s descended below the line of writing.

26.3 Titles

Italicize the titles of books, plays, magazines, newspapers, ships, and artistic works:

Webster's New World Dictionary	*The Free Press*
Othello	SS. *Pennsylvania*
Scientific American	*Mona Lisa*

27. CAPITALIZATION (CAP)

We provide the more important rules of capitalization. For a complete rundown see your college dictionary.

27.1 Proper Nouns

Capitalize all proper nouns and their derivatives:

Places
America   American   Americanize   Americanism

Days of the Week and Months
Monday   Tuesday   January   February

But not the seasons:

winter   spring   summer   fall

Organizations and Their Abbreviations
American Kennel Club (AKC)

United States Air Force (USAF)

Capitalize *geographic areas* when you refer to them as areas:

The Andersons toured the Southwest.

But do not capitalize words that merely indicate direction:

We flew west over the Pacific.

Capitalize the names of *studies* in a curriculum only if the names are already proper nouns or derivatives of proper nouns or if they are part of the official title of a department or course:

Department of Geology

English Literature 25

the study of literature

the study of English literature

Note: Many nouns (and their derivatives) that were originally proper have been so broadened in application and have become so familiar that they are no longer capitalized: *boycott, macadam, spoonerism, italicize, platonic, chinaware, quixotic.*

27.2 Literary Titles
Capitalize the first word, the last word, and every important word in literary titles:

But What's a Dictionary For

The Meaning of Ethics

How to Write and Be Read

27.3 Rank, Position, Family Relationships

Capitalize the titles of rank, position, and family relationship unless they are preceded by *my, his, their,* or similar possessive pronouns:

Professor J. E. Higgins

I visited Uncle Timothy.

I visited my uncle Timothy.

Dr. Milton Weller, Head, Department of Entomology

28. NUMBERS (NUM)

There is a good deal of inconsistency in the rules for handling numbers. We will give you the general rules. Your instructor or your organization may give you others. As in all matters of format, you must satisfy whomever you are working for at the moment. Do, however, be internally consistent within your reports. Do not handle numbers differently from page to page of a report.

28.1 As Words

Generally, you write all numbers ten and under, and rounded-off large numbers, as words:

six generators

about a million dollars

However, when you are writing a series of numbers, do not mix up figures and words. Let the larger numbers determine the form used:

five boys and six girls

But:

It took us 6 months and 25 days to complete the experiment.

28.1.1 Beginning a Sentence

Do not begin sentences with a figure. If you can, write the number as a word. If this would be cumbersome, write the sentence so as to get the figure out of the beginning position:

Fifteen months ago, we saw the new wheat for the first time.

We found 35 deficient steering systems.

28.1.2 Compound Number Adjectives

When you write two numbers together in a compound number adjective spell out the first one or the shorter one to avoid confusing the reader:

> twenty 10-inch trout

> 100 twelve-volt batteries

28.1.3 Hyphens

Two-word numbers are hyphenated:

> Eighty-five boxes

or:

> Eighty-five should be enough.

28.2 Figures

The general rule here is to write all exact numbers over ten as figures. This rule probably holds more true in technical writing with its heavy reliance on numbers than it does in general writing. However, as we noted in 28.1, rounded-off numbers are commonly written as words. The precise figure could give the reader an impression of exactness that might not be called for.

Certain conventional uses call for figures at all times.

28.2.1 Dates, Exact Sums of Money, Time, Address

> 1 January 1984 *or* January 1, 1984

> $3,422.67 *but* about three thousand dollars

> 1:57 P.M. *but* two o'clock

> 660 Fuller Road

28.2.2 Technical Units of Measurement

> 6 cu ft

> 4,000 rpm

28.2.3 Cross-References

> See page 22.

> Refer to Figure 2.

28.3 Fractions

When a fraction stands alone, write it an an unhyphenated compound:

two thirds

fifteen thousandths

When a fraction is used as an adjective you may write it as a hyphenated compound. But if either the numerator or the denominator is hyphenated, do not hyphenate the compound. More commonly, fractions used as adjectives are written as figures.

two-thirds engine speed

twenty-five thousandths

3/4 rpm

29. ABBREVIATIONS (AB)

Although most people are familiar with the kinds of abbreviations we encounter in everyday conversation and written material, from *mph* to *Mon.* to *Dr.*, technical abbreviations are something else again. Each scientific and professional field generates hundreds of specialized terms, and many of these terms are often abbreviated for the sake of conciseness and simplicity.

Thus, before deciding to use technical abbreviations in an article or report, you must first consider your audience—lay people or scientists, executives or engineers? Only readers with a technical background will be able to interpret the specialized shorthand for the field in question. When in doubt, then, avoid all but the most common abbreviations. If you must use a technical term, spell it out in full the first time it appears and include both the abbreviation and a definition in parentheses after it. You can then safely use the abbreviation if the term crops up again in your report.

Standard technical and scientific abbreviations include the following:

absolute	abs
acre or acres	(spell out)
alternating current (as adjective)	a-c
atomic weight	at. wt
barometer	bar.

Brinell hardness number	Bhn
British thermal units	Btu or B
meter	m
square meter	m^2
microwatt or microwatts	mu w
miles per hour	mph
National Electric Code	NEC
per	(spell out)
revolutions per minute	rpm
rod	(spell out)
ton	(spell out)

The system implied by these illustrative abbreviations can be described by a brief set of rules.

1. Use the same (singular) form of abbreviation for both singular and plural terms:

cu ft	either cubic foot or cubic feet
cm	either centimeter or centimeters

But there are some common exceptions:

no.	number
nos.	numbers
p.	page
pp.	pages
ms.	manuscript
mss.	manuscripts

2. Use small (lowercase) letters except for letters standing for proper nouns or proper adjectives:

abs	*but*	Btu or B
mph	*but*	Bhn

3. For technical terms, use periods only after abbreviations that spell complete words. For example, *in* is a word, and the abbreviation for inches could be confused with it. Therefore, use a period:

ft	*but*	in.
abs	*but*	bar.
cu ft	*but*	at. wt

4. Remember the hyphen in the abbreviations a-c and d-c when you use them as adjectives:

This a-c motor can be converted to 28 volts dc.

5. Spell out many short and common words:

acre   rod   per   ton

6. In compound abbreviations, use internal spacing only if the first word is represented by more than its first letter:

rpm	*but*	cu ft
mph	*but*	mu w

7. With few exceptions, form the abbreviations of organization names without periods or spacing:

NEC   ASA

8. Abbreviate terms of measurement only if they are preceded by an arabic expression of exact quantity:

55 mph	*and*	20-lb anchor

But:

We will need an engine of greater horsepower.

APPENDIXES

Appendix # A

A Student Report

930 Cleveland Avenue, South
St. Paul, MN 55116
March 10, 1983

Dr. Milton W. Weller
Professor and Head
Department of Entomology, Fisheries, and Wildlife
University of Minnesota
St. Paul, MN 55108

Dear Dr. Weller:

Enclosed is my Feasibility Report on a Portable Cassette
Tape Recorder for Field Research. It is submitted for your
consideration and approval.

The report recommends that the department purchase either a
Nakamichi 550 portable cassette tape recorder or, as a
close second choice, the AIWA TPR-945. All recorders
considered in the report meet the minimum standards set by
the department. Final conclusions and recommendations are
based on a comparison using frequency response, cost, size
and weight, and battery requirements.

Dr. Milton W. Weller   -2-   March 10, 1983

Dr. Milton W. Weller
March 10, 1983

I hope this report will help you decide on a portable
cassette tape recorder for your research projects. If you
have any questions or problems concerning the report,
please write or call (690-5849).

Sincerely yours,

David M. Zellar

Enclosure

FEASIBILITY REPORT ON A PORTABLE CASSETTE

TAPE RECORDER FOR FIELD RESEARCH

Prepared for

Dr. Milton W. Weller

Professor and Head

Department of Entomology, Fisheries, and Wildlife

University of Minnesota

by

David M. Zellar

Abstract

This report presents the results of a study that considered
the purchase by the Department of Entomology, Fisheries,
and Wildlife of a portable cassette recorder for field
research. The report discusses the minimum specifications
this tape recorder must meet and compares tape recorders
meeting these minimum specifications on the basis of
frequency response, cost, size and weight, and battery
requirements. Conclusions and recommendations are reached.

March 10, 1983

CONTENTS

FEASIBILITY REPORT ON A PORTABLE CASSETTE

TAPE RECORDER FOR FIELD RESEARCH

INTRODUCTION

This report presents the results of a study that

considered the purchase by the Department of Entomology,

Fisheries, and Wildlife (EFW) of a portable cassette tape

recorder for the Department's field research projects.

The study was designed to determine, first, if a

portable cassette tape recorder was available that could

meet EFW's minimum specifications for field research

applications, as follows: (1) weigh 20 pounds or less; (2)

be powered by disposable, reasonably priced, readily

available batteries; (3) have a frequency response of 100-

12,000 Hz or better; (4) cost $750 or less; and (5) be

purchased and serviced locally. I found four such

recorders: the Marantz Superscope CD-330, the RMF 740 AV

Stereo Recorder, the AIWA TPR-945, and the Nakamichi 550.

Second, I compared these four on the basis of frequency

response, cost, size and weight, and battery requirements.

(All technical information concerning the tape recorders

compared comes from the respective company publications

1

listed in the References.) Included in each of these comparisons are a definition and a rating of <u>good</u> or <u>acceptable</u> for each of the tape recorders according to the definition. Because EFW considers frequency response and cost the most important criteria, my recommendations have come mostly from the comparisons based on them.

FACTUAL SUMMARY

Because the tape recorder is being purchased primarily to record bird vocalizations, the frequency response should match such vocalizations as closely as possible. Because bird vocalizations may range from 100-18,000 Hz, ideally, the recorder response should match that range. Cost is important because only a limited amount of money is available.

Following is a table that summarizes ratings and key data for the four tape recorders that meet the minimum specifications. Because battery life was not provided in the information available for the recorders considered, satisfaction of the battery requirements criterion is not rated.

	Frequency Response	Cost	Size and Weight
Marantz Superscope CD-330	acceptable (40-13,000 Hz)	good ($249)	good (7.8 lb)
RMF 740 AV Stereo Recorder	acceptable (40-12,500 Hz)	acceptable ($579)	good (8.0 lb)
AIWA TPR-945	good (50-15,000 Hz)	good ($350)	acceptable (12.5 lb)
Nakamichi 550	good (40-17,000 Hz)	acceptable ($640)	good (8.3 lb)

As the data show, the Marantz, RMF, and Nakamichi machines are approximately the same weight. The Aiwa is more than four pounds heavier than the others and is also larger in all its dimensions.

CONCLUSIONS

1. Four portable cassette tape recorders are available that meet minimum specifications.

2. On the basis of the most important criterion, frequency response, the Nakamichi is the best choice.

3. On the basis of cost, the Marantz is the best choice.

4. On the basis of cost and frequency response, the AIWA may be a good compromise choice.

3

5. On the basis of size and weight, all the recorders are at least acceptable, but on this criterion, the AIWA is the least desirable.

6. The results of the battery requirements comparison are too ambiguous to use as a basis of comparison for this study.

7. None of the recorders is clearly superior, but because frequency response is considered most important in this study, the Nakamichi 550 may be the best buy. However, if cost and frequency response are considered equally important, and size and weight are not considered, the AIWA TPR-945 would be the best buy.

RECOMMENDATIONS

On the basis of this study I recommend the following to EFW:

1. If sufficient money is available, buy the Nakamichi 550 portable cassette tape recorder.

2. If the Nakamichi is considered too expensive, buy the AIWA TPR-945.

4

A.   FREQUENCY RESPONSE

EFW will purchase the tape recorder primarily to record bird vocalizations. The department may later use these recordings to construct sonagrams for further studies. The frequency response of the tape recorder used is the single most important factor in making sonagrams from tape recordings.

Definition

The cassette recorder purchased should, ideally, have a frequency response that would be able to record all bird vocalizations. Robbins et al. (1966) include sonagrams for a variety of birds in their book Birds of North America. These sonagrams indicate that bird vocalizations may range from 100-18,000 Hz. Ideally, therefore, the recorder purchased should have a frequency response to match that range. However, very few portable cassette tape recorders are capable of such a frequency response, and the few that are cost too much to be considered. Further study of the sonagrams, however, reveals that the majority of bird vocalizations fall within the 100-12,000 Hz range with only a rare few exceeding 15,000 Hz. Therefore, for this study a

5

frequency response of 100-15,000 Hz will be rated

acceptable, and a frequency response that exceeds this

range will be rated good. These ranges may limit the

cassette tape recorder's research application for a few

species, but will allow the purchase of a very acceptable

machine for most studies.

Sensitivity of Recording Tapes

The frequency response depends somewhat on the type of

recording tape used. Three basic types are currently

available:

1.   Standard tape

2. CrO_2 tape

3.   FeCr tape

The construction of these tapes makes a difference in the

sensitivity of the sound they will reproduce. FeCr tapes

are the most sensitive of the three and also the most

expensive. Also, they are not always readily available,

which makes them somwhat impractical for use. The CrO_2 tapes

are not quite as sensitive as the FeCr tapes, but they are

better than the standard tapes. They are slightly more

expensive than standard tapes, but they are readily

6

available. Because frequency response is so important, CrO_2 tape will be used for all frequency response data in this study. Standard tapes will reduce the response slightly, and FeCr tapes will improve it slightly.

Comparison

Two of the four cassette tape recorders in this study are rated good. Of the two, the Nakamichi 550 has the better frequency response: 40-17,000 Hz. This comes very close to the ideal frequency response of 100-18,000 Hz and is exceptionally good for a portable cassette tape recorder. The AIWA TPR-945 is also rated good, having a frequency response of 50-15,000 Hz. Both of these machines would be good choices if only frequency response is considered. The other two machines are not as good, but both are rated acceptable. The Marantz Superscope CD-330 has a frequency response of 40-13,000 Hz, slightly better than the 40-12,500 Hz range of the RMF 740 AV Stereo Recorder.

B.   COST

Definition

The original proposal set $500 as the maximum amount to be paid for the tape recorder. After further discussion, however, Dr. Weller agreed to $750 as a more realistic

limit. This amount is for the purchase of the tape recorder

only. Microphones and other accessories are already

available for use with the machine. Quoted prices were

accurate as of March 1 but may change before the machine is

actually purchased. For this study a cost of less than $500

is rated good; a cost of between $500 and $750 is rated

acceptable.

Comparison

Two of the four recorders cost less than $500 and are

rated good. The Marantz Superscope CD-330, at $249, costs

the least of the machines compared. The AIWA TPR-945, at

$350, is still well below the original $500 maximum. For

machines that meet the specifications of this study, the

prices of both these machines are quite low. The other two

machines cost considerably more but are still in the

acceptable range. The RMF 740 AV Stereo Recorder is $579;

the Nakamichi 550 is $640.

C.   SIZE AND WEIGHT

Definition

Although frequency response and cost are the two major

8

criteria, size and weight cannot be overlooked in a machine

to be used in field research. Size and weight become

especially important if the researcher must carry the

machine long distances over rough terrain. Three of the

four recorders are very similar in size and weight. The

fourth is considerably larger and heavier. For comparison

the three similarly sized recorders are rated good, and the

fourth is rated acceptable.

Comparison

The following table summarizes the dimensions (in

inches) and the weights (in pounds) of the four recorders

compared:

	Length	Width	Height	Weight
Marantz Superscope CD-330	11.6	7.75	3.25	7.8
RMF 740 AV Stereo Recorder	11.3	8.6	4.25	8.0
AIWA TPR-945	19.3	14.6	5.2	12.5
Nakamichi 550	11.5	9.6	4.5	8.3

The three recorders rated good are the Marantz, the RMF, and

the Nakamichi. The Marantz is the smallest and lightest of

the three followed closely by the RMF and the Nakamichi. All

9

three machines are very small and light for quality

portable cassette tape recorders and, compared by size and

weight only, would be good choices. The AIWA, although

larger and heavier than the other three, is rated

acceptable. Its weight of 12.5 pounds is well within the

minimum specification of 20 pounds set by EFW.

D.   BATTERY REQUIREMENTS

Definition

The original proposal stated that the cassette

recorder must be powered by batteries that are disposable,

reasonably priced, and readily available. The only

cassette recorders that meet this requirement are powered

by disposable size C and size D dry-cell batteries.

Therefore, all the machines considered are powered by

similar batteries, and the only important difference is the

number of batteries needed. The number may be especially

important if the field study is to be of extended length and

replacement batteries must be carried.

Comparison

The following table summarizes the number and size of

batteries needed by each machine:

10

	Number	Size
Marantz Superscope CD-330	4	D
RMF 740 AV Stereo Recorder	5	C
AIWA TPR-945	6	C
Nakamichi 550	8	C

The Marantz, using the fewest batteries, would seem to be the best choice for an extended field trip if only battery requirements are considered. But the length of operation on a set of batteries is also a factor. According to dealers, the Nakamichi's batteries will record for 15 continuous hours. However, no definite evidence is available in the literature to prove this point one way or another. Actually, none of the literature for any of the recorders is clear about how long batteries will last in operation. Because, on an extended field trip, length of operation could be as important as number of batteries required, which machine satisfies this criterion the best remains somewhat ambiguous. Therefore, I have not rated any of the machines good or acceptable in this category.

11

REFERENCES

N.d. <u>Catalog.</u> New York: RMF Products, Inc.

N.d. <u>Catalogue of portable cassette products.</u> Moonachie,

   N.J.: AIWA America, Inc.

N.d. <u>Nakamichi product directory.</u> New York: Nakamichi

   Research.

N.d. <u>Portable stereo cassette decks.</u> Chatsworth, Calif.:

   Superscope, Inc.

Robbins, Chandler S. et al. 1966. <u>Birds of North America.</u>

   New York: Golden Press.

Prepared by Donald J. Barrett,
Chief Reference Librarian, United
States Air Force Academy

Appendix

Technical Reference Guides

BOOK GUIDES

Information on books in science and technology is generally found
in standard guides to book publication and general reviewing tools.
A few tools devoted specifically to technical publications do exist, but
most cover only a small portion of each year's book production, so
the more general tools must be used. Several of the major guides are
listed here and will be found in most libraries. Each should be examined for its inclusion policy and limitations.

Book review digest. 1905–.

(Index to selected book reviews, mostly from general periodicals.
Gives publication data, brief descriptive notes, exact citations to
reviews in about 75 periodicals. Subject and title indexes.)

Books in print and *Subject guide to books in print.* Annual.

(Guide to book availability [in print] from 3,600 American publishers. *Books in print:* Author and title lists—give author, title, usually
date of publication, edition, price, and publisher. *Subject guide to
books in print:* Books entered under Library of Congress subject
heading, then by author, with title and other bibliographic data.)

Cumulative book index. 1928–.

(World list of all books published in the English language. Author,
title, and subject listings arranged in a dictionary sequence. Main
entry [fullest information] under author. Give author, title, edition,
series, pagination, price, publisher, date, Library of Congress card
number.)

Library of Congress catalog—Books: Subjects. 1950–.
(Cumulative list by subject of works represented by Library of Congress printed cards for publications printed 1945 or later.)

New technical books. 1915–.
(Selectively annotated titles by New York Public Library staff. Classed subject arrangement, includes table of contents, annotation for each book.)

Proceedings in print. 1964–.
(Announcement journal for availability of proceedings of conferences. Full citation, price, subject and agency indexes.)

Technical book review index. 1935–.
(Very good guide to reviews appearing in scientific, technical, and trade journals, with bibliographic data and exact references to sources of reviews.)

United Nations documents index. 1950–.
(Contains checklists of documents and publications issued by various U.N. agencies, with subject and author indexes cumulated annually.)

REFERENCE BOOKS

Each subject or academic field frequently has its own literature, often ranging from encyclopedias to indexes, dictionaries, biographical tools, etc. The amount of literature specific to one subject may vary greatly. When you first approach a field, one principal guide, Sheehy's, is available to assist you in becoming familiar with its literature. The current edition of this guide should be found in every major collection.

Guide to reference books. Eugene P. Sheehy, ed. 9th ed. Chicago: American Library Association, 1976. Supplements issued biennially.

Encyclopedias

The encyclopedia, although considered too general by some specialists, is often extremely useful to the researcher in getting under way and learning a field. Several of the encyclopedias for specific subjects are in fact quite detailed and scholarly. The editors and contributors to a well-written special encyclopedia will often be experts in their fields. Of course, the latest developments in a subject would be available only in periodical and report literature. Still, the general works are of great value toward understanding a subject, and they often include bibliographic citations to aid in further research.

Dictionary of organic compounds: The constitution and physical, chemical, and other properties of the principal carbon compounds and their derivatives, together with relevant literature references. Ian M.

Heilbron, ed. 4th rev. ed. New York: Oxford University Press, 1965. 5 volumes, 2 cumulative supplements, 1965–73, plus annual supplements. 1975–.

(Alphabetical list of compounds, large number of cross-references.)

Encyclopaedic dictionary of physics. New York: Pergamon Press, 1961–. 9 volumes, 5 supplements.

(Scholarly work, alphabetically arranged, articles generally under 3,000 words, most with bibliographies. Includes articles on general, nuclear, solid state, molecular, chemical, metal, and vacuum physics. Index, plus a multilingual glossary in six languages.)

Encyclopedia of the biological sciences. Peter Gray, ed. 2d ed. New York: Van Nostrand Reinhold Co., 1970.

(Contains 800 articles covering the broad field of the biological sciences as viewed by experts in their developmental, ecological, functional, genetic, structural, and taxonomic aspects. Bibliographies, biographical articles, illustrations, and diagrams are helpful features.)

Encyclopedia of chemical technology. Raymond Kirk and Donald Othmer, eds. 2d ed. New York: Wiley-Interscience, 1963–71. 22 volumes, index and supplement. 3d ed. began publication in 1978.

(Main subject is chemical technology; about half the articles deal with chemical substances. There are also articles on industrial processes. A bibliography is included for each product, as well as information on properties, sources, manufacture, and uses.)

Encyclopedia of engineering signs and symbols. New York: Odyssey Press, 1965.

(Compilation of all graphic symbols used on engineering drawings that are most widely accepted in their respective engineering fields. Includes military standards, industry standards established by associations of the particular industries concerned, and engineering standards established by professional societies.)

Encyclopedia of polymer science and technology: plastics, resins, rubbers, fibers. New York: Wiley-Interscience, 1964–72. 15 volumes, plus index, supplement.

(Articles designed to present a balanced account of all aspects of polymer science and technology, with bibliographies included.)

McGraw-Hill encyclopedia of science and technology. 5th ed. New York: McGraw-Hill Book Co., 1982. 15 volumes, annual supplements.

(Main set includes 7,600 articles, kept current by annual supplements. Covers the basic subject matter of all the sciences and their major applications in engineering, agriculture, and other technologies. Separate index volume. Has many diagrams and charts, and complicated subjects are treated in clear and readable language. Contributors identified in index volume.)

Van Nostrand's scientific encyclopedia. 6th ed. New York: Van Nostrand Reinhold Co., 1983.

> (Includes articles on both basic and applied sciences. Defines and explains over 16,500 terms, arranged alphabetically with extensive cross-references.)

Subject Guides

Bibliographers and librarians have gathered research suggestions and bibliographies for many fields into published guides. It should be remembered that the rapidly changing literature in many subjects partially outdates any guide. Therefore, Sheehy's *Guide to Reference Books* and other tools listing current books and indexes should be consulted to supplement these guides.

Current information sources in mathematics; an annotated guide to books and periodicals, 1960–72. Elie M. Dick. Littleton, Colo.: Libraries Unlimited, 1973.

> (Classed arrangement, contents of books noted, with subject and author indexes.)

Geologic reference sources: A subject and regional bibliography to publications and maps in the geological sciences. Dederick Ward and Marjorie Wheeler. 2d ed. Metuchen, N.J.: Scarecrow Press, 1981.

> (Subject bibliography, most items not annotated. Subject and geographic indexes. Section on geologic maps.)

Guide to basic information sources in engineering. Ellis Mount. New York: John Wiley & Sons, 1976.

> (Annotated list by type material and subject. Author and title index.)

Guide to the literature of the life sciences. Roger C. Smith, W. M. Reid, and A. E. Luchsinger. 9th ed. Minneapolis: Burgess Publishing Co., 1980.

> (Stresses the importance of knowing the literature on a subject and using that knowledge effectively.)

Guide to scientific and technical journals in translation. Carl E. Himmelsbach and Grace Brociner. 2d ed. New York: Special Libraries Association, 1972.

> (Lists 278 cover-to-cover English translations of foreign-language journals and their availability.)

Mechanical engineering: The sources of information. Bernard Houghton. Hamden, Conn.: Archon Books, 1970.

> (Some British emphasis, but useful to students of the subject.)

Reference sources in science and technology. Earl J. Lasworth. Metuchen, N.J.: Scarecrow Press, 1972.

> (Guide to performing literature searches. Bibliographical references, author and title indexes.)

Science and engineering literature: A guide to reference sources. Harold R. Malinowsky. 3d ed. Littleton, Colo.: Libraries Unlimited, 1980.
> (Selected evaluative list of basic reference sources, arranged by major subjects such as physics, chemistry, astronomy, etc.)

Science information sources: A universal and international guide. Herman de Jaeger. 2d ed. Nijmegen, Holland: Association Scientifique et Technique, 1974.
> (Selective list by subject of scientific and technical information sources.)

Science and technology: An introduction to the literature. Denis Grogan. 3d ed. Hamden, Conn.: Linnet Books, 1976.
> (Student guide to structure of the literature of science and technology.)

Scientific and technical information sources. Krishna Subramanyam. New York: Marcel Dekker, 1981.
> (Discussions of all types of literature, access to sources and bibliographic control.)

Technical information sources: A guide to patent specifications, standards, and technical reports literature. Bernard Houghton. 2d ed. Hamden, Conn.: Linnet Books, 1972.
> (British emphasis, covers use of patents as source of technical information, use of specifications and reports.)

Use of chemical literature. R. T. Bottle, 3d ed. London: Butterworths, 1979.
> (Subject arrangement, guides to resources and services evaluated.)

The use of earth sciences literature. David Wood, ed. Hamden, Conn.: Archon Books, 1973.
> (British author, but with emphasis on international literature sources. Includes bibliographies of basic works and sources of abstracts for each subject area.)

Use of engineering literature. K. W. Mildren, ed. London: Butterworths, 1976.
> (Discusses forms of literature and services with 22 chapters on various engineering fields.)

Use of mathematical literature. A. R. Dorling, ed. London: Butterworths, 1977.
> (Describes general literature and use of particular tools, then provides critical accounts by subject experts in their fields. Author and subject indexes.)

Use of physics literature. Herbert Coblans, ed. London: Butterworths, 1975.
> (Subject chapter bibliographic essays, with selective indexes.)

Using the biological literature: A practical guide. Elisabeth B. Davis. New York: Marcel Dekker, 1981.

(Broad subject chapters, bibliographic lists with brief subject content notes.)

Using the chemical literature: A practical guide. Henry Woodburn. New York: Marcel Dekker, 1974.

(Practical guide to use of chemical literature, not a comprehensive bibliography of sources.)

Bibliographies

The literature of a field may often be compiled into bibliographies in connection with other publications, and in some cases as an indication of the work of an agency or company. A guide to bibliographies and examples of other types of compilations are given here.

Bibliographic index: A cumulative bibliography of bibliographies. 1937–.

(Alphabetical subject list of separately published bibliographies and bibliographies appearing in books, pamphlets, and periodicals.)

Bibliography of agriculture. 1942–.

(Classified bibliography of current literature received in the National Agricultural Library, with cumulative annual subject and author indexes.)

Bibliography and index of geology. 1969–.

(Index produced by Geological Society of America. Arranged in broad subject categories, with author and subject indexes. Formerly *Bibliography and index of geology exclusive of North Ameirca* and *Bibliography of North American geology.)*

Chemical titles. 1960–.

(Author and keyword indexes to titles from 700 journals in pure and applied chemistry. A computer-produced bibliography.)

Dissertation abstracts international; abstracts of dissertations available on microfilm or as xerographic reproductions. 1952–.

(Compilation of abstracts of doctoral dissertations from most American universities. Since 1966, Part B has been devoted to the sciences and engineering.)

Index of selected publications of the RAND Corporation. 1946–.

(Coverage includes unclassified publications of the corporation. Abstracts, listed by subject and author, describe content and indexes.)

Science citation index. 1961–.

(Computer-produced index that provides access to related articles by indicating sources in which a known article by an author has been cited. Not ideally suited to subject searching.)

Translations register-index. 1967–.

(Lists new translations received in the Special Libraries Association translations center, plus those available from government and commercial sources.)

Vertical file index. 1932–.

(Subject and title index to selected pamphlet material in all subjects of interest to the general library.)

Biographies

Identification of authors and significant figures in the scientific and technical areas is frequently a problem. Some of the most notable biographical sources are commented on here. A check of Sheehy's *Guide to Reference Books* will reveal many more directories in almost every major subject field.

American men and women of science. 15th ed. New York: Jaques Cattel/R. R. Bowker, 1982. 7 volumes.

(Standard biographical set for personages in the sciences.)

Biography index. 1946–.

(Index to biographical material in books and magazines. Alphabetical, with index by profesison and occupation.)

Dictionary of scientific biography. New York: Charles Scribner's Sons, 1970–80. 14 volumes, supplement and index.

(Comprehensive; covers historical and current persons in science field.)

Who's who in America: A biographical dictionary of notable living men and women. Chicago: Marquis, 1899–. Biennial.

(The standard dictionary of contemporary biographical data. Regional volumes cover persons not of national prominence.)

Who's who in science in Europe: A new reference guide to West European scientists. 3d ed. New York: International Publications Service, 1978. 4 volumes.

(More than 40,000 entries including natural and physical sciences.)

Dictionaries

Definition of terms for the student and scholar is a problem in the sciences, as in any field. A few general guides are available in addition to glossaries for a single field.

Chambers dictionary of science and technology. T. G. Collocott, ed. New York: Barnes & Noble, 1972.

(Successor to *Chamber's technical dictionary;* completely revised.)

A dictionary of physical sciences. John Daintith, ed. New York: Pica Press, 1977.

(Includes some diagrams and cross-references.)

Dictionary of technical terms. Frederic S. Crispin, ed. 11th ed. rev. New York: Bruce, 1970.

(Definitions of commonly used terms; first published in 1929.)

McGraw-Hill dictionary of the life sciences. Daniel N. Lapedes, ed. New York: McGraw-Hill Book Co., 1976.

(Provides vocabulary of the biological sciences and related disciplines. Over 20,000 Terms, useful appendixes.)

McGraw-Hill dictionary of physics and mathematics. Daniel N. Lapedes, ed. New York: McGraw-Hill Book Co., 1978.

(More than 20,000 terms, containing both basic vocabulary and current specialized terminology. Illustrated.)

McGraw-Hill dictionary of scientific and technical terms. Daniel N. Lapedes, ed. 2d. ed. New York: McGraw-Hill Book Co., 1978.

(Gives almost 100,000 definitions, amplified by 2800 illustrations. Each definition identified with the field of science in which it is primarily used.)

Modern science dictionary. A. Hechtlinger. 2d ed. Palisade, N.J.: Franklin, 1975.

(Briefly defined terms for beginning science students.)

Commercial Guides

Access to materials from companies working on a specific product can be facilitated by the use of product association and company address information. The standards are an example of tools that many industries must use to satisfy a contractor's requirements.

Annual book of ASTM standards, with related material. Philadelphia: American Society for Testing and Materials. Annual.

(Approximately 50 parts, including index. Contains 4,900 ASTM Standards and Tentatives in effect at the time of publication. An example of an essential reference book in industrial technology.)

MacRae's blue book. Chicago: MacRae's Blue Book. Annual. 5 volumes.

(Buying directory for engineering products from over 50,000 companies.)

Thomas register of American manufacturers and Thomas register catalog. New York: Thomas Publishing Co. Annual. 17 volumes.

(Products lists, alphabetical listing of manufacturers, trade names.)

PERIODICALS

In almost any current research, the latest developments in a field will be published in the current periodicals and professional journals. Your first task as a researcher may be to determine what periodicals are published in a given field. Next, you may want to determine what indexing or abstracting services give access to a specific journal. The guides are the most significant in assisting you to locate this type of information.

Gebbie Press house magazine directory: A public relations and free lance guide to the nation's leading house magazines. New York: Gebbie Press, 1952–. Triennial.

(Alphabetical list of magazines published by industrial organizations.)

Irregular serials and annuals; an international directory. 7th ed. New York: R. R. Bowker Co., 1982.

(Publication information on over 35,000 serials issued annually or irregularly. Lists many yearbooks, proceedings, transactions, etc. Alphabetical subject arrangement, title index.)

Ulrich's international periodicals directory; a classified guide to current periodicals, foreign and domestic. New York: R. R. Bowker Co. Biennial, with supplement.

(Alphabetical subject list of over 55,000 periodicals published in all languages. Alphabetical title index. Indicates coverage of titles in periodical indexes and abstracting services.)

World list of scientific periodicals published in the years 1900–1960. 4th ed. London: Butterworths, 1963–65. 3 volumes. Supplement, 1960–68, 1970.

(Lists more than 60,000 titles of periodicals concerned with the natural sciences and technology.)

Periodical Indexes

A periodical index provides ready access to articles appearing in professional journals and general periodicals. Each index should be examined for inclusion principles, entry format, and any peculiarities unique to that index. The principal indexes for your consideration cover both general and specific fields, and the primary newspaper index to the *New York Times* is also listed.

Agricultural index, subject index to a selected list of agricultural periodicals and bulletins. 1916–64.

(Detailed alphabetical subject index. Continued as *Biological and agricultural index.)*

Air University Library index to military periodicals. 1949–.

(Subject index to significant articles in approximately 70 military and aeronautical periodicals not covered in readily available commercial indexes.)

Applied science and technology index. 1958–.

(Subject index to periodicals in aeronautics, automation, chemistry, construction, electricity and electrical communication, engineering, geology and metallurgy, industrial and mechanical arts, machinery, physics, transportation and related subjects. Formerly part of the *Industrial arts index.)*

Art Index. 1929–.
(Author and subject index to fine-arts periodicals. Also includes coverage of architecture, graphic arts, industrial design, planning, and landscape design.)

Biological and agricultural index. 1964–.
(Detailed subject index to approximately 190 English-language periodicals. Reports, bulletins, and other agricultural agency publications formerly covered in the *Agricultural index* are no longer covered.)

British technology index. 1962–.
(Current subject guide to articles in British technical journals, with author index.)

Business periodicals index. 1958–.
(Subject index to business, financial, and management periodicals and specific industry and trade journals. Formerly part of the *Industrial arts index.)*

General science index. 1978–.
(Subject index to 90 periodicals in fields including astronomy, atmospheric sciences, biological sciences, earth sciences, environment and conservation, genetics, oceanography, physics, physiology, and zoology.)

Index to U.S. Government periodicals. 1974–.
(Computer-generated guide to 139 selected titles by author and subject.)

Industrial arts index. 1913–57.
(Subject index, split into *Applied science and technology index* and the *Business periodicals index.*)

New York Times index. 1851–.
(Subject index with precise reference to date, page, and column for each article. Well cross-referenced, brief synopses of articles. Can serve as a guide to locating articles in other unindexed papers.)

Public affairs information service bulletin (PAIS). 1915–.
(Very useful index to government, economics, sociology, etc., covering books, periodicals, documents, and reports. Includes selective indexing of over 1,000 periodicals.)

Readers' guide to periodical literature. 1900–.
(Best-known periodical index. Covers U.S. periodicals of a broad, general nature in all subjects and scientific fields.)

Selected references on environment quality as it relates to health. 1971–.
(Indexing service from the National Library of Medicine. Subject and author sections. Covers U.S. and foreign titles.)

Social sciences index. 1974–.
(Author and subject index to periodicals in fields including eco-

nomics, environmental science, psychology, and public administration. Covers 263 titles on a more scholarly level than the *Readers' guide*. Formerly part of the *Social sciences and humanities index* and the *International index to periodicals.*)

Abstract Services

The abstracting journals assist access to periodical, book, and report literature as does an index. The significant difference is the abstract itself, which frequently gives a better indication of article content and hence makes the abstracting journal significantly more useful than the periodical index.

Abstracts and indexes in science and technology: A descriptive guide. Dolores B. Owen. Metuchen, N.J.: Scarecrow Press, 1974.
(Describes approximately 125 abstract and indexing services by subject coverage, arrangement, indexes, abstracts, and other features.)

Abstracts of North American geology. 1966–71.
   (Abstracts of books, technical papers, maps on the geology of North America. Complements the *Bibliography of North American Geology.*)

Air pollution abstracts. 1970–76.
   (Produced by the Air Pollution Technical Information Center of the Environmental Protection Agency. Broad subject arrangement with specific author-subject indexes. Covers periodicals, books, proceedings, legislation, and standards.)

ASM review of metal literature. 1944–67.
   (American Society for Metals abstracting journal of the world's literature concerned with the production, properties, fabrication, and application of metals, their alloys and compounds. Combined to form part of *Metal abstracts.)*

Biological abstracts. 1926–.
   (Broad subject coverage of periodicals, books, and papers in all biological fields. Author, keyword, and systematic indexes.)

Ceramic abstracts. 1922–.
   (Published as a section of the *American Ceramic Society journal.)*

Chemical abstracts. 1908–.
   (Covers chemical periodicals in all languages. Arranged in 80 subject sections; entry includes title, author, publication date, and abstract. Index sections by chemical substance, formula, numbered patent, patent concordance, author, and keyword.)

Energy research abstracts. 1976–.
   (Covers all scientific and technical reports, journal articles, conference papers and proceedings, books, patents, theses originated by the Department of Energy, its laboratories and contractors.

Classed subject arrangement, with author, title, subject, report, and contract indexes. Succeeds *Nuclear science abstracts.)*

Engineering index. 1906–.

(Abstracting journal includes coverage of serial publications, papers of conferences and symposia, separates and nonserial publications, some books. Excludes patents. Entries arranged by subject. Separate author index.)

Geophysical Abstracts. 1929–71.

(Abstracts of current literature pertaining to the physics of the solid earth and to geophysical exploration. Annual author and subject indexes.)

A guide to the world's abstracting and indexing services in science and technology. Washington, D.C.: National Federation of Science Abstracting and Indexing Services, 1963.

(List of 1,855 titles originating in 40 countries. Covers the pure and applied sciences including medicine, agriculture, etc. Includes country and subject indexes.)

Information science abstracts. 1966–.

(Classified subject arrangement listing books and periodicals, international in scope. Formerly *Documentation abstracts.)*

International aerospace abstracts. 1961–.

(Covers published literature in aeronautics and space science and technology. Companion publication to *Scientific and technical aerospace reports.* Arranged in 75 subject categories; each entry gives accession number, title, author, source (book, conference, periodical, etc.), date, pagination, and abstract. Indexed by specific subject, personal author, contract number, meeting paper, accession number.)

Mathematical reviews. 1940–.

(Subject index to mathematical periodicals and books. Arranged by broad subject, with abstracts for most entries. Separate author index, subject classification.)

Metallurgical abstracts. 1934–67.

(British publication, combined to form part of *Metals abstracts.)*

Metals abstracts. 1968–.

(Covers all aspects of the science and practice of metallurgy and related fields. Classed subject arrangement with author index. Publication is a merger of *Metallurgical abstracts* and the *ASM review of metal literature.)*

Meterological and geoastrophysical abstracts. 1950–.

(Includes foreign publications, arranged by Universal Decimal Classification subjects. Has separate author, subject, and geographical location indexes.)

Mineralogical abstracts. 1920–.

(Classified list of abstracts covering current international literature, including books, periodicals, pamphlets, reports.)

Nuclear science abstracts. 1947–76.

(Covers reports of the U.S. Energy Research and Development Administration [formerly the Atomic Energy Commission], government agencies, universities, industrial and independent research organizations, and worldwide book, journal, and patent literature dealing with nuclear science and technology. Arranged by subject field. Indexes by corporate author, personal author, subject, and report number. Continued by *Energy Research Abstracts*.)

Oceanic abstracts. 1964–.

(Covers the worldwide book, periodical, and report literature on the oceans, including pollution, engineering, geology, and oceanography.)

Pollution abstracts. 1970–.

(Classed subject listing of books, periodicals, reports, documents.)

Psychological abstracts, 1927–.

(Covers books, periodicals, and reports; arranged by subject with full author and subject indexes; cumulated indexes for 1927–74.)

Science abstracts. Section A–Physics abstracts. 1898–. *Section B–Electrical and electronic abstracts.* 1898–. *Section C–Computer and control abstracts.* 1966–.

(Covers books, periodicals, papers in all languages; sections do not overlap. All sections arranged by subjects. Separate author and conference indexes.)

Selected water resources abstracts. 1968–.

(Covers water in respect to quality, resources, engineering, and related aspects from books, journals, and reports. Subject and author indexes.)

REPORT LITERATURE

The publication of reports by academic, industrial, and government agencies has been a major development since World War II. Many contracts funded by government agencies have required publication of such reports. The report is now frequently the most recent information on a subject. The bibliographical guides to this type of literature are only recently being developed.

Dictionary of report series codes. Lois Godfrey and Helen Redman, eds. 2nd ed. New York: Special Libraries Association, 1973.

(Explains and identifies letter and number codes used in issuing report literature. Reference notes explain some of the systems used by various agencies in assigning report codes. Alphabetical list of designations related to issuing agency; alphabetical agency list to related series codes.)

Government reports announcements and index. 1964–.

(Formerly titled *Government reports announcements* (1971–75), *U.S. government research and development reports* (1965–71), and *U.S. government research reports* (1954–64). Covers new reports of U.S. Government-sponsored research and development released by the Department of Defense and other federal agencies. Arranged by broad subject areas, each entry gives complete bibliographical citation, descriptors and availability data, and usually an abstract.)

Government reports index. 1965–75.

(Continued as part of *Government reports announcements and index.* Previously issued under other titles. Indexes reports by subject, author, corporate source, and report number.)

Scientific and technical aerospace reports. 1963–.

(Comprehensive abstracting journal covering worldwide report literature on the science and technology of space and aeronautics. Companion publication to *International aerospace abstracts.* Arranged in 74 subject categories. Indexed by subject, corporate source, individual author, contract number, report number, and accession number.)

Technical abstract bulletin. 1953–.

(Primary guide to report literature until the mid-1960s, published by Defense Technical Information Center, formerly the Defense Documentation Center. Covers technical reports of U.S. and foreign governments and commercial contractors as submitted to DTIC. Now covers material with distribution limitations; abstract section issues before 1978 are security classified until six years after issue. All index sections are declassified. Both sections are unclassified after July 1978.)

Use of reports literature. Charles P. Auger, ed. Hamden, Conn.: Archon Books, 1975.

(Discusses report literature nature, control, and value. Evaluates sources.)

ATLASES AND STATISTICAL GUIDES

Basic data of interest to the technical researcher on many subjects are found in reliable and frequently updated standard guides. The quality atlas generally contains much more than maps of geographical locations. Statistical guides are of great reference importance to original research in economic, industrial, and social questions.

Commerical atlas and marketing guide. Annual. Chicago: Rand McNally & Co.

(In addition to maps, contains much statistical data on trade, manufacturing, business, population, and transportation.)

The national atlas of the United States of America. Washington, D.C.: U.S. Department of the Interior Geological Survey, 1970.

(Outstanding collection of 765 maps, many in color. Covers general reference and special subjects, including landforms, geophysical forces, geology, marine features, soils, climate, water, history, economic, sociocultural, administrative, with maps, data tables, and diagrams.)

Statistical abstract of the United States. 1878–. Annual. Washington, D.C.: Bureau of the Census.

(Official government standard summary of statistics on the social, political, and economic organization of the United States. Excellent source citations, index.)

The world almanac and book of facts. 1868–. Annual. New York: Newspaper Enterprise Association.

(The most comprehensive and generally useful of the almanacs of miscellaneous information. Excellent statistical and news summary coverage.)

COMPUTERIZED INFORMATION RETRIEVAL

Since the 1970s, a growing number of bibliographic files have been made available for on-line interactive searching and information retrieval. Normal access is from a local computer terminal to a firm offering access to data base or a system of bases. Charges are calculated on the number of minutes a file is used and the number of citations received, usually by off-line printing and airmail delivery. Over 14 million citations beginning in about 1970 are now available for searching. Three commercial services offer access to a wide spectrum of data bases. They are ORBIT II, from Systems Development Corporation, Santa Monica, California; DIALOG, Dialog Information Retrieval Services, Palo Alto, California; and BRS, Bibliographic Retrieval Services, Latham, New York. An average search can be expected to cost a minimum of $20–$50, with a key to the cost being good preplanning of search terms and search strategy by an experienced operator. Many major colleges, universities, and information centers offer these services through service bureaus or libraries, with some libraries paying part of all of the costs. Individual access directly through originating firms is also possible. Custom searches of over 800,000 reports from federal agencies and federally sponsored research are available from National Technical Information Service, U.S. Department of Commerce, Springfield, Virginia 22161. The cost is about $100 for up to 100 abstracts. Some of the current subjects available in the scientific and technical areas are given here.

Agriculture: AGRICOLA, developed by the National Library of Agriculture.

Biological sciences: BIOSIS, prepared by Biological Sciences Information Service.

Business: ABI/INFORM, produced by Data Courier, Inc.

Chemistry: CA CONDENSATES/CASIA, produced by Chemical Abstracts Service of the American Chemical Society.

Education: ERIC (Educational Resources Information Center), developed by the National Institute of Education.

Engineering: COMPENDEX, produced by Engineering Index, Inc.

Environmental studies: ENVIROLINE, prepared by Environment Information Center, Inc.

Geosciences: GEO-REF, produced by the American Geological Institute.

Government research and development reports: NTIS, produced by the National Technical Information Service of the Department of Commerce.

Mechanical engineering: ISMEC, prepared by Data Courier, Inc.

Physics: SPIN (Searchable Physics Information Notices), produced by American Institute of Physics.

Pollution and environment: POLLUTION, produced by the publishers of *Pollution abstracts.*

Psychology: Psychological Abstracts, produced by the American Psychological Association.

Science abstracts: INSPEC, produced by the Institution of Electrical Engineers.

Science and technology: SCISEARCH, produced by the Institute for Scientific Information.

Many other data bases are available for searching, but some have special use restrictions such as being limited to specific industrial group members.

U.S. GOVERNMENT PUBLICATIONS

Access to government publications, releases, and directives is possible only from a series of federally produced and commercially supplemented catalogs and indexes. The nature of the subject indexing is generally less specific than with nongovernmental periodical indexes and abstracting services. Therefore, more ingenuity on the part of the researcher is generally needed to find pertinent resources.

American statistics index and abstracts, annual and retrospective edition. A comprehensive guide and index to the statistical publications of the U.S. Government. Washington, D.C.: Congressional Information Service.

(Aims to be a master guide and index to all federally produced statistical data. Does not contain the data but describes data and identifies sources.)

Bureau of the Census catalog of publications, 1790–1972.

(An example of a departmental catalog, kept up-to-date by frequent supplements. Many agencies issue such retrospective catalogs.)

CIS/Index to publications of the United States Congress. Washington, D.C.: Congressional Information Service, 1970–.

(Part one: Abstracts of congressional publications and legislative histories; Part two: Index of congressional publications and public laws. Commercial index giving significant insight into contents of congressional publications, with a detailed subject index.)

Code of federal regulations. 1949–.

(Contains codifications of general and permanent administrative rules and regulations of general applicability and future effect.)

Cumulative subject index to the monthly catalog of United States Government publications, 1900–1971. Washington, D.C.: Carrollton Press, 1973–75. 15 volumes.

(Covers about 800,000 publications. Be sure to read the introduction for exclusions and entry policies.)

Federal register. 1936–.

(Daily publication of executive orders, presidential proclamations, and announcements of important rules and regulations of the federal government. Indexed by agency and significant subjects.)

Monthly catalog of United States Government publications. 1895–.

(List by agency of publications, printed and processed, issued each month. Subject index only until 1973, then added author and title indexes cumulated annually. Entry gives title, author, publication data, price or availability indication. Superintendent of Documents classification number.)

Monthly checklist of state publications. 1910–.
> (Records those documents and publications issued by the various states and received in the Library of Congress.)

Subject bibliographies. 1975–.
> (Lists of publications available in specified subject areas from the Government Printing Office.)

United States Government manual. 1935–. Annual. Washington, D.C.: U.S. Government Printing Office.
> (The official handbook of the federal government; describes the purposes and programs of most official and quasi-official agencies, with lists of current officials.)

Weekly compilation of presidential documents. 1965–.
> (Makes available transcripts of the president's news conferences, messages to Congress, public speeches and statements, and other presidential materials. Indexed.)

Code of federal regulations. 1949–.
> (Contains codifications of general and permanent administrative rules and regulations of general applicability and future effect.)

Monthly checklist of state publications. 1910–.
> (Records those documents and publications issued by the various states and received in the Library of Congress.)

Subject bibliographies. 1975–.
> (Lists of publications available in specified subject areas from the Government Printing Office.)

American statistics index and abstracts, annual and retrospective edition. A comprehensive guide and index to the statistical publications of the U.S. Government. Washington, D.C.: Congressional Information Service.
> (Aims to be a master guide and index to all federally produced statistical data. Does not contain the data but describes data and identifies sources.)

Bureau of the Census catalog of publications, 1790–1972.
> (An example of the departmental catalog, kept up-to-date by frequent supplements. Many agencies issue such retrospective catalogs.)

Appendix C

A Selected Bibliography

TECHNICAL WRITING

BLICQ, RON S. *Technically Write!* 2d ed. Englewood Cliffs, N.J.: Prentice-Hall, Inc., 1981.

BRUSAW, CHARLES, ALRED, GERALD J., and OLIU, WALTER E. *Handbook of Technical Writing.* 2d ed. New York: St. Martin's Press, 1982.

DAY, ROBERT A. *How to Write and Publish a Scientific Paper.* Philadelphia: ISI Press, 1979.

FEAR, DAVID E. *Technical Communication.* 2d ed. Glenview, Ill.: Scott, Foresman & Company, 1981.

JORDAN, STELLO, KLEINMAN, JOSEPH, M., and SHIMBERG, H. LEE, eds. *Handbook of Technical Writing Practices.* 2 vols. New York: Wiley-Interscience, 1971.

LANNON, JOHN. *Technical Writing.* 2d ed. Boston: Little, Brown and Company, 1982.

MATHES, J. C., and STEVENSON, DWIGHT. *Designing Technical Reports.* Indianapolis: The Bobbs-Merrill Company, Inc., 1976.

MICHAELSON, HERBERT B. *How to Write and Publish Engineering Papers and Reports.* Philadelphia: ISI Press, 1982.

MILLS, GORDON H., and WALTER, JOHN A. *Technical Writing.* 4th ed. New York: Holt, Rinehart and Winston, 1978.

PEARSALL, THOMAS E., and CUNNINGHAM, DONALD H. *How to Write for the World of Work.* 2d ed. New York: Holt, Rinehart and Winston, 1982.

PICKETT, NELL A., and LASTER, ANN A. *Technical English.* 3d ed. New York: Harper & Row, Publishers, Inc., 1980.

SOUTHER, JAMES W., and WHITE, MYRON L. *Technical Report Writing.* 2d ed. New York: John Wiley & Sons, Inc., 1977.

WARREN, THOMAS. *Technical Communication: An Outline.* Totowa, N.J.: Littlefield, Adams and Co., 1978.

WEISMAN, HERMAN M. *Basic Technical Writing.* 4th ed. Columbus, Ohio: Charles E. Merrill Books, Inc., 1980.

BUSINESS COMMUNICATION

LEONARD, DONALD. *Shurter's Communication in Business.* 4th ed. New York: McGraw-Hill Book Company, 1979.

MURPHY, HERTA A., and PECK, CHARLES E. *Effective Business Communications.* 3d ed. New York: McGraw-Hill Book Company, 1980.

TREECE, MALRA. *Successful Business Writing.* Boston: Allyn and Bacon, Inc., 1980.

WHALEN, DORIS H. *Handbook of Business English.* New York: Harcourt Brace Jovanovich, Inc. 1980.

WILKINSON, C. W., CLARKE, PETER B., and WILKINSON, DOROTHY C. M. *Communicating Through Letters and Reports.* 7th ed. Homewood Ill.: Richard D. Irwin, Inc., 1980.

Writing Better Letters, Reports and Memos. New York: American Management Association, Inc., 1976.

WRITING IN GENERAL

FELKER, DANIEL B., et al. *Guidelines for Document Designers.* Washington, D.C.: American Institutes for Research, 1981.

FLESCH, RUDOLF. *Art of Readable Writing.* Rev. ed. New York: Harper & Row, Publishers, Inc., 1974.

FLOWER, LINDA. *Problem Solving Strategies for Writing.* New York: Harcourt Brace Jovanovich, Inc., 1981.

KINNEAVY, JAMES L. *A Theory of Discourse.* New York: W. W. Norton & Company, 1980.

LAMBUTH, DAVID, et al. *The Golden Book on Writing.* New York: Penguin Books, Inc., 1976.

QUILLER-COUCH, SIR ARTHUR. *On the Art of Writing.* New York: G. P. Putnam's Sons, 1916.

VAN BUREN, ROBERT, and BUEHLER, MARY FRAN. *The Levels of Edit.* 2d ed. Pasadena, Calif.: Jet Propulsion Laboratory, California Institute of Technology, 1980.

STYLE AND USAGE

CBE STYLE MANUAL COMMITTEE, *Council of Biology Editors Style Manual.* 4th ed. Arlington, Va.: Council of Biology Editors, 1978.

The Chicago Manual of Style. 13th ed. Chicago: The University of Chicago Press, 1982.

EBBITT, WILMA R., and EBBITT, DAVID R. *Writer's Guide and Index to English.* 7th ed. Glenview, Ill.: Scott, Foresman & Company, 1981.

FOLLETT, WILSON. *Modern American Usage.* Edited and completed by Jacques Barzun. New York: Hill and Wang, 1966.

FOWLER, HENRY W. *A Dictionary of Modern English Usage.* 2d ed. Revised by Sir Ernest Gowers. New York: Oxford University Press, 1965.

GOVERNMENT PRINTING OFFICE. *Style Manual.* Washington, D.C.: U.S. Government Printing Office, 1982.

GRAVES, ROBERT, and HODGE, ALAN. *The Reader Over Your Shoulder.* 2nd ed. New York: Vintage Books, 1979.

WILLIAMS, JOSEPH M. *Style: Ten Lessons in Clarity and Grace.* Glenview, Ill.: Scott, Foresman & Company, 1981.

SPEECH

BRYANT, DONALD C., et al. *Oral Communication.* 5th ed. Englewood Cliffs, N.J.: Prentice-Hall, Inc., 1982.

CONNOLLY, JAMES. *Effective Technical Presentations.* St. Paul, Minn.: 3M Business Press, 1968.

EHNINGER, DOUGLAS, et al. *Principles of Speech Communication.* 8th ed. Glenview, Ill.: Scott, Foresman & Company, 1980.

TACEY, WILLIAM S. *Business and Professional Speaking.* 3rd ed. Dubuque, Iowa: Wm. C. Brown Company Publishers, 1980.

TIMM, PAUL R. *Functional Business Presentations.* Englewood Cliffs, N.J.: Prentice-Hall, Inc., 1981.

SEMANTICS

CONDON, JOHN C., JR. *Semantics and Communication.* 2d ed. New York: Macmillan Publishing Co., Inc., 1975.

HAYAKAWA, S. I. *Language in Thought and Action.* 4th ed. New York: Harcourt Brace Jovanovich, Inc., 1978.

LOGIC

COHEN, MORRIS R., and NAGEL, ERNEST. *Introduction to Logic.* New York: Harcourt Brace Jovanovich, Inc., 1962.

COPI, IRVING M. *Introduction to Logic.* 6th ed. New York: Macmillan Publishing Co., Inc., 1982.

GRAPHICS

MACGREGOR, A. J. *Graphics Simplified: How to Plan and Prepare Effective Charts, Graphs, Illustrations, and Other Visual Aids.* Toronto: University of Toronto Press, 1979.

TURNBULL, ARTHUR T., and BAIRD, RUSSELL N. *The Graphics of Communication.* 4th ed. New York: Holt, Rinehart and Winston, 1980.

LIBRARY RESEARCH

BARZUN, JACQUES, and GRAFF, HENRY F. *The Modern Researcher.* 3d ed. New York: Harcourt Brace Jovanovich, Inc., 1977.

DOWNS, ROBERT B., and KELLER, CLARA D. *How to Do Library Research.* 2d ed. Urbana, Ill.: University of Illinois Press, 1975.

HERNER, SAUL. *A Brief Guide to Sources of Scientific and Technical Information.* 2d ed. Arlington, Va.: Information Resources Press, 1980.

JACKSON, ELLEN. *Subject Guide to Major United States Government Publications.* Chicago: American Library Association, 1968.

McCORMICK, MONA. *The New York Times Guide to Reference Materials.* New York: Popular Library, 1982.

MOREHEAD, JOE. *Introduction to United States Public Documents.* 2d ed. Littleton, Colo.: Libraries Unlimited, Inc., 1978.

PALIC, VLADIMIR M. *Government Publications: A Guide to Bibliographic Tools.* Washington, D.C.: Library of Congress, 1975.

SCHMECKEBIER, LAURENCE F., and EASTIN, ROY B. *Government Publications and Their Use.* Rev. ed. Washington, D.C.: Brookings Institution, 1969.

Chapter Notes

Preface

1. Thomas L. Warren, "Style in Technical Writing," *The Technical Writing Teacher* 6, no. 2 (1979): 49.
2. Richard M. Davis, "How Important is Technical Writing?—A Survey of the Opinions of Successful Engineers," *Journal of Technical Writing and Communication* 4, no. 3 (1977): 83–88.
3. John Naisbitt, "The New Economic and Political Order of the 1980's," *Vital Speeches* 46, no. 18 (1980): 557–62. See also John Naisbitt, *Megatrends* (New York: Warner Books, 1982).
4. Philip W. Swain, "Giving Power to Words," *American Journal of Physics* 13, no. 5 (1945): 320.

Chapter 2: Getting Started

1. Suggested by Terry C. Smith, "What Bugs Engineers Most About Report Writing?" *Technical Communication* 23, no. 4 (1976): 2–6.
2. Adapted from a form created by Professor John S. Harris of Brigham Young University.

Chapter 3: Analyzing Your Audience

1. Walter James Miller, "What Can the Technical Writer of the Past Teach the Technical Writer of Today?" *IRE Transactions on Engineering Writing and Speech* 4, no. 3 (1961): 69–76.
2. Philip W. Swain, "Giving Power to Words," *American Journal of Physics* 13, no. 5 (1945): 318.
3. U.S., Bureau of the Census, *Statistical Abstract of the United States*, 102d ed. (Washington, D.C.: U.S. Government Printing Office, 1981), 141.
4. Richard Conniff, "Eye on the Storm," *Raytheon Magazine*, Fall 1982, 21.

5. Norman Cousins, "Growing Your Own Fuel," *Saturday Review*, 30 September 1978, 24.

6. Michael Ryan and James Tankard, Jr., "Problem Areas in Science News Writing," *Journal of Technical Writing and Communication* 4, no. 3 (1974): 233.

7. *Dietary Salt* (Chicago: Institute of Food Technologists, 1980), 85.

8. Joseph Callanan and Cameron Foote, "The Light Fantastic," *Raytheon Magazine*, Fall 1981, 16.

9. Excerpted from the July 1978 *The Harvard Medical School Health Letter*. Copyright © 1978 President and Fellows of Harvard College.

10. Ibid., 1–2.

11. Ibid, 1.

12. Ryan and Tankard, "Problem Areas in Science News Writing," 230.

13. Daniel B. Felker et al., *Guidelines for Document Designers* (Washington, D.C.: American Institutes for Research, 1981), 41–48.

14. Mary Fran Buehler and Andrea Stein, "History, Science, and People: Shaping Today's Technical Message with Yesterday's Art," presented at the 28th International Technical Communication Conference, Pittsburgh, Pennsylvania, May 20–23, 1981, and published in the *Proceedings* of that conference.

15. Umberto F. Gianola and Richard R. Shively, "Signal Processor Sorts Sounds from the Sea," *Bell Laboratories Record*, May 1980, 167–68.

16. James W. Souther, "What Management Wants in the Technical Report," *Journal of Engineering Education* 52, no. 8 (1962): 500.

17. Richard W. Dodge, "What to Report," *Westinghouse Engineer* 22, no. 4–5 (1962): 108–111. The information in "What to Report" is based upon a study made at Westinghouse in 1959–60 by Professor James W. Souther of the University of Washington. All material reprinted from "What to Report" is done so through the courtesy of the Westinghouse Electric Corporation.

18. Reported by Mary B. Coney, "The Use of the Reader in Technical Writing," *Journal of Technical Writing and Communication* 8, no. 2 (1978): 104.

19. Ibid.

20. Souther, "What Management Wants," p. 501.

21. David F. Cope, "Nuclear Power: A Basic Briefing," *Mechanical Engineering* 89, no. 6 (1967): 50.

22. M. Daily et al., *Application of Multispectal Radar and LANDSAT Imagery to Geologic Mapping in Death Valley* (Pasadena, Calif.: Jet Propulsion Laboratory, California Institute of Technology, 1978), 44.

23. J. J. Degan, "Microwave Resonance Isolators," *Bell Laboratories Record*, April 1966, 123.

24. Ibid., 125

Chapter 5: Gathering and Checking Information

1. William H. Whyte, "Small Space Is Beautiful: Design as if People Mattered," *Technology Review* 85, no. 5 (1982): 40.

2. Ibid., 39.

3. For detailed treatment we suggest you read Douglas R. Berdie and John F. Anderson's *Questionnaires: Design and Use* (Metuchen, N.J.: The Scarecrow Press, Inc., 1974).

Chapter 6: Technical Exposition

1. For a series of interesting articles on this point, see *Technical Communication*, 4th Quarter, 1978.
2. Breyne Arlene Moskowitz, "The Acquisition of Language," *Scientific American*, November 1978, 92.
3. Samuel Clemens [Mark Twain], *A Tramp Abroad* (Hartford, Conn., 1880), 329–30.
4. U.S., Department of the Interior, *The Severity of an Earthquake* (Washington, D.C.: U.S. Government Printing Office, 1979), 3–7.
5. D. W. Phillipson, "The Spread of the Bantu Language, *Scientific American*, April 1977, 106.
6. For this idea we are indebted to W. Earl Britton, and we suggest you read his "What to Do About Hard Words," *STWP Review*, October 1964, 13–16.
7. F. Richard Stephenson and David H. Clark, "Historical Supernovas," *Scientific American*, June 1976, 100.
8. John A. Lofgren, *Controlling Insect Pests of Shade and Ornamental Trees*, Agricultural Extension Service of the University of Minnesota Fact Sheet no. 28 (St. Paul, Minn., 1973).
9. Sir James Jeans, *Stars in Their Courses* (Cambridge: Cambridge University Press, 1931), 23–24. Copyright © 1931. Reprinted by permission of the publisher.
10. Joel Gurin, "In the Beginning," *Science 80*, July/August 1980, 50.
11. V. Thomas Mawhinney, "Mommy, Can I Leave the Lights On?" *Family Health*, October 1978, 33.
12. Willard W. Cochane, "U.S. Farm Policy, Collision in the 70's," *Minnesota Science*, Fall 1970, 4.
13. Jet Propulsion Laboratory, "Satellites of the Outer Planets," *Mission to Jupiter/Saturn* 2 (November 1977), 1. Published by U.S. National Aeronautics and Space Administration.
14. U.S., National Aeronautics and Space Administration, *The Planet Venus* (Washington, D.C.: U.S. Government Printing Office, 1978), 2–4.

Chapter 7: Technical Narration, Description, and Argumentation

1. George Cowan, "A Natural Fission Reactor," *Scientific American*, July 1976, 36.
2. U.S., National Aeronautics and Space Administration, *Voyager at Saturn: 1981*. Prepared by the Jet Propulsion Laboratory (Pasadena, Calif., 1981), 1.
3. Ibid., 2.
4. Ibid., 3.
5. John Strait, "Haying Systems Using Cubic Bales," *Minnesota Science* 27, no. 4 (1971): 11–12.

6. U.S., Department of Agriculture, *Wood Handbook*, prepared by the Forest Products Laboratory (Washington, D.C., 1974), 17.

7. Ibid., 20–24.

8. John S. McNown, "Canals in America," *Scientific American*, July 1976, 117.

9. *The New Columbia Encyclopedia*, 4th ed., s.v. "Camera."

10. *Collier's Encyclopedia*, s.v. "Camera." Reprinted with permission from *Collier's Encyclopedia*, copyright © 1969, Crowell-Collier Educational Corporation.

11. Ibid., s.v. "Engine."

12. *Wood Handbook*, 2–5.

13. *The Mead Short Course in the Graphic Arts*, vol. 4, *Capabilities Booklets* (Dayton, Ohio: Mead Paper, 1974), 10. Copyright © 1974 by Mead Paper. All rights reserved.

14. Robin Birley, "A Frontier Post in Roman Britain," *Scientific American*, February 1977, 39.

15. U.S., National Aeronautics and Space Administration, *The Planet Venus* (Washington, D.C.: U.S. Government Printing Office, 1978), 4.

16. U.S., National Aeronautics and Space Administration, *The Voyager Flights to Jupiter and Saturn*. Prepared by the Jet Propulsion Laboratory (Pasadena, Calif.: 1982), 11.

17. Victor Gilinsky, "Washington, Report: Full Ahead for Nuclear Power?" *Technology Review* 85, no. 2 (1982): 10.

18. Z. A. Zasada and John W. Benzie, *Mechanized Harvesting for Thinning Sawtimber Red Pine*, Miscellaneous Report 99, Forestry Series 9 (St. Paul, Minn.: University of Minnesota Agricultural Experiment Station, 1970), 13–14.

Chapter 8: Organizing Your Report

1. Terry C. Smith, "What Bugs Engineers Most About Report Writing?" *Technical Communication* 23, no. 4 (1976): 2.

2. Ernest Hemingway, *A Moveable Feast* (New York: Bantam Books, 1965), 12.

3. If you *would* like to know more about this subject, we suggest you read Walter James Miller's "What Can the Technical Writier of the Past Teach the Technical Writer of Today," *IRE Transactions on Engineering and Speech* 4, no. 3 (1961): 69–76.

4. Blaine McKee, "Do Professional Writers Use an Outline When They Write?" *Technical Communication* 19, no. 1 (1972): 10–13.

Chapter 9: Achieving a Clear Style

1. U.S., National Aeronautics and Space Administration, *The New Frontier: Linking Earth and Planets*, Issue 5 (Pasadena, Calif: Jet Propulsion Laboratory, 1974), 2.

2. *Diet and Hyperactivity: Any Connection* (Chicago: Institute of Food Technologists, 1976), 1–2.

3. Rudolph Flesch, *The Art of Plain Talk* (New York: Harper & Brothers, 1946), 38.

4. Porter G. Perrin and George H. Smith, *Handbook of Current English* (New York: Scott, Foresman and Company, 1955) 211.

5. George R. Klare, "Assessing Readability," *Reading Research Quarterly* 10, no. 1 (1974–1975): 97.

6. Francis Christensen, "Notes Toward a New Rhetoric," *College English*, October 1963, 7–18.

7. Daniel B. Felker et al., *Guidelines for Document Designers* (Washington, D.C.: American Institutes for Research, 1981), 47–48.

8. As quoted in *Guidelines*, 64.

9. As revised in *Guidelines*, 65.

10. As quoted in Janice C. Redish, *The Language of Bureaucracy* (Washington, D.C.: American Institutes for Research, 1981), 1.

11. CBE Style Manual Committee, *Council of Biology Editors Style Manual: A Guide for Authors, Editors, and Publishers in the Biological Sciences,* 4th ed. (Arlington Va.: Council of Biology Editors, 1978), 21.

12. The excerpts from the St. Paul Fire and Marine Insurance Company's old and new Personal Liability Catastrophe Policy are reprinted with the permission of the St. Paul Companies, St. Paul, Minn. 55102.

13. CBE Style Manual Committee, *Style Manual*, 19–23. Reproduced with permission from *Council of Biology Editors Style Manual*, 4th edition. CBE Style Manual Committee. Council of Biology Editors, 1978.

14. "Planners Outlaw Jargon," *Plain English*, April 1981, 1.

Chapter 11: Formal Elements of Reports

1. William E. Splinter, "Center-Pivot Irrigation," *Scientific American*, June 1976, 90.

2. Carolyn J. Mullins, "Once Is Not Enough: WP Files Can Do Extra Duty," *Technical Communication* 29, no. 1 (1982): 20.

3. Max Weber, "Human Engineering and Its Application to Technical Writing, *Technical Communication* 19, no. 3 (1972): 2.

4. *Phthalates in Food* (Chicago: Institute of Food Technologists, 1974), 1.

5. Carolyn Krause, "Carbon Dioxide and Climate," *Oak Ridge National Laboratory Review* 10 (Fall 1977): 40–41.

6. *Phthalates in Food*, 2.

7. Krause, "Carbon Dioxide and Climate," 47.

8. J. J. Bull, R. C. Vogt, and C. J. McCoy, "Sex Determining Temperatures in Turtles: A Geographic Comparison," *Evolution* 36 (1982): 331.

9. *The Chicago Manual of Style*, 13th ed. (Chicago: University of Chicago Press, 1982).

10. *MLA Handbook* (New York: Modern Language Association, 1977).

11. Institute of Food Technologists' Expert Panel on Food Safety and Nutrition and the Committee on Public Information, *Mercury in Food* (Chicago: Institute of Food Technologists, 1973), 1–2.

Chapter 13: Correspondence

1. Avery Comarow, "Tracking the Elusive Job," *The Graduate* (Knoxville, Tenn.: Approach 13–30 Corporation, 1977), 42.

2. Jane L. Anton, Michael L. Russell, and the Research Committee of the Western College Placement Association, *Employer Attitudes and Opinions Regarding Potential College Graduate Employees* (Hayward, Calif.: Western College Placement Association, 1974), 10.

3. Rosemary Ullrich, "The Power of a Positive Job Search," *Business World Women*, Fall 1977, 11.

4. Office of Career Development, *Writing Resumes* (Boston: Harvard University Graduate School of Business Administration, n.d.), 3.

Chapter 14: Instructions for Performing a Process

1. G. B. Harrison, *Profession of English* (New York: Harcourt, Brace and World, 1962), 149.

2. Richard Rideout, *Planting Landscape Trees* (St. Paul, Minn.: University of Minnesota, Agricultural Extension Service, 1978), 2.

3. *The Mead Short Course in the Graphic Arts* (Dayton, Ohio: Mead Paper, 1974), vol. 4, *Capabilities Booklets*, 2. Copyright © 1974 by Mead Paper. All rights reserved.

4. U.S., General Services Administration, *Paint and Painting* (Washington, D.C.: U.S. Government Printing Office, 1977), p. 15.

5. *Fundamentals of Machine Operation: Hay and Forage Harvesting* (Moline, Ill.: Deere and Company, 1976), 88–89. Copyright © 1976 by Deere and Company. All rights reserved.

6. Ibid., 100.

7. *Capabilities Booklets*, 4–7.

8. General Services Administration, *Paint and Painting*, 24.

9. *Fundamentals of Machine Operation*, 92–94.

10. Ibid., 102.

11. *Operator's Manual: 850 and 950 Tractors* (Moline, Ill,: Deere and Company, n.d.), 43.

12. *Capabilities Booklets*, 26–27.

13. Roger E. Machmeier, *Driving a Wellpoint* (St. Paul, Minn.: University of Minnesota, Agricultural Extension Service, 1972).

Chapter 17: Feasibility Reports

1. Excerpts from "The Feasibility of Removing Sediment Deposits from Wyoming Lake" reprinted with the permission of Barbara J. Buschatz.

Chapter 18: Empirical Research Reports

1. J. J. Bull, R. C. Vogt, and C. J. McCoy, "Sex Determining Temperatures in Turtles: A Geographic Comparison," *Evolution* 36 (1982): 326.

2. Lynn S. Best and Paulette Bierzychudek, "Pollinator Foraging on Foxglove *(Digitalis purpurea):* A Test of a New Model," *Evolution* 36 (1982): 70.

3. Leslie K. Johnson, "Sexual Selection in a Brentid Weevil," *Evolution* 36 (1982): 251.

4. Ibid.

5. Hugh M. Robertson and Hugh E. H. Paterson, "Mate Recognition and Mechanical Isolation in *Enallagma* Damselflies *(Odonata: Coenagrionidae)*," *Evolution* 36, no. 2 (1982): 243.

6. CBE Style Manual Committee, *Council of Biology Editors Style Manual: A Guide for Authors, Editors and Publishers in the Biological Sciences,* 4th ed. Arlington, Va.: Council of Biology Editors, 1978), 10.

7. Bull, Vogt, and McCoy, "Sex Determining Temperatures," 326.

8. David M. Craig, "Group Selection Versus Individual Selection: An Experimental Analysis," *Evolution* 36 (1982): 272.

9. Gerald L. Chan and John B. Little, "Further Studies on the Survival of Non-Proliferating Human Diploid Fibroblasts Irradiated with Ultraviolet Light," *International Journal of Radiation Biology* 41 (1982): 360.

10. Craig, "Group Selection," 272.

11. Bull, Vogt, and McCoy, "Sex Determining Temperatures," 326–27.

12. Johnson, "Sexual Selection," 253.

13. Bull, Vogt, and McCoy, "Sex Determining Temperatures," 327–29.

14. Ibid., 329–30.

Chapter 19: Oral Reports and Group Conferences

1. These introspections were compiled at a session of the National Training Laboratory in Group Development that one of the authors attended in Bethel, Maine. Many of the ideas expressed in this chapter were first developed in the National Training Laboratory.

2. The material in this section has been especially prepared for this chapter by Professor James Connolly of the University of Minnesota, based upon his book *Effective Technical Presentations* (St. Paul, Minn.: 3M Business Press, 1968.)

Index

MARKING SYMBOLS

This list of marking symbols refers you to Part 3, the "Handbook," where comments about style and form can be found. The list furnishes you with a caption and a page reference.